Carbon Criminals,
Climate Crimes

Library of Congress Cataloging-in-Publication Data

Names: Kramer, Ronald C., author.
Title: Carbon criminals, climate crimes / Ronald C. Kramer.
Description: New Brunswick, New Jersey : Rutgers University Press, [2020] |
    Includes bibliographical references and index.
Identifiers: LCCN 2019025770 | ISBN 9781978805590 (hardback) |
    ISBN 9781978805583 (paperback) | ISBN 9781978805606 (epub) |
    ISBN 9781978807631 (mobi) | ISBN 9781978807648 (pdf)
Subjects: LCSH: Offenses against the environment. |
    Criminology—Environmental aspects. | Climatic changes—Moral and ethical
    aspects. | Global warming—Moral and ethical aspects. | Climatic changes—
    Government policy. | Global warming—Government policy. | Corporations—
    Corrupt practices. | Environmental justice.
Classification: LCC HV6401 .K73 2020 | DDC 364.1/45—dc23
LC record available at https://lccn.loc.gov/2019025770

A British Cataloging-in-Publication record for this book is available
from the British Library.

♾ The paper used in this publication meets the requirements of the American National
Standard for Information Sciences—Permanence of Paper for Printed Library Materials,
ANSI Z39.48-1992.

www.rutgersuniversitypress.org

Manufactured in the United States of America

For my grandsons, Truman, Malcolm, and Calvin

CONTENTS

FOREWORD

OUR LEADERS HAVE GONE MAD, and they have taken the world with them. This is most evident in the way in which global warming is distorting what used to be the familiar or ordinary patterns of weather. Extreme weather is increasingly becoming the new "norm" and is evident in extreme temperatures, intense precipitation, cyclones, storms and hurricanes, droughts, and forest fires.

More and more things are out of kilter today as the climate is changing and unusual weather events are proliferating worldwide. This is *climate disruption*, one of the key outcomes of global warming and overall changes in climatic conditions.

Rises in global average temperatures do not translate into uniform warming. Rather, they trigger a whole series of changes in existing weather patterns, for example, bringing into the equation excessive precipitation in some parts of the world, extreme drought in others. The point is that climate change disrupts normal weather patterns in various and unexpected ways.

The number of climate-related natural catastrophes has been rising steeply since the 1980s, and the most damaging are often the result of combinations of variables—a phenomenon called compound extremes. These include geophysical events such as landslides through to weather phenomena like storms. Lives both human and nonhuman are being lost; lands are being devastated and waters polluted. The earth and its inhabitants are suffering.

So where does criminology fit into this picture?

The answer is simple. The harms associated with climate disruption are and always have been preventable. Ideas about the "greenhouse gas effect" have been around for well over a century and certainly widely known since the United Nations Rio Summit in 1992. The problem and its consequences have been known for decades.

The systematic destruction and degradation of environments at the planetary level and the continued pumping of greenhouse gases into the atmosphere constitute *ecocide*. It is harmful, and it is wrong, and it is the most significant crime of the twenty-first century.

Enter Ron Kramer. For many years now, Ron has been at the forefront of criminological work that casts a critical eye over global warming and its consequences. This book is the culmination of part of his overall project. It is both an exposition and intervention.

At the heart of Ron's argument is the observation that climate disruption does not happen by chance, accident, or simply because of human activities in general. Rather, it is corporate–state collusion that is mostly to blame for perpetuating global warming and for delaying action to prevent or forestall further climate change.

Global warming is accelerated by the activities of governments, corporations, and individuals that rely upon or involve pumping greenhouse gases into the atmosphere. It is also fostered by the failure of governments to regulate carbon emissions, for example, letting the dirty industries continue to do what they do best—which is to continue to profit from irresponsible and destructive behaviors.

The harms are demonstrable and the culprits identifiable. For Ron Kramer this is a clear example of state–corporate crime of the highest order. This book tells us how and why this is the case. This is public criminology at its best.

The book provides an exploration of how the perspective of state–corporate crime can be utilized to unpack and interpret crimes of the powerful in relation to climate change. As the book unfolds, we can see that this analysis has many different layers and dimensions, which allows the reader to fully appreciate the terms and concepts that best explain the travesty that is climate injustice in the U.S. context (but which nevertheless has global implications and consequences). Moreover, in taking the reader on this journey, Ron Kramer draws inspiration and material from a wide range of sources (academic, government, United Nations reports, journalists and other commentators, activists), all of which contribute to the complex mosaic that portrays the dynamics of climate change state–corporate crime.

The carbon criminals are those who pretend that climate change is not happening or who believe that climate policy should not take precedence over immediate economic gain. Many are contrarians—eschewing scientific evidence in favor of bias and ill-informed opinion. Nothing will convince them otherwise because their specific sectoral interests override universal human and ecological interests.

The carbon criminals are those who continue to facilitate carbon emissions: governments that foster deforestation and massive oil, gas, and coal projects; corporations that construe energy policy as fundamentally about fossil fuels rather than alternative sources—these are the purveyors of future costs that already affect us in the here and now.

The carbon criminals are those who fail to prevent and stop the activities and policies that are killing the planet and life as we know it. Delayed action is in effect a green light to even greater climate disruption happening at an even greater pace.

This is a powerful book, which is timely given the counternarratives of the Trump administration and the prevalence and continuance of acts and omissions that contribute further to the phenomenon of climate change. For criminology it is a much-needed call to arms. In this, the more voices that speak out, the more systematic the analysis, and the more condemning of the perpetrators of harm, the more pressure that can be built in responding to the most pressing issue of our era.

I have known for some time about Ron's passion to pursue issues pertaining to climate change and why it has been the driver for the present work. He has a burgeoning "history" in this area and as such brings to the enterprise considerable knowledge and commitment. The motivation is both academic (insofar as the criminological framework of state–corporate crime "fits" this like a glove) and personal (in that Ron has children and grandchildren). There is an urgency about Ron's writing that is important and that reflects the actual urgency of global warming and climate change as it rapidly affects all our lives.

Failure to act, now, can only be described as criminal.

Rob White
University of Tasmania, Australia

Carbon Criminals,
Climate Crimes

CHAPTER 1

"This Was a Crime"

CLIMATE CHANGE AS A CRIMINOLOGICAL CONCERN

THE CLIMATE CRISIS IS the most urgent problem facing humanity. The purpose of this book is to analyze the critical issue of climate change from a criminological perspective. Before I turn to an examination of carbon criminals and climate crimes, however, it is important for me to sketch out how I came to focus on the issue of global warming, and in the process became a "green" criminologist.

I was born in a small town in Ohio and raised in a working-class family. I grew up in the post–World War II era when economic growth and environmental destruction accelerated and the global American Empire fully emerged. In high school I began to develop a political consciousness due to the Civil Rights struggle and mounting protests against the War on Vietnam. As a college student, I joined in the protest movement against that war, which I considered immoral and illegal. In May 1971, my friends Rick and Dean and I drove all night from the University of Toledo in Ohio to Washington, DC, to attend a large antiwar rally. That was the first of many trips I would make to the Capital to engage in various protest activities. It was the beginning of my political engagement with serious social problems, and it came about at the same time that I started to study criminology as an undergraduate student.

From the University of Toledo, I went on to graduate school at the Ohio State University and became a sociological criminologist. As I was finishing my doctoral dissertation, I developed an interest in white-collar crime due to the Ford Pinto scandal, which was much in the news at that time. After joining the faculty of the Sociology Department at Western Michigan University in 1978, I started to teach about and conduct research on white-collar, corporate, and state crimes; but environmental harms were not yet on my criminological radar. When I became a parent in the early 1980s, I developed a deep concern about the nuclear threat and wrote about the "sociology of nuclear weapons" (Kramer and Marullo, 1985) and eventually the "crimes of the American nuclear state" (Kauzlarich and Kramer, 1998). As I carried out academic studies of the crimes of the powerful, I also continued to be involved in a variety of peace and social justice organizations and movements. While I was certainly in favor of protecting the environment, I still did not yet see myself as an environmentalist, and I knew little about the emerging field of green criminology (the study of environmental crimes).

As I worked on war and peace issues over the years, both as a scholar and an activist, I gradually became aware of something called the greenhouse effect. I knew it was an important problem but due to other commitments did not pursue it. As time went on, however, I kept hearing more and more about greenhouse gas emissions, global warming, and climate change, and I knew I had to educate myself about these topics. Finally, during the summer of 2010, I took a fateful plunge as I dove into a pile of books about global warming and associated climate disruption. Little did I know, I was about to become a green criminologist engaged in the analysis of environmental harms.

Before starting this reading project, I knew generally that anthropogenic (human-caused) global warming and the climatic changes that result were grave social and political problems, but I knew little about the specifics concerning these issues. What I read that summer and beyond about greenhouse gas emissions and the heating of the planet would have a profound personal impact on me and lead me to take on the study of climate change as a criminological concern, the focus of this book.

When I started to read the global warming literature, two things stood out. First, I was struck by how grave the situation really is. Of course, I already knew that climate change was a looming threat. But until then I did not fully appreciate the gravity of the problem. I had not fully appreciated that the climate crisis threatened human civilization itself, what the elegant scholar Robert Heilbroner (1974) called the "human prospect." As American journalist Mark Hertsgaard (2011b: 247) observes in *Hot: Living through the Next Fifty Years on Earth*, everyone who finally "gets it" about climate change has an "oh, shit" moment: "an instant when the pieces all fall into place, the full implications of the science at last become clear, and you are left staring in horror at the monstrous situation humanity has created for itself." My "oh, shit" moment came that summer of 2010, and I concluded that systemic climate change induced by human-caused global warming is *the most urgent threat, the most important political issue, and the greatest moral challenge* that the world faces today.

The second thing that struck me from the readings I did that summer (and subsequently) was how often those who commented on climate change used the word "crime" in discussing the topic. The final chapter in Hertsgaard's book is titled "This Was a Crime," the declaration of Nobel Prize–winning German physicist Hans Joachim Schellnhuber about the United States government's failure to address global warming in the early years of the twenty-first century. According to Schellnhuber, the former chief climate advisor to the government of Germany, the current inability of the international political community to respond effectively to the alarming climate crisis caused by global greenhouse gas emissions "is the result of the lost decade under George W. Bush, the *crime* [emphasis added] of not taking action these past ten years" (Hertsgaard, 2011b: 254).

As a criminologist who studies corporate crime, state crime, and wrongdoing at the intersection of business and government (what Ray Michalowski and I have called "state–corporate crime"), I was very intrigued by Schellnhuber's remark. And, as I soon discovered, he was not alone in using the language of crime and criminality to refer to global warming and climate disruption. In the climate change literature, references to social phenomena—like state failures to mitigate greenhouse gas emissions, socially organized denial of climate change, and the continued extraction of oil, gas, and coal by fossil fuel corporations—as "crimes" were so frequent and striking that I eventually started compiling a list (a list that I add to on a regular basis). Here are some samples (emphasis added throughout).

John Sauven, who was the executive director of Greenpeace UK, stated after the failure of the 2009 United Nations conference on climate change in Copenhagen that "The city of Copenhagen is a *crime scene* tonight, with the *guilty* men and women fleeing to the airport. There are no targets for carbon cuts and no agreement on a legally binding treaty" (quoted in Dyer, 2010: 208).

Penn State University climate scientist Donald Brown (2010: 2), speaking of those who reject or deny climate science, argued that "We may not have a word for this *type of crime* yet, but the international community should find a way of classifying extraordinarily irresponsible scientific claims that could lead to mass suffering as some type of *crime against humanity*."

Journalist Wen Stephenson (2015: xiii), in *What We're Fighting for Now Is Each Other: Dispatches from the Front Lines of Climate Justice*, asserts that "to deny the science, deceive the public, and willfully obstruct any serious response to the climate catastrophe is to allow entire countries and cultures to disappear. It is to *rob* people, starting with the poorest and most vulnerable on the planet, of their land, their homes, their livelihoods, even their lives—and their children's lives, and their children's children's lives. For profit. And for political power. There's a word for this: these are *crimes*. They are *crimes against the earth*, and they are *crimes against humanity*."

Mark Karlin (2014: 1), editor of the website Buzzflash, states that "the fossil fuel companies commit so many egregious *crimes against life and the planet*—and the federal government is so *complicit* in playing down the damage—that a numbing sets in for most of the United States population."

Tom Engelhardt (2014: 142), editor of TomDispatch.com, argues, "If the oil execs aren't terrarists [*terra* is the Latin word for Earth], then who is? And if that doesn't make the big energy companies *criminal enterprises*, then how would you define that term? To destroy our planet with malice afore-thought, with only the most immediate profits on the brain, with only your own comfort and well-being (and those of your shareholders) in mind: Isn't that the *ultimate crime?*"

In *This Changes Everything: Capitalism vs. the Climate*, Canadian journal-ist Naomi Klein (2014: 456) proclaims that "Heading an oil company that actively sabotages climate science, lobbies aggressively against emission con-trols while laying claim to enough interred carbon to drown populous nations like Bangladesh and boil sub-Saharan Africa is indeed a *heinous moral crime.*"

In *Shadow Sovereigns: How Global Corporations Are Seizing Power*, Susan George (2015: 154) asserts, "If the *climate criminals* were only out to destroy America and America's children, we could say that was the Americans' prob-lem, but they are holding the entire human family to hostage, not to men-tion every other living thing on earth. I don't know what can be done about them, but at the very least they must be designated as *criminals* and we need an International Court for *crimes against the environment and the human race.*"

Canadian environmental scientist and climate activist David Suzuki, in an interview with American journalist Bill Moyers, stated that "Our politi-cians should be thrown in the slammer for willful blindness. . . . I think that we are being willfully blind to the consequences for our children and grandchildren. It's an *intergenerational crime*" (B. Baker, 2014: 1).

Hertsgaard (2017: 11), commenting on President Trump's decision to withdraw from the Paris Agreement, observed: "*Crimes against Humanity* is a phrase to use with caution, but it fits Trump's repudiation of the accord—and, indeed, his entire climate policy."

After hurricanes Harvey and Irma struck Texas and Florida in late sum-mer 2017, May Boeve, executive director of the climate activist organ-ization 350.org, argued in a press release that "Right now, the people who are paying for the climate crisis are those who have done the least to cause it. Instead of putting the burden on low-income and communities of color, we should be holding these fossil fuel billionaires accountable for the dam-age they've done. Climate change isn't just a crisis, it's a *crime*. It's high time to hold the *criminals* accountable."

To a criminologist, the language of crime in these statements com-mands attention. Yet, none of these people are themselves criminologists. None of them engage in the systematic or scientific study of crime and criminals. Still, all of them felt compelled to use the language of crime and criminality when commenting on the issues of global warming and climate change. In each of these statements, the concept of "crime" is used as a symbolic device or expressive term to convey a deep concern over actions regarded as moral transgressions or blameworthy harms (acts that are evil, shameful, or wrong). And, in the opinion of these writers, those harms should be publicly condemned and perhaps legally sanctioned. As criminologist Robert Reiner (2016: 5) observes, "To call something a crime is to register disapproval, fear, disgust or condemnation in the strongest possible terms and to demand urgent remedies—but not necessarily the pain of criminal penalties." Concerning climate change, the quoted comments can be viewed as rhetorical moves within the larger social and political effort to mitigate carbon emissions, limit global warming, and respond to climate disruption.

As a criminologist, I was both impressed and galvanized by this use of the language of crime. I started to read about climate change because I wanted to better educate myself as a concerned citizen. I did not set out to study global warming as a criminological topic. But the more I came across this language in the climate change literature, the more I started to think about it as a criminological question. That inclination was strengthened when I reflected on the specific content of the comments. Many of them focus on the failure of the United States government, or the international political community, to mitigate global warming, that is, to effectively reduce the greenhouse gas emissions that are responsible for the warming of the planet. Some of them target the organized climate change denial countermovement that has played a powerful role in blocking effective mitigation efforts. Many of them point out the central role that corporations in the fossil fuel industry play in funding and organizing the denial movement. Many also note that these same companies continue to extract and market fossil fuels even in the face of the overwhelming scientific evidence of the catastrophic consequences of the greenhouse gas emissions that result from their consumption. Some of the commentators argue that these corporate and state actions (or omissions, that is, the failure to act when one has a legal or moral obligation to do so) constitute what the international political community designates as "crimes against humanity" or represent "intergenerational crimes" (serious threats to today's children and future generations).

After reading these books and articles, coming to appreciate the existential threat to the human prospect posed by global warming, and being struck by the fact that this literature is replete with references to corporate and state criminality, I decided to investigate global warming and climate change from a criminological perspective—and with that I became a green criminologist. Green criminology, a term coined by criminologist Michael J. Lynch, emerged in the early 1990s (Lynch, 1990; Lynch and Stretesky, 2003;

Lynch, Long, Stretesky, and Barrett, 2017) as "a critical and sustained approach to the study of environmental crimes and harms" (R. White and Heckenberg, 2014: 7). My objective within this field, and within this book, is to identify, describe, and explain environmental and social harms that I call *climate crimes*: organizational acts (and omissions) that are responsible for causing global warming, that deny climate change is real or humanly caused, that fail to mitigate greenhouse gas emissions, or that adapt to climate disruptions in militaristic and unjust ways. I will also offer some ideas for the extremely difficult task of responding to and reducing these various blameworthy harms and holding accountable the *carbon criminals* who commit them. As a scholar who has spent most of my career analyzing "the crimes of the powerful" more generally, I decided to examine this topic through the lens of the concepts of corporate crime, state crime, and the intersection of the two, *state–corporate crime* (Michalowski and Kramer, 2006). As criminologists Rob White and Diane Heckenberg (2014: 112) have argued, "The question of justice in relation to climate change inevitably leads one to consider the nature and dynamics of state–corporate crime." Since most corporate and state crimes related to climate change are the result of the symbiotic nexus of relationships between business corporations and government entities, I will generally use the term *state–corporate crime* to refer to all of these forms of organizational crime.

CLIMATE CHANGE AS STATE–CORPORATE CRIME

The concept of state–corporate crime refers to serious social harms that result from the interaction of political and economic organizations. The idea emerged out of the recognition that some organizational crimes are the collective product of interactions between a business corporation and a state agency engaged in a joint endeavor. State–corporate crime analysis seeks to breach the conceptual wall between economic crimes and political crimes to create a new lens through which we can examine the ways serious illegal acts and profound social injuries often emerge from intersections of economic and political power. As Ray Michalowski and I (Michalowski and Kramer, 2007: 201) have noted, "Contemporary social scientists have largely forgotten what our 19th century counterparts knew so well. There is neither economics nor politics; there is only *political-economy* [emphasis added]."

In *State–Corporate Crime: Wrongdoing at the Intersection of Business and Government* (Michalowski and Kramer, 2006: 15), we formally defined state–corporate crime as "illegal or socially injurious actions that result from a mutually reinforcing interaction between (1) policies and/or practices in pursuit of the goals of one or more institutions of political governance and (2) policies and/or practices in pursuit of the goals of one or more institutions of economic production and distribution." As this definition makes clear, we proposed extending the scope of criminology beyond legal definitions (an issue I will address later in this chapter) to incorporate harmful blameworthy social actions that may not currently violate either criminal or regulatory laws at the state level. While the concept of state–corporate crime could be applied to illegal or other socially injurious actions in societies ranging from private production systems to centrally planned political economies, most of the research to date has focused on state–corporate crimes within the private production system of modern neoliberal capitalism (Tombs, 2014). State–corporate crimes within a global capitalist political economy involve the active participation of two or more organizations, at least one of which is in the civil sector and one of which is in the state sector. This book will extend the state–corporate crime framework to the study of the critical role of corporations and political states in both promoting the release of greenhouse gases and refusing to seriously address the resulting consequences of global warming and planetary climate change.

As I continued to read the climate change literature after that summer of 2010, I did so with an eye for empirical evidence on what corporations in the fossil fuel industry, the U.S. government, and the international political community did to cause global warming or did not do to mitigate carbon emissions or adapt in just and progressive ways to emerging climate disruptions. I discovered that several sociologists and criminologists had already started to focus on global warming and climate change (Agnew, 2011a; Derber, 2010; Dryzek, Norgaard, and Schlosberg, 2011; Giddens, 2011; McNall, 2011). This literature analyzed the social causes, social consequences, and politics of climate change, with criminologists focusing on the relationship between global warming, violence, and crime. Reading the literature on green criminology and environmental crime further revealed that a few green criminologists (Burns, Lynch, and Stretesky, 2008; Lynch and Stretesky, 2010; R. White, 2010, 2011) had begun to study climate change from a criminological perspective and that some of them had even argued that the concept of state–corporate crime "provides a useful tool for examining" crimes related to global warming (Lynch, Burns, and Stretesky, 2010: 215).

After reviewing the sociological and criminological literature on climate change, I presented a series of academic papers on global warming as state–corporate crime at various professional conferences that were subsequently published as book chapters or journal articles. One paper, (Kramer and Michalowski, 2012), titled "Is Global Warming a State–Corporate Crime?," was published in a volume edited by Rob White, *Climate Change from a Criminological Perspective* (2012), which demonstrated that criminologists do indeed have much to contribute to the overall analysis of the issue of climate change on a wide range of topics. In 2013, White and I organized a session at the annual meeting of the American Society of Criminology in

Chicago and then co-edited a special issue of the *Journal of Critical Criminology* (R. White and Kramer, 2015a), both with that same title.

Criminological interest in global warming continues to grow. New books and journal articles on green criminology offering insights on climate crimes appear on a regular basis, as Rob White (2018) reports in a major new work, *Climate Change Criminology*. Outside of the discipline of criminology, a group of scholars and activists issued a manifesto in summer 2015 titled "Act to Stop Climate Crimes," which called for mass action in Paris during the December 2015 climate conference "to declare our determination to stop climate crimes and keep fossil fuels in the ground." Peter Carter and Elizabeth Woodworth (2018) have published a book about climate change denial as "an unprecedented crime." One objective of this book is to blend the emerging criminological scholarship on global warming with the research and analysis carried out by other scholars and environmental organizations in the public sphere (such as 350.org, Greenpeace, and the Union of Concerned Scientists) and, in doing so, to establish that climate change is a form of crime, describe the nature and extent of specific climate crimes, explain the social structural causes of these crimes, and explore some ideas about social, political, and legal actions that can be undertaken to prevent or control (stop) climate crimes.

CONTRIBUTIONS OF THE BOOK

A systematic examination of "climate crimes" can make three important contributions. First, it can contribute to the discipline of criminology by enlarging its conceptualization and enhancing its analysis of "the crimes of the powerful" (Barak, 2015; Rothe and Kauzlarich, 2016). As criminologist David Friedrichs (2010) points out, traditionally within the field there has been an "inverse relationship" between harm and criminological work; that is, the greater the harm of an act, the less research criminologists conduct on that topic. While the discipline is still dominated by the study of traditional "street" crimes, the crimes of the powerful have been an important topic of criminological inquiry since sociologist Edwin H. Sutherland (1940) first advanced the concept of white-collar crime in his 1939 Presidential Address to the American Sociological Society in Philadelphia. Scholarship on various forms of white-collar crimes, including organizational, corporate, state, political, and state–corporate crimes, has grown tremendously since the 1970s (Friedrichs, 2010; Michalowski and Kramer, 2006; Rothe and Kauzlarich, 2016). By enlarging the substantive content of this important academic subfield within the discipline of criminology, research and theory on global warming and climate change as forms of corporate, state, or state–corporate crime can make a significant theoretical contribution.

Creating theoretical frameworks to recognize and analyze organizational actions related to global warming and climate change as state–corporate crimes can also create a systematic foundation for addressing them through public policy. Thus, a second contribution that can be made by an analysis of climate crimes is that by focusing on the systems of heretofore unaccountable power (some private, some governmental) that are responsible for the serious harms of global warming, we can spur public conversation and political debate about how society might finally hold carbon criminals (corporations in the fossil fuel industry and complicit government agencies) legally and morally accountable for these crimes and achieve forms of climate justice for victims. Journalist Kate Aronoff (2019), in "It's Time to Try Fossil-Fuel Executives for Crimes against Humanity," argues that these executives are "mass murderers." While this is an understandable position, most criminologists contend that these crimes are engaged in by organizational actors, and they do not necessarily advocate for the criminal prosecution and imprisonment of specific fossil fuel corporate executives or negligent government officials who occupy positions within these organizational structures. But as Australian criminologist John Braithwaite (1989: 4) has argued, "potent shaming" is "the essential necessary condition" for lowering crime. Thus, the proper shaming of carbon criminals (organizations or individual corporate executives and government officials) may be essential and necessary to reduce climate crimes and achieve climate justice. Using the concepts of carbon criminals, climate crimes, and climate justice can be an important way to frame the broader issue of global warming and promote strategies of political change (Hadden, 2017). Labeling the shameful actions of these corporations and states that cause climate disruption and threaten human civilization as *crimes* may generate moral outrage and, as American climate activist Bill McKibben (2012a: 9) observes, moral outrage may "spark a transformative challenge to fossil fuel" and just might "give rise to a real movement."

A third and related contribution is that an examination of the social structural causes of climate crimes can help a real climate movement identify and promote specific political actions and policies that are necessary to mitigate greenhouse gas emissions and construct just forms of adaptation to climate disruptions. Responding to the climate crisis requires both an understanding of the institutional forces that drive these crimes and, as McKibben (2012a) points out, a broad-based global political movement that can address these social forces and produce the necessary structural changes. Swedish ecologist Andreas Malm (2016: 394) argues that such a global climate movement should be "the movement of movements, at the top of the food chain, on a mission to protect the very existence of the terrain on which all others operate." Naomi Klein (2014: 10) notes that a mass social movement to address climate change can be a catalyzing force for advancing a broader progressive political agenda: "As part of the project of getting our emissions down to the levels many scientists recommend, we once again have the chance to advance policies that dramatically improve lives, close the gap between rich and poor, create huge numbers of good jobs, and reinvigorate democracy from the ground up." My hope is that the concept of carbon criminals and the analysis of climate crimes may be important catalysts for building such a social movement. And, with the presidency of Donald Trump, his ignorant rejection of climate science, his irresponsible appointment of climate deniers to key cabinet positions, and his reckless dismantling of existing efforts to mitigate carbon emissions, a mass political movement working for climate justice is even more critical now if we are to avoid climate catastrophe.

PUBLIC CRIMINOLOGY AND
THE PROPHETIC VOICE

Holding carbon criminals (organizations or individuals) accountable for their climate crimes, generating moral outrage at the social and environmental harms caused by the continued emission of greenhouse gases by the fossil fuel industry, and working for social movements and public policies that reduce global warming and promote climate justice can be viewed as forms of *public criminology*. According to American criminologist Elliott Currie (2007: 175), a public criminology "takes as part of its defining mission a more vigorous, systematic and effective intervention in the world of social policy and social action." Sociologist Michael Burawoy (2007: 28) has argued that public sociology (including criminology) "brings sociology into a conversation with publics, understood as people who are themselves involved in a conversation." Following these conceptualizations, I argue that a public criminology of state–corporate crimes (such as climate crimes) would attempt to systematically and effectively intervene in the world of social policy and social action by seeking out nonacademic audiences and entering into conversations with various publics. These publics could include the victims of state–corporate crimes, environmental organizations, the international political community, nongovernmental organizations (NGOs), state agents, media organizations, and other more generalized, amorphous "publics," such as the generic collective known as "the American people" that politicians love to address. The content of these conversations will be quite varied but would be, at the more abstract level, a dialogue about the moral and political implications of criminological research findings and theoretical explanations concerning state–corporate crimes related to climate change. More concretely, these conversations may help to overcome the fact that, as British environmentalist George Marshall (2014) points out in *Don't Even Think about It*, for a variety of reasons like our common psychology, our perception of risk, and our deepest instincts to defend our family and tribe, "our brains are wired to ignore climate change."

A public criminology of state–corporate crime can take several different forms. Again, following Burawoy (2007), I distinguish between *traditional* public criminology and *organic* public criminology. Traditional public criminology attempts to effectively intervene in the world of social policy and action by initiating a conversation, instigating a debate, or provoking a critical questioning within or between publics through the publication of books and articles addressed to audiences outside the academy or through opinion pieces in national or international newspapers or social media that identify and analyze state–corporate crimes or comment on important public issues related to such crimes. This book is one such example. Organic public criminology, on the other hand, involves criminologists working directly with specific groups, organizations, movements, or state officials and engaging in a dialogue or a process of mutual education that may or may not lead to specific political policies or actions related to the prevention or control of state–corporate crimes.

This book expands on these conceptions of public criminology by using a slightly different language. I argue that criminologists, as part of their professional role, can and should assume two important responsibilities in the larger struggle to respond to the climate crisis and other state–corporate crimes and the serious harms they cause. First, criminologists should take seriously the responsibility to *speak in the prophetic voice* concerning state and corporate crimes (traditional public criminology); and second, they should take the responsibility where they can to *engage in social and political actions* to reduce or prevent these harms (organic public criminology). The language here comes from American journalist and scholar Robert Jensen's *All My Bones Shake: Seeking a Progressive Path to the Prophetic Voice*. As Jensen (2009: 162–163) argues, given the "cascading crises" humanity faces, "It is time to recognize that we all must strive to be prophets now. It is time for each of us to take responsibility for speaking in the prophetic voice."

"Prophecy" in this sense does not mean predicting the future. To speak prophetically regarding state and corporate crimes in general and to climate crimes specifically is to identify or name the harms committed by corporations and states as "criminal," call out the social injustices that these organizations produce or tolerate, confront the abuses of powerful officials, and analyze how political and economic systems cause destruction, devastation, and untold suffering. By speaking in the prophetic voice, criminologists attempt to break through the denial and normalization of state and corporate crimes, critique the structural and organizational forces that give rise to them, and create political or "deliberative frames" (Wilson, 2009) that can orient debate on these issues.

To speak in the prophetic voice carries a responsibility to then intervene or act in the political arena to effect structural or policy changes. To engage in progressive political action involves, among other things, organizing or participating in peace, environmental, or transitional justice movements, but it also involves challenging empire, contesting the power of the corporate state, working to reinvigorate democratic governance, and enhancing the power and control of international political and legal institutions. By engaging in such "political" actions (regarded as heresy by conventional guidelines for conducting criminology), criminologists can raise awareness of the obstacles to and possibilities for progressive social change and perhaps contribute to the construction of collective solutions to state–corporate criminality.

By speaking in the prophetic voice to publicly identify, sociologically analyze, and politically frame major forms of climate crimes, criminologists can assist public and legal efforts to prevent and control harmful state and corporate actions related to global warming. As I will analyze throughout this book, these corporate and state actions threaten to destroy the biosphere and annihilate most forms of life on earth. Speaking in the prophetic voice would require criminologists to identify and describe the actions of states and corporations that are driving us to this dystopian future as "criminal" and then explain the structural and cultural forces that drive them. By speaking in the prophetic voice about climate change in this way, criminologists can help orient public debate and generate effective political action to limit global warming and the associated environmental and social harms resulting from climate disruption. We can, as Bill McKibben (2019b: 17) puts it, help keep the "human game going."

As the quotes I have offered indicate, people often use the concept of crime as a symbolic device to call attention to a grievous moral harm or an evil, shameful act that they think should be publicly condemned and legally sanctioned. The people quoted all believe that something—some systematic and effective intervention in the world of social policy and social action— needs to be done about the climate crisis, and they are each engaged in a symbolic crusade to help bring about a transformative international social movement with the political power to drastically reduce greenhouse gas emissions, limit global warming, transition to clean, renewable energy, and adapt in just ways to the climate disruptions already under way. This book argues that criminologists can play an important role in assisting this political movement by using the powerful conceptual and symbolic language of "crime" as well as the analytic tools of criminology to bring greater public attention to the climate crisis and help generate moral outrage at the relationship between governments and the fossil fuel industry that allows the destructive forces of catastrophic climate change to continue. By analyzing these state–corporate actions as "climate crimes," we can attempt to, in the words of Adrian Parr (2012: 2), "interrupt and contest the historical and institutional conditions that regulate and organize the frames of reference through which we think and act."

Research shows that humans give the greatest salience to threats that are concrete, immediate, and indisputable (Kahneman, 2011). By labeling and analyzing climate change as a crime, we may perhaps be giving greater salience to the threat of climate change than it might normally have for the public. As a precedent for this effort, criminologists can reflect on Sutherland's pathbreaking work on white-collar crime. Gil Geis and Colin Goff (1982: 18) have observed that "On the podium in Philadelphia in 1939, what Sutherland really said in his key speech as President of the American Sociological Society—once the camouflage is removed—is that white-collar crime is wrong—indeed, that often it is despicable—and that sociologists and economists ought to pay close attention to such matters and join with him in a crusade to do something about them." Following Sutherland, I argue that criminologists today must act as public criminologists; that is, we must speak in the prophetic voice about climate crimes, generate moral outrage about these existential threats, provide a critical frame to analyze these destructive corporate and state actions, and join in the global political crusade "to do something about them."

CRIMINOLOGISTS AND THE DEBATE OVER THE DEFINITION OF CRIME

If climate change is to be considered a criminological concern, we need to explore what criminologists mean by the concept of crime and set out a working definition of the term. This is not an easy task. As Reiner (2016: 3) notes, "crime is an essentially contested, conflict-ridden concept that is treated as if it were essentially uncontested and consensual." Most conventional *Introduction to Criminology* textbooks define crime simply as behavior that violates the criminal law, and then they give no further consideration to what can be an enormously complex issue. In the academic literature on criminology, there is a long-standing and controversial "debate" over the definition of crime. This debate is about several different things. First, it is important to note that two different abstract, fundamental images or definitions of crime (paradigms) guide criminological inquiry. The most common image or definition of crime views it as a distinctive form of harmful behavior. The major debate within this "behavioral" or "realist" framework is whether legal or social norms or standards should be used to classify behavior as criminal for the purpose of doing criminological work. I will return to this important debate shortly, but it is instructive to start with the lesser used "labeling," "constructionist," or "social definitional" framework.

The social definitional approach starts with the premise that no behavior is inherently criminal. As the labeling theorists in criminology argued, there is no act that is, in and of itself, criminal. Criminality is not a quality that resides within behavior or persons. If crime is not intrinsically real and objectively given, then it follows that in any arena we can identify (political or scholarly, for example), some subjective, social definitional process must be used to construct the existence of *crime*. And since "criminal" behavior is not pre-existent, there must be some procedure that can be used to identify acts that are to be considered criminal (for some purpose) and acts that are not.

Most political jurisdictions, for example, must develop some legislative or juridical process to select out certain behaviors (and persons) to define as criminal and subject to formal social control. Thus, legislative bodies pass criminal laws (thereby creating crime or criminal labels). Police agencies enforce the laws and usually initiate the process of applying criminal labels to specific individuals by making arrests. Prosecutors bring formal charges against those arrested on the basis of these laws and take accused offenders into criminal courts where they can be officially convicted (effectively creating criminals by applying the legal label to them). Judges interpret the laws and hand down legal sanctions. Importantly, these laws and their supporting legal institutions are not value-free. They are instead rooted in the moral values and concrete interests of those who create them or of their political and economic supporters. It is this overall criminalization process that is the subject of much of the work carried out within the social definitional or political labeling framework (Hartjen, 1978; Kramer, 1982). The primary focus of this research is on the discretionary decisions of the legal agents who create laws and apply criminal labels to people, not on criminals or criminal behavior.

Most criminologists, however—as well as most concerned citizens—are more interested in focusing on "bad guys" and the harmful "criminal" acts they engage in, such as murder, rape, and robbery (and, in this book, the corporate and state acts that I call climate crimes). Returning to the more traditional behavioral or realist framework, we find that just as state authorities follow a subjective, value- and interest-driven social process to classify behavior as criminal for the purposes of formal social control, so too must criminologists who work within the behavioral approach follow a similar process to classify behavior as criminal for the purposes of scientific study. Criminologists possess no value-free mechanism to create these definitions of crime. Thus, criminologists face a dilemma. They must choose which set of standards they are going to use or privilege to classify behavior as criminal. It is this dilemma within the behavioral framework that has generated the long-standing debate over the definition of crime. And white-collar crime, in its various forms, has often been at the heart of that debate.

Within the discipline of criminology, the traditional and easiest way to resolve the dilemma over which standards or norms of conduct should be used to classify behavior as criminal for the purpose of study is to utilize the set of legal standards created by the state; thus, most conventional criminologists define crime as behavior that violates the criminal law. And there is usually no specification that this classification requires an actual enforcement action (arrest or conviction). Furthermore, as criminologist Robert Agnew (a past president of the American Society of Criminology) points out, this critical decision that "determines the scope of the discipline" and "has a major effect on the content of the discipline as well" is usually "taken for granted" (Agnew, 2011b: 13–14). Conventional criminologists, Agnew (2011b: 13–14) argues, "spend surprisingly little time discussing the actual definition of crime," and he notes that "the legal definition . . . is so pervasive in mainstream criminology that it is no longer necessary to present or defend it."

Those criminologists who see themselves as critical criminologists (myself included) have long challenged this approach to defining the concept of crime. In an earlier work (Kramer, 1982), I argued that by choosing the legal definition of crime, criminologists tacitly agree to use the moral values and interests (usually of dominant political and economic groups) that are encoded in state criminal law. The traditional criminal law under capitalism contains a distinct class bias (Michalowski, 1985; Reiner, 2016; Schwendinger and Schwendinger, 1977). This not only results in the greater criminalization of conventional "street" crimes and the working- and lower-class individuals who tend to be disproportionately involved in these crimes, but such a definition of crime also restricts criminologists from studying other types of socially harmful and morally blameworthy acts, especially those perpetrated by the powerful—by corporations and capitalist states (such as climate crimes). As criminologists Larry Tifft and Dennis Sullivan (1980: 6) have noted, "By assuming definitions of crime within the framework of law, by insisting on legal assumptions as sacred, criminologists comply in the concealment and distortion of the reality of social harms inflicted by persons with power."

The critical criminological proposition that the existing criminal law reflects the values and interests of dominant economic and political groups under capitalism has generated much controversy over the years. After carefully reviewing the arguments and evidence for and against this proposition, Agnew (2011b: 18) draws the cautious conclusion that even though "parts of the criminal law do appear to reflect the interests of all people," at the same time, "many harmful acts are not defined as crime, and certain relatively harmless acts are defined as crimes." He then goes on to support what I think is the most important point that critical criminologists make in this debate: "If the criminal law partly reflects the interests and values of dominant groups, it follows that certain harmful behaviors may not be defined as crimes, particularly those behaviors that serve the interests of dominant groups. Those *harmful acts committed by corporations and states* figure prominently here" (Agnew, 2011b: 18, emphasis added). Critical criminologists recognize that this reality requires them to construct and utilize alternative standards.

Before examining some of the more recent alternative standards proposed by critical criminologists to classify behavior as crime, I want to briefly note the critique of the traditional legal definition of crime offered by Edwin H. Sutherland in his famous work on white-collar crime. Like a few criminologists before him, Sutherland came to reject the criminal law as the sole basis for defining the concept of crime. His rejection of this traditional definition was not based on scientific or methodological criteria, as were those of his predecessors, but by his moral concern over the issue of white-collar crime and his desire to reform the legal system. In his attempt to bring the socially harmful and morally blameworthy behavior of businessmen and corporations within the boundaries of criminology, Sutherland felt compelled to define white-collar crime as "real crime" and to apply legal standards to such behavior. He argued that the conventional criminal law definition of crime must be supplemented, contending that "other agencies than the criminal court must be included, for the criminal court is not the only agency which makes official decisions regarding violations of the criminal law" (Sutherland, 1940: 6). Note that Sutherland was relying on actual legal decisions (enforcement actions) here and not just the judgment of the criminologist as to whether the laws under examination apply to the behavior in question. In his view, criminal convictions must be supplemented by data on decisions against businessmen and corporations rendered by civil courts and regulatory agencies.

In his empirical study of the seventy largest corporations in the country at that time, Sutherland found 779 civil and administrative decisions that he concluded were also "crimes." To justify his conclusions concerning these civil and administrative actions, he reformulated the definition of crime, writing that "the essential characteristic of crime is that it is behavior which is prohibited by the State as an injury to the State and against which the State may react, at least as a last resort, by punishment. The two abstract criteria generally regarded by legal scholars as necessary elements in a definition of crime are legal descriptions of an act as socially harmful and legal provision of a penalty for the act" (Sutherland, 1949: 32). Note that this is still a state-based "legal" definition of crime. But it is an expanded legal definition of crime—expanded to include civil and regulatory violations in addition to violations of the traditional criminal law—and does not require that an actual legal decision be made.

In his consideration of a set of standards to classify behavior as criminal for the purpose of study, Sutherland adheres to legal boundaries but expands the nature of the legal norms that can be used. Sutherland was attacked for his attempt to expand the legal definition of crime, most notably by lawyer and sociologist Paul Tappan (1947) in his famous article, "Who Is the Criminal?" Tappan, by the way, insisted that crime could be defined only in reference to an actual criminal court conviction. At the time, most conventional criminologists (like Tappan) remained unconvinced by Sutherland's attempts to expand the definition of crime outside the realm of criminal law. Within the study of white-collar crime, however, Sutherland's use of civil and regulatory law to classify behavior as crime eventually came to be widely accepted. Most criminologists who study white-collar crime today accept the idea that civil and regulatory legal standards can be used to define crime for the purpose of scientific study, often without the additional requirement that an actual legal decision be made or state punishment inflicted.

With the rise of a radical/critical criminology in the 1960s and 1970s came renewed attacks on the traditional legal definition of crime. Radical criminologists such as Herman and Julia Schwendinger (1977: 8) argued that "Legal definitions of crime are ideological instrumentalities which shape and develop the language and objectives of science in such a way as to strengthen class domination." Radical criminologists pointed out that legal definitions under capitalism confine criminologists to study only those behaviors sanctioned by the capitalist state as criminal, while excluding other types of socially harmful behaviors from analysis, especially those committed by the state itself and other powerful actors. Thus, American criminologist Tony Platt (1974: 5) asserted: "A radical criminology requires a re-definition of subject matter, concerns and commitments. . . . We need a definition of crime which reflects the reality of a legal system based on power and privilege; to accept the legal definition of crime is to accept the fiction of neutral law."

The most noteworthy attempt to redefine crime from the perspective of radical criminology from the Schwendingers (1970), who proposed a social definition of crime based on the existence of fundamental, historically determined human rights. They proclaimed that "All persons must be guaranteed the fundamental prerequisites for well being, including food, shelter, clothing, medical services, challenging work, and recreational experiences, as well as security from predatory individuals and repressive and imperialistic social elites" (Schwendinger and Schwendinger, 1970: 145). To the Schwendingers, these material requirements and basic services are not to be regarded as rewards or privileges but as basic human rights whose deprivation is a crime.

A few criminologists who followed the Schwendingers down the path toward a human rights definition of crime did so by making explicit reference to specific forms of public international law in their work (something the Schwendingers did not do). John Galliher (1991: 2–3), for example, suggested that the essential human rights to be protected could be found in the Charter of the United Nations, the Declaration of Independence, and the United Nation's Universal Declaration of Human Rights. Gregg Barak (1991: 8), concerned with gross human rights violations by the capitalist state, noted that "higher criteria for establishing state criminality exist in various international treaties and laws." Two colleagues and I (Kramer, Kauzlarich, and Smith, 1992; see also Kramer, 2013b) argued that to make the violence of states a more central focus of criminology, we should use the standards found in both international humanitarian law (the laws of war) and human rights law.

In addition to the efforts of the Schwendingers and others to ground the concept of crime in some broad notion of human rights, there have been several other noteworthy attempts by critical criminologists to reformulate the definition of crime outside of legal boundaries. For example, in their discussion of defining the state as criminal, Penny Green and Tony Ward (2004) use the term *organizational deviance* and stress the important role of various "social audiences." Michalowski developed an important social definition of crime with his consideration of "crimes of capital" (illegal or harmful acts that arise from the ownership or management of capital); he included not only acts prohibited under criminal and regulatory law but also what he called "analogous forms of social injury" (Michalowski, 1985: 317). Analogous social injuries refer to "legally permissible acts or sets of conditions whose consequences are similar to those of illegal acts." Thus, Michalowski's (1985: 317–318) examination of the crimes of capital focused on behaviors or conditions that "arise in connection with the process of capital accumulation and that, regardless of whether or not they are prohibited by law," result in "violent or untimely death, illness or disease, deprivation of adequate food, clothing, shelter, or medical care, or reduction or elimination of the opportunity for individuals to participate effectively in the political decision-making processes that affect their lives."

While Michalowski favors the concept of social injury, other scholars have utilized the notion of "social harm" in their efforts to create a broader social definition of behavior as crime. Tifft and Sullivan (2001: 191) discuss a "needs based, social harms definition" that "extends our definition of crime to social conditions, social arrangements, or actions of intent or indifference that interfere with the fulfillment of fundamental needs and obstruct the spontaneous unfolding of human potential." This definition allows them to focus attention on social structural harms, what Norwegian sociologist Johan Galtung (1969) called "structural violence," or preventable suffering and premature death that arises from structural inequalities. Green criminologists are among those who argue that the legal definition of crime is unnecessarily restrictive and needs to be supplemented by a broader harm-based approach. As Rob White (2011: 21) asserts, "A basic premise of green criminology is that we need to take environmental harm seriously, and in order to do this we need a conceptualization of harms that goes beyond conventional understandings." Following the lead of these scholars, I argue that to examine state–corporate crimes in general, and climate crimes in particular, criminologists must step outside the conventional understanding of crime and utilize a broader harms-based definition.

Defining Climate Crimes

To classify organizational behavior that causes global warming and climate change, or acts of political omission related to the climate crisis as crime, we need a broad and flexible definition that can incorporate many of the significant elements that have been part of the broader debate over the definition of crime within the field of criminology (domestic legal standards, international law, social injury and harm, human rights, social audience definitions). This book will use Agnew's "integrated" definition of crime, an important effort that combines norms of conduct, reactions of social audiences, and legal enforcement actions. As he notes, while "the integrated definition assigns a central place to violations of the criminal law and street crimes . . . it also focuses on a range of harmful acts that are not legally defined as crimes, *including acts committed by states and corporations*" (Agnew, 2011b: 30, emphasis added).

According to Agnew (2011b: 37), the three general characteristics that should be used to classify behavior as criminal are "the extent to which they are (1) blameworthy harms; (2) condemned by the public; and (3) sanctioned by the state." A morally blameworthy harm is an act that threatens physical security (or a failure to act to prevent harm when one has a moral or legal duty to intervene), is voluntary and intentional behavior, and is unjustifiable and inexcusable. Public condemnation means that there is an emotional reaction to the blameworthy harm and a desire by citizens to sanction the act in some way. Finally, the notion of being sanctioned by the state means that some type of legal action, broadly defined, is undertaken in response to the harmful act. Thus, Agnew (2011b: 38) argues that "any behavior classified as a blameworthy harm, subject to at least modest condemnation by a significant portion of the public, or classified as a crime or 'crime-like' civil violation by the state should be viewed as a proper part of the subject matter of criminology."

This integrated definition initially yields a list of what Agnew calls "core crimes," which are mostly traditional forms of street crime. Conventional criminologists generally confine themselves to the study of such behaviors. Among several other forms of crime that Agnew discusses in relationship to the integrated definition of crime, the most important category in my view is "unrecognized blameworthy harms," a crucial distinction that allows him to incorporate much of the work of those who advocate for a broader social definition of crime. Unrecognized blameworthy harms are behaviors that are not yet strongly condemned by the public at large and not sanctioned strongly by the state. Many climate crimes appear to fit this description. According to Agnew (2011b: 38), in general, "Much state and corporate harm falls into this category, since the power of state and corporate actors makes it easier for them to justify and excuse harm, hide harm, hide blameworthiness, and prevent state sanction." Bringing these harms within the boundaries of criminology is an important step toward Agnew's goal of expanding the core of the discipline. Yet, since they are often unrecognized by social audiences, bringing them within the boundaries of criminology requires that criminologists make an independent judgment or determination that such behaviors are morally blameworthy harms. But Agnew does not only think unrecognized blameworthy harms should be made part of the subject matter of criminology; he also thinks that criminologists, acting as public intellectuals, can provide an important service by bringing these crimes to the attention of social audiences. As Agnew (2011b: 38) argues, "In such cases, criminologists can play an important role in making the harm and/or blameworthiness apparent through their research and *advocacy* [emphasis added]." This seems to be a perfect illustration of criminologists speaking in the prophetic voice.

Speaking prophetically about climate crimes requires that we identify, describe, and explain a variety of blameworthy harms related to global warming and climate change and offer some ideas about how to respond to these crimes. The climate crimes I will focus on are primarily committed by corporations in the fossil fuel industry, nation-states operating within the international political community, military institutions, various other government agencies within states, conservative foundations, and ideologically oriented think tanks.

Following the integrated definition of crime developed by Agnew, I will first document the blameworthy harms that these organizational actors engage in related to global warming and climate change and then, where possible, assess the level of public condemnation of these harms and analyze the nature and extent of legal actions states and others have undertaken in response to the commission of these harms. While the level of public condemnation of climate harms has been rising, and more and more legal actions of various kinds have emerged in reaction, there are still many blameworthy harms related to global warming and climate change that, following Agnew, can be characterized as unrecognized. One of the objectives of this book is to help bring these unrecognized blameworthy harms to the attention of significant social audiences so that they might be subjected to stronger public condemnation and more effective legal sanctioning.

Four specific forms of state–corporate climate crime are identified and examined in the following chapters: crimes of continued extraction and rising emissions; crimes of political omission; crimes of socially organized denial; and climate crimes of empire, which include wars for oil, military greenhouse gas emissions, and forms of unjust and militaristic adaptation to climate disruptions. For each climate crime, the blameworthy harms will be described, the level of public concern will be assessed, and the social and legal actions that have been taken in response will be explored. And for each of these crimes I will also ask the theoretical question: Why? A theoretical analysis of climate crimes must examine the broader social structural forces—the political and economic factors—that explain why these crimes occur. Human economic systems of extraction and industrial production, and the political economy of global capitalism and the process of capital accumulation that cause ecological destruction, will therefore be explored in relation to each climate crime identified.

Overview of the Book

The climate crimes of continuing fossil fuel extraction and rising carbon emissions, political omission (the failure to reduce emissions), organized climate change denial, and climate crimes of empire will be examined in the following chapters. Before I analyze these particular state–corporate crimes, however, it is important to briefly describe the major impacts of climate change and identify the specific social actors that are responsible for those blameworthy harms. I address these issues in chapter 2, which begins with the argument that 1988 was a benchmark year for the climate movement: despite the fact that the greenhouse effect had been the subject of scientific inquiry for more than a century, it was not until 1988 that the issue of anthropogenic global warming was elevated to a new level of public awareness and political concern. I then briefly review the scientific research on global warming and describe the ecological and social impacts of climate disruption—both of which can be characterized as "beyond catastrophic." These impacts are also increasingly conceptualized as various forms of violence (interpersonal, structural, or slow). The term *ecocide*, for example, is frequently used to characterize the environmental harm resulting from climate change.

In chapter 2, I also examine the question of whether the world has entered into a new geological epoch known as the Anthropocene and consider whether the "Great Acceleration" of economic growth, carbon emissions, and human ecological impact that took place after World War II is the causative factor for this geological development. Following Agnew's discussion of blameworthiness, I also raise the question of who is responsible for the ecological and social destruction of climate change and identify the corporations that interact and compete within the fossil fuel industry, military institutions, and other government agencies as the organizational actors that are primarily involved in perpetrating climate crimes. These "elite-dominated and undemocratic organizations" (Downey, 2015) will be examined as independent social actors, but it is also extremely important to place them within the broader structural context—the institutional and industry environment—in which they operate. A critical component of that larger structural context is the historical global capitalist system and the process of capital accumulation. Treadmill of production theory will be introduced in chapter 2, and I will use this approach in later chapters to provide a theoretical explanation of the political and economic forces that drive state–corporate climate crimes.

In chapter 3, I focus on the blameworthy harms of extracting fossil fuels such as oil, gas, and coal—*crimes of ecological withdrawal*—and the marketing and subsequent burning of these fuels that sends heat-trapping greenhouse gases into the atmosphere—*crimes of ecological addition* (Stretesky, Long, and Lynch, 2014). The central question animating this chapter is, what did the fossil fuel industry know about the greenhouse effect, and when did they know it? Scientists have understood—and have been warning—that the emission of carbon dioxide into the air causes global temperatures to rise since at least 1896 when Swedish chemist (and future Nobel laureate) Svante Arrhenius first proposed the idea (McKibben, 2011). Both the science and the warnings grew stronger in the second half of the twentieth century. And we now know that scientists working for corporations in the fossil fuel industry had substantial evidence of the greenhouse effect in the 1960s and 1970s and issued dire internal warnings about the threat that global warming posed to society and the bottom line of the companies they worked for. The specific case of what ExxonMobil knew about carbon emissions, global warming, and climate change, and what the company did or did not do with that knowledge, has been well documented by investigative reporters and will be presented in this chapter. Despite the strong evidence that burning fossil fuels is causing the earth to warm with potentially catastrophic environmental and social effects, as well as increasing demands from the climate justice movement to "leave it in the ground," the oil, gas, and coal corporations have continued to extract (sometimes using more extreme extraction methods) and market these fuels, causing global warming to continue unabated with increasingly severe consequences.

The interrelated crimes of political omission (failure to act when one is legally or morally required to act to prevent harm) and socially organized climate change denial are examined in chapters 4 and 5. The larger political failure to mitigate greenhouse gas emissions (which could have slowed global warming and limited climate disruptions) is a critical blameworthy harm that can be classified as a climate crime—and this failure has occurred at the level of both the international political community and the individual nation-state. As Schellnhuber asserted, the failure of the George W. Bush administration to do anything about climate change for the eight years it was in office was a crime (Hertsgaard, 2011b; Lynch, Burns, and Stretesky, 2010), and earlier presidential administrations in the United States are also responsible for political omission on the issue of climate change. In chapter 4, I document what the Ronald Reagan, George Herbert Walker Bush, Bill Clinton, and George W. Bush administrations did or did not do with regards to global warming. I also examine the repeated failure of the international political community to negotiate an effective, binding treaty to reduce greenhouse gas emissions over the years as an ongoing climate crime (although the Paris Agreement of 2015 may start to reverse that pattern).

Additionally, as the evidence for human-caused global warming and associated climate disruption continued to mount in the 1980s, there emerged a socially organized ideological denial of climate science (Hoffman, 2015; Kramer, 2013a). Through a variety of deceptive tactics and narratives, this conservative political countermovement has denied that climate change is occurring at all or, if it concedes that global warming is occurring, denies that it is human-caused. These social and political efforts have been very effective in creating doubt and reshaping public opinion concerning global warming, and this highly organized disinformation campaign has been a key factor in the Republican Party's increasing obstructionism on the issue in the United States, successfully blocking actions that might mitigate greenhouse gas emissions. Climate change denial efforts are largely carried out by right-wing think tanks, such as the Heartland Institute, which are funded by corporations in the fossil fuel industry, such as Exxon-Mobil and Koch Industries, as well as more than 140 conservative foundations (Brulle, 2013). While climate change denial can also be analyzed at the social psychological level by examining ideological worldviews, cultural cognition, and social identity, it is the organized structural resistance to climate science by fossil fuel companies and conservative think tanks that has prevented strong political action to reduce emissions. This is another

blameworthy harm that can be identified as a climate crime and is also analyzed in chapter 4.

In chapter 5, I focus on the climate crimes of political omission and
denial that occurred during the eight years of the Barack Obama administration and the first three years of the Donald Trump administration. After
getting off to a stumbling start on the issue of global warming during his
first term with the failure of the 2009 UN climate meeting in Copenhagen
and the congressional failure to pass the cap-and-trade bill in 2010, Obama
could point to some important accomplishments during his second term.
He developed new fuel efficiency standards for the automobile industry,
signed a historic agreement to reduce emissions with China, developed the
Clean Power Plan to reduce carbon emissions from coal-fired power plants,
rejected the Keystone XL pipeline, and negotiated the historic Paris Agreement with the international political community in 2015. While these
limited achievements did not go far enough to address the climate crisis,
they were signs of progress. That progress would be systematically undone
by the Donald Trump administration that took office in 2017.

Trump believes that climate change is a hoax and has appointed climate
change deniers to leadership positions within his cabinet, including the
State Department, the Department of the Interior, the Department of
Energy, and the Environmental Protection Agency. He has also taken steps
to withdraw the United States from the Paris Agreement and scrap Obama's
Clean Power Plan. During the 2016 presidential campaign, Noam Chomsky warned in an interview that Trump's goal to maximize fossil fuel
extraction and consumption at all costs would constitute "almost a death
knell for the human species" were he to be elected (Benedictus, 2016: 3).
After the election, Engelhardt (2018: 61) predicted that the Trump administration would likely become involved in many political and economic
crimes in the coming years but that the new administration's suppression of
scientific knowledge about climate change and refusal to take any steps
whatsoever to reduce greenhouse gas emissions will prove them to be "the
greatest criminals of all time."

The climate crimes of empire are examined in chapter 6, which begins
by describing a major transformative process that altered the twentieth
century: the full emergence of the "global American empire," accompanied
by an influential "culture of militarism" in U.S. society (Boggs, 2017; Freeman, 2012). I make the argument that the U.S. global empire, American
militarism, and the increasing production of greenhouse gas emissions
through the burning of fossil fuels during the Great Acceleration all became
"a way of life" during this half-century and served to not only sustain fossil
capitalism at home and abroad but to aggravate the problem of climate disruption as well.

After presenting a short history of the American Empire, I analyze three specific, interrelated blameworthy harms that I call climate crimes of empire. These crimes include (1) a large-scale pattern of illegal imperial interventions and wars, many undertaken to secure access to and control over oil by the U.S. "warfare state," which in turn facilitated the greater extraction of fossil fuels and ensured rising carbon emissions during the Great Acceleration; (2) the deployment and operation of the American military during these imperial actions and wars, coupled with a supporting domestic and foreign military base structure, resulting in the release of huge quantities of greenhouse gases by the Pentagon; and (3) the planning for and beginning execution of a "militarized" form of adaptation to climate disruption that involves responding to the problem with "the politics of the armed lifeboat," that is, "by arming, excluding, forgetting, repressing, policing, and killing" vulnerable people, most often climate refugees from the Global South (Parenti, 2011: 11). I argue that these state–corporate harms are climate crimes that can best be understood within the context of the development of the post–World War II global U.S. Empire, the normalization of a culture of militarism within American society, and the resulting creation of a "treadmill of destruction" (Clark and Jorgenson, 2012; Hooks and Smith, 2005).

The analysis of specific forms of climate crimes in chapters 3 to 6 is an attempt to speak in the prophetic voice about grave threats to the human prospect. The serious blameworthy harms identified and analyzed in these chapters raise the critical question of whether climate catastrophe can be averted. While some thoughtful analysts contend that society's efforts to respond to the climate crisis are doomed to failure, I hold out what historian Howard Zinn (1994) calls the "possibility of hope" that the most extreme forms of climate disruption can still be avoided through "a state of engagement" (McKibben, 2019: 3) with the issue. In chapter 7, I analyze ongoing and potential public sphere efforts and state actions to implement the elements of a broad climate action plan and control or prevent climate crimes—and argue that a broad-based, integrated global social movement can foster hope, resist climate crimes, and achieve climate justice.

My use of the concept of crime throughout this book is an effort at "persuasive definition" (Jamieson, 2014). It is not only an important academic criminological exercise; it also constitutes a symbolic and political effort to make this a salient public issue, create moral outrage at and demand legal accountability from the carbon criminals, and mobilize people to respond to the existential threat of climate change. As a society, we must respond by taking actions to drastically reduce greenhouse gas emissions, making systemic changes to reduce our dependence on fossil fuels, developing clean, alternative sources of energy, and creating progressive and just adaptations to the climate disruptions that are already under way. In order to undertake these actions, we will have to directly counter the powerful denialist movement and expose its deceptive tactics, work with progressive social movements to challenge the corporate capitalist state, foster structural changes in the political economy, rethink our consumptive lifestyles, and work within the international political community to implement the Paris Agreement. It is also imperative that we go beyond Paris and negotiate an even stronger international treaty to keep fossil fuels in the ground and block unjust, militaristic forms of adaptation to climate change that would constitute still another destructive climate crime.

Implementing this progressive political agenda to mitigate global warming and adapt to climate disruptions was already a daunting challenge—a long shot if you will—prior to November 2016. But Trump's presidency has now magnified that challenge exponentially. His election was, in the words of McKibben (2017c: 1), "a stunning blow to hopes for avoiding the worst impacts of global warming." But American psychiatrist Robert Jay Lifton (2017) argues that despite Trump's ascension to power, his reckless destruction of Obama's climate policies, and his immoral promotion of fossil fuels, a "climate swerve" is under way, a more "formed awareness" of ourselves as a single threatened species that provides Lifton with hope that constructive action might still be taken before it is too late. I conclude the book by examining a wide variety of public sphere efforts and state actions that provide concrete evidence for the climate swerve and, if brought together in a mass social movement, may still be able to stop climate crimes, slow global warming, resist the destruction of the human species, and achieve climate justice.

CHAPTER 2

"Beyond Catastrophic"

THE CLIMATE CRISIS, CARBON CRIMINALS, AND FOSSIL CAPITALISM

IN THE UNITED STATES, the summer of 1988 was hot and dry. Persistent heat waves and intensifying drought gripped much of the country. On June 23, the thermometer hit 98 degrees Fahrenheit in Washington, DC, as Dr. James E. Hansen of the National Aeronautics and Space Administration (NASA) sat down in Room 366 of the Dirksen Senate Office Building to testify about global warming before the Senate's Committee on Energy and Natural Resources. Hansen, director of NASA's Goddard Institute for Space Studies and a leading expert on climate science, had been invited to appear before the committee by Senator Timothy E. Wirth, a Democrat from Colorado. This was not the first time that Hansen had testified in front of Congress about the "greenhouse effect," but his statement on this occasion would come to have a far more dramatic impact on the climate change debate. In his testimony to the Senate on that sweltering June day as the building's air conditioning system failed to function, Hansen drew three main conclusions. "Number one," he said, "the earth is warmer in 1988 than at any time in the history of instrumental measurements. Number two, the global warming is now large enough that we can ascribe with a high degree of confidence that a cause and effect relationship to the greenhouse effect. And number three, our computer climate simulations indicate that the greenhouse effect is already large enough to begin to effect the probability of extreme events such as summer heat waves" (Shabecoff, 1988: 1).

Hansen's dramatic testimony, against the backdrop of just such an early summer heat wave, had an immediate impact. The next day an above-the-fold headline on the front page of the *New York Times* blared, "Global Warming Has Begun, Expert Tells Senate." Writing for the *Times*, Philip Shabecoff (1988: 1) pointed out that "If Dr. Hansen and the other scientists are correct, then humans, by burning of fossil fuels and other activities, have altered the global climate in a manner that will affect life on earth for centuries to come." Shabecoff then quoted Hansen as saying: "It is time to stop waffling so much and say that the evidence is pretty strong that the greenhouse effect is here." Wirth, who presided at the hearing, added, "As I read it, the scientific evidence is compelling: the global climate is changing as the earth's atmosphere gets warmer. Now, the Congress must begin to consider how we are going to slow or halt that warming trend and how we are going to cope with the changes that may already be inevitable." Later that year in Congress, thirty-two climate bills, including Wirth's National Energy Policy Act of 1988, were introduced, and soon thereafter funding for scientific work on climate change substantially increased. As climate activist Bill McKibben (2011: 46) later noted, "Hansen's testimony had ignited the most intense period of scientific investigation of any topic ever."

But, again, it is important to note that the summer of 1988 was not the first time that scientists like Hansen addressed what came to be called the "natural greenhouse effect" and sounded the alarm about the connection between carbon dioxide emissions and global warming. As far back as 1824, French mathematician and physicist Joseph Fourier suggested that gases in the earth's atmosphere might cause it to act as an insulator and keep the planet warmer than should be expected. He reasoned that the atmosphere impedes outgoing heat (infrared or long-wave) radiation, the way a glass cover keeps air heated by the sun from escaping a box. Eunice Foote, an American scientist, discovered the warming properties of what was then called "carbonic acid" (CO_2) and in an 1856 publication was the first to describe the so-called greenhouse effect (when gases such as carbon dioxide absorb and trap heat in the atmosphere and re-emit it, thus warming the earth). In 1859, Irish physicist John Tyndall also began conducting research on the absorptive property of atmospheric gases that allowed him to suggest (without citing Foote's research) that variations in the composition of the atmosphere could result in changes to Earth's climate over time. And in the late nineteenth century, Swedish chemist (and future Nobel laureate) Svante Arrhenius became the first scientist to propose that it was the burning of fossil fuels (primarily coal at that time) that released carbonic acid into the air and would eventually heat the planet by changing the earth's radiative balance, "the relationship between incoming solar radiation and outgoing terrestrial energy" (Leichenko and O'Brien, 2019: 27). Without benefit of a computer, he calculated that an approximate doubling of carbon dioxide in the atmosphere could raise the earth's temperature by 5–6 degrees Celsius, a remarkably accurate prediction.

In the 1930s, British engineer Guy Stewart Callendar compiled the first empirical evidence that greenhouse gases released by human activities were responsible for the warming of the earth. By burning fossil fuels, Callendar noted, humans had become "able to speed up the process of nature" (Rich, 2018a: 14). In 1957, American oceanographer Roger Revelle and his colleague Hans Suess published an important paper suggesting that, by burning fossil fuels and releasing carbon dioxide into the atmosphere, humans may be "carrying out a large-scale geophysical experiment of a kind that could not have happened in the past nor could be reproduced in the future" (McKibben, 2011: 38). Working at the Scripps Institution of Oceanography in San Diego, Revelle hired Charles David Keeling, a young American chemist who developed a new and more precise way to measure atmospheric concentrations of carbon dioxide. In 1958, Keeling and Revelle helped set up a new U.S. Weather Bureau observatory, high on the Mauna Loa volcanic mountain on the Big Island of Hawaii, that deployed this new technique. The resulting measurement chart, known as the "Keeling Curve" (for which Keeling would receive the National Medal of Science), famously shows that carbon dioxide levels have been climbing steadily since these measurements began. McKibben (2011: 44) suggests that this chart "may be the most famous image of the climate change era." Revelle would later attempt to influence government policy on climate change and also teach at Harvard University, where his lectures on global warming and presentation of the data in the Keeling Curve made a huge impression on a young undergraduate student (and future politician) by the name of Al Gore.

Keeling, Revelle, and other scientists recognized that the inexorable rise in carbon dioxide levels, which could lead to a rise in global temperatures, was not just an academic research issue but would also have major public policy implications. In the early 1960s, Revelle chaired the environmental pollution subcommittee of President John F. Kennedy's Science Advisory Committee, which would, a few years later, influence President Lyndon B. Johnson to include a warning about carbon dioxide emissions in his "Special Message to the Congress on Conservation and Restoration of Natural Beauty" on February 8, 1965. In a section of the message on pollution Johnson (1965: 6) noted that "Air pollution is no longer confined to isolated places. This generation has altered the composition of the atmosphere on a global scale through radioactive materials and a *steady increase in carbon dioxide from the burning of fossil fuels* [emphasis added]." As Ken Caldeira, an atmospheric scientist at the Carnegie Institution for Science, has observed: "To the best of my knowledge, 1965 was the first time that a U.S. President was ever officially warned of environmental risks from the accumulation of fossil-fuel carbon dioxide in the atmosphere" (Lavelle, 2015: 1). In November 1965, the President's Science Advisory Committee went on to issue a report formally warning that the continued burning of fossil fuels could produce "marked changes in climate, not controllable through local or even national efforts" (Otto, 2016: 271).

In the 1960s and 1970s, scientists working for corporations in the fossil fuel industry, influenced by the findings of Keeling, Revelle, and other academic scientists, also carried out research that documented the greenhouse effect and, as I will analyze in chapter 3, issued dire internal warnings about climate disruption. In 1974, a classified report by the Central Intelligence Agency warned that future impacts of a warming world would be "almost beyond comprehension" (Rich, 2018a: 16). Given the growing concern among industry and government scientists, the Jimmy Carter administration ordered the prestigious National Academy of Sciences (NAS) to assemble a group of experts and assess the issue. Chaired by atmospheric physicist Jule Charney (the father of modern meteorology), the group met at Woods Holes, Massachusetts, in July 1979. Their report, *Carbon Dioxide and Climate: A Scientific Assessment*, examined the impact that rising carbon dioxide levels would have on global temperatures, concluding: "If carbon dioxide continues to increase, the study group finds no reason to doubt that climate changes will result and no reason to believe that these changes will be negligible" (Charney, 1979: 2). Yet, even though scientists were sounding the alarm, numerous presidential administrations in the United States attempted to dampen concerns over global warming and took almost no significant policy actions to mitigate greenhouse gas emissions (what I call the crime of political omission).

While the executive branch of government gave little more than lip service to the issue of climate change for many years, some members of Congress, in both the House and the Senate, expressed grave concerns, held hearings, and attempted to at least promote research on the problem. In September 1978, Congress passed the National Climate Program Act, which established a federal program, "a first effort" to "assist the Nation and the world to understand and respond to natural and human-induced climate processes and their implications" (Pielke, 2000: 12). Three of the federal agencies participating in the Climate Program—NASA, the National Oceanic and Atmospheric Administration (NOAA), and the National Science Foundation (NSF)—would become major players in climate change science in the 1980s and 1990s. The work of the scientists in these agencies would eventually lead to the creation of the important Global Change Research Program in 1990 (Pielke, 2000), a program that would issue periodic National Climate Assessment (NCA) reports in the future.

Congressional hearings on global warming began in the early 1980s with little media attention. Senator Paul Tsongas, a Democrat from Massachusetts, held the first Senate hearing on the buildup of carbon dioxide in the atmosphere in April 1980. Gore, then a congressman from Tennessee, held House hearings on rising levels of atmospheric carbon dioxide in 1981, 1982, and 1984. The Senate again held hearings on global warming in the fall of 1985. Then, in June 1986, a group of scientists appeared in front of the Senate Committee on Environment and Public Works for two days of hearings about "Ozone Depletion, the Greenhouse Effect, and Climate Change." These hearings, convened by Republican Senator John H. Chafee of Rhode Island, garnered considerable media coverage but positioned the ozone hole as the more immediate threat than predictions of future climate disruptions due to global warming. NASA scientist Robert Watson, however, caused a bit of a stir at the Chafee hearings when he testified, "I believe global warming is inevitable. It is only a question of the magnitude and the timing" (Pielke, 2000: 17). Watson's statement was picked up by the *New York Times* and other major newspapers, but during the second day of hearings, officials from the Ronald Reagan administration's Environmental Protection Agency (EPA) and the Commerce, State, and Energy Departments all downplayed the significance of his testimony. This led Chafee to notably observe: "It was the scientists yesterday who sounded the alarm, and it was the politicians, or the government witnesses, who put the damper on it" (Pielke, 2000: 17). It was a prescient comment, as the subsequent history of the U.S. government's handling of the climate crisis would show.

By the late 1980s, more and more members of Congress were becoming concerned about climate change caused by rising greenhouse gas emissions. In 1987, Congress passed the Global Climate Protection Act, which directed the National Climate Program to carry out "research into methods to control [greenhouse gas] emissions and possible cooperation in international efforts to control climate change" (Keenan, 2008: 170). The Act also gave the EPA and the State Department the authority to develop climate change policy. Although the Reagan administration opposed the legislation, the president still signed the measure but then worked behind the scenes to maintain political control over the troublesome issue. As Roger Pielke (2000: 18) observes, "The Reagan Administration used such tactics (i.e., formal acceptance but actual opposition) to effectively thwart the intent of the Global Climate Protection Act by retaining control over climate change policy at the highest levels." Governmental failure to respond effectively to the global warming issue would become a persistent problem in the decades to come.

While the early congressional hearings on the greenhouse effect and calls for climate research did gain some media coverage at the time (on the CBS Evening News in March 1982, anchor Dan Rather devoted three minutes to the topic), press and public attention was "characteristically short-lived" (Pielke, 2000: 17). This was due in part to a false summary that was put out by oceanographer William Nierenberg as chair of a blue-ribbon NAS panel on climate change (Rich, 2018a). In his press interviews concerning the October 1983 NAS report *Changing Climate*, Nierenberg downplayed the problem and said that there was no urgent need for action, which directly contradicted the actual conclusions of the report. "Haste on Global Warming Trend Is Opposed" read a *New York Times* headline (Rich, 2018a: 38).

One year later, Nierenberg helped found the George C. Marshall Institute, a conservative think tank that would go on to become a major player in the climate change denial countermovement in the 1990s.

While Nierenberg (and the Reagan administration) tempered the issue in 1983, Hansen's June 1988 Senate testimony would resurrect and amplify the "political story" of climate change, elevating the greenhouse effect to the level of a critical national public issue. Hansen had a "greater public impact when it came to consciousness-raising—in part because at that point, he said that the warming of the globe caused by humans was already detectable" (Mooney, 2016: 2). As Naomi Klein (2014: 73) puts it: "If the climate movement had a birthday, a moment when the issue pierced the public consciousness and could no longer be ignored, it would have to be June 23, 1988." Hansen's testimony also frightened the fossil fuel industry. As Nathaniel Rich (2018a: 64) reports, "It was James Hansen's testimony before Congress in 1988 that, for the first time since the [NAS] 'Changing Climate' report [1983], made oil-and-gas executives begin to consider the issue's potential to hurt their profits."

In addition to Hansen's Senate testimony and its political consciousness-raising impact in the United States, other landmark events concerning climate change also took place in the pivotal year of 1988. In June, a combination of scientists and policymakers (a first) gathered in Toronto for an international conference on the changing atmosphere, a "Woodstock" for climate change, according to Rich (2018a: 50). The four-day World Conference on the Changing Atmosphere: Implications for Global Security highlighted atmospheric issues such as global warming in a comprehensive way, and for the first time emission reductions were discussed as a policy issue. Even more important, late in 1988 the United Nations Environment Programme (UNEP) and the World Meteorological Organization (WMO) created the Intergovernmental Panel on Climate Change (IPCC) to "assess on a comprehensive, open and transparent basis the scientific, technical and socioeconomic information relevant to understanding the scientific basis of risk of human-induced climate change, its potential impacts and options for adaptation and mitigation" (M. Mann and Kump, 2015: 8). The IPCC, which includes most of the world's climate scientists and a wide variety of other experts in related fields, reviews and summarizes the current state of scientific knowledge about climate change and issues periodic assessment reports. In 2007, the IPCC was awarded the Nobel Peace Prize for its work on climate change, along with that Harvard undergraduate influenced by Roger Revelle, former U.S. vice president Al Gore, whose book and film An Inconvenient Truth (2006) had done much to publicize the "Keeling Curve" and the overall "planetary emergency of global warming."

Whatever scientific findings and warnings about global warming and climate change had been issued before (and there were many), the events of 1988 served to spotlight in a new way the grave risk posed by these phenomena to all life on earth and the conclusion that these existential dangers were being caused by the burning of fossil fuels. The issue had become so significant that *Time* magazine, rather than announcing its traditional "Man of the Year" in 1988, proclaimed a rather shocking choice: "Planet of the Year: Endangered Earth," raising the issue of anthropogenic global climate change to a new level of public awareness and political concern. After 1988, it should have been politically difficult and morally questionable to deny that global warming was occurring, and it should have been socially and legally irresponsible to fail to act to mitigate fossil fuel emissions and plan for just adaptations to climate disruption. But corporations in the fossil fuel industry, despite their own prior knowledge of the problem, were able to stop or slow any political efforts to reduce carbon emissions and rein in global warming while they continued to extract the last drop of profit from what is ultimately a terminal industry. And this "predatory delay" (Steffen, 2016), a blameworthy harm, allowed dangerous greenhouse gas emissions to continue to rise. Similarly, despite their knowledge of the problem and their responsibility to act, governments—both individually and collectively as an international political community—also failed to take the necessary actions after 1988 to mitigate carbon emissions and adapt to climate disruptions caused by global warming (another significant blameworthy harm). The year 1988 therefore serves as a historic benchmark to judge when organizational actions like these constitute morally blameworthy harms that can be identified and analyzed as climate crimes.

The Impacts of Climate Change: "Beyond Catastrophic"

As Hansen's 1988 testimony made clear, the emission of heat-trapping greenhouse gases into the atmosphere, primarily through the burning of fossil fuels, causes the planet to warm as the atmosphere that blankets the earth holds heat in (the greenhouse effect). As this human-caused global warming increases, it has severely harmful effects—preventable effects that should be considered morally blameworthy harms and, more specifically, climate crimes. Scientific research on global warming and climate change, much of which is reviewed and summarized in the IPCC's Fifth Assessment Report (2013), demonstrates the catastrophic nature of the harms that are being inflicted on the environment and human societies. And research published since 2013 suggests that climate change is happening even faster than predicted by the IPCC and is causing even more extreme harms than expected (Berwyn, 2017; Romm, 2019). An alarming study in 2019 found that scientists have been consistently underestimating the pace of climate change (Oppenheimer, Oreskes, et al., 2019). New evidence shows that climate disruption has now become such a great risk that a paper published in the *Proceedings of the National Academy of Sciences* in 2017 argued that a new category of climate change is necessary, one that goes beyond "dangerous" and even "catastrophic" to "unknown" (Xu and Ramanthan, 2017).

One study supporting a categorization of "beyond catastrophic" was published in the journal *Nature Climate Change* in 2018 and found that global warming poses such wide-ranging risks to humanity that some parts of the world could experience as many as six climate-related threats simultaneously (Mora, Spirandelli, et al., 2018). Another article in *Nature Geoscience* in 2018 warned that global warming could be much worse than current climate models project, since these models tend to underestimate long-term warming by as much as a factor of two (Fischer, Meissner, and Zhou, 2018). And a study published in *Nature* in 2017 suggests that the worst-case predictions of climate disruption turn out to be the most accurate (P. Brown and Caldeira, 2017). These more realistic and increasingly dire scientific predictions about the effects of global warming are starting to reach a broader audience. In July 2017, American journalist David Wallace-Wells published a provocative cover story on climate change in *New York Magazine* titled "The Uninhabitable Earth." The widely read article starts by asserting: "It is, I promise worse than you think." Wallace-Wells goes on to review recent research on global warming and describe in detail some of the worst-case scenarios that could result from climate disruption. The picture he painted was so bleak and alarming that even some well-known climate scientists like Michael E. Mann (2017: 2) felt the need to point out some errors and overstatements in the article and complain that such a stark portrait of climate catastrophe "feeds a paralyzing narrative of doom and hopelessness." But Margaret Klein Salamon (2017), a clinical psychologist, disagreed with Mann and defended Wallace-Wells's gloomy portrayal, arguing that "climate truth" is a strategic asset that possesses a "transformative power" that could spark climate mobilization. In 2019, Wallace-Wells expanded the article into a best-selling book, *The Uninhabitable Earth: Life after Warming*. In the book, he asserts that "The path we are on as a planet should terrify anyone living on it, but . . . all the relevant inputs are within our control, and there is no mysticism required to interpret or command the fate of the earth. Only an acceptance of responsibility" (Wallace-Wells, 2019: 226).

More "climate truth" was dispensed to the public in November 2017 when the United States Global Change Research Program (2017) released the first volume of its Fourth National Climate Assessment (NCA4). The NCA4, mandated by Congress and assembled by thirteen federal agencies, concluded that global warming is affecting the United States more than ever before and that these negative impacts are expected to increase in the future (Fountain and Plumer, 2017). This official U.S. government report (which, surprisingly, the science-denying Trump administration allowed to be released) points out that since NCA3 (four years earlier), stronger evidence has emerged to document continuing, rapid, human-caused warming of the earth's atmosphere and oceans. In a *New York Times* opinion piece, many of the climate scientists who wrote the report stated that "even though we cannot quantify all of the possible changes to every element of the climate system, the risks to things we care about—from the health of our children, to the future economic viability of our low-lying coastal cities and infrastructure—are real and growing" (Horton, Hayhoe, Kopp, and Doherty, 2017: 3).

In November 2018, the second volume of the NCA4 (U.S. Global Change Research Program, 2018) was released by the Trump administration, which tried to bury the report by issuing it on "Black Friday," the day after Thanksgiving when many Americans are busy shopping or traveling and media attention was expected to be low. Volume 2, a 1,656-page assessment, "lays out the devastating effects of a changing climate on the economy, health and environment, including record wildfires in California, crop failures in the Midwest and crumbling infrastructure in the South" (Davenport and Pierre-Louis, 2018: 1). The report describes how climate change will result in trade disruptions, agricultural risks, and other economic impacts. It puts precise price tags on the cost of climate impacts to the United States economy: $141 billion from heat-related deaths, $118 billion from sea-level rise, and $32 billion from infrastructure damage by the end of the century (Davenport and Pierre-Louis, 2018). NCA4 emphasizes that the occurrence of these disastrous outcomes and enormous costs depends on how quickly and decisively the world acts to mitigate greenhouse gas emissions through policies such as carbon pricing, government regulations, or investments in renewable energy. This emphasis on policy is ironic since the Trump administration, despite allowing the report to be released, rejects all efforts to mitigate carbon emissions. When asked about the report's conclusions concerning the potential economic impact of climate change, Trump declared, "I don't believe it." This prompted meteorologist Eric Holthaus to assert, "In all seriousness, the willful denial and obfuscation by Trump on climate change is a crime against humanity. Billions of people will bear incalculable harm for generations to come" (Germanos, 2018a: 1).

In addition to these general public reports, more specific climate research details the increasingly disastrous ecological and social impacts of global warming. The evidence shows that the burning of fossil fuels (and to a lesser extent deforestation) has already raised the temperature of the planet by one degree Celsius (1.8 degrees Fahrenheit) over the pre-industrial average, and some scientists estimate it could go as high as an alarming 5 or 6 degrees Celsius, or 9 to 11 degrees Fahrenheit (Lynas, 2008; Hamilton, 2010; M. Mann and Kump, 2015; Goodell, 2019). The planet is warming faster than it has in the last 2,000 years, and there are no natural causes for this unprecedented rise in temperatures (Neukom, Barboza, et al., 2019). As officials from 150 countries gathered at the United Nations on April 22, 2016 (Earth Day), to sign the Paris Agreement—an accord aimed at reducing greenhouse gas emissions to keep the rise in global temperatures below 2 degrees Celsius, and preferably 1.5 degrees Celsius—the planet had just experienced seven straight months of record-breaking heat and 2016 was on track for the largest increase in global temperatures ever in a single year (Jamail, 2016). 2016 would go on to become the hottest year on record (National Oceanic and Atmospheric Administration, 2017), and 2015, 2016, and 2017 have been determined to be the three hottest years that have ever been measured (Kusnetz, 2018b). The average surface temperature of the earth in 2018 "was the fourth-highest since 1880, when record-keeping began," June 2019 "was the hottest June ever recorded," and "astonishingly, July [2019] was the hottest month in human history"(Goodell, 2019: 79). McKibben (2018b: 49) points out that twenty of the hottest years ever recorded have occurred in the past thirty years. As one climate scientist put it, such warming is "our new normal" (Milman, 2017: 3).

The concentration of carbon dioxide in the atmosphere has increased from 275 parts per million (ppm) at the dawn of the extractive industrial age to more than 410 ppm today. Shortly before the Conference of the Parties (COP 23) climate negotiations in Bonn, Germany, in November 2017, the *Greenhouse Gas Bulletin*, published by the Global Atmosphere Watch program of the WMO, found that globally averaged carbon dioxide concentrations had surged to 403.3 ppm in 2016—the highest level in approximately three million years, when natural factors caused a huge release of carbon dioxide. The WMO report raised alarm among climate scientists and prompted calls for the consideration of more drastic cuts in carbon emissions at COP 23. In 2017 global atmospheric carbon dioxide levels hit a new record high (Watts, 2017b). Then, in April 2018, still a new climate change threshold was crossed when, for the first time since humans have been monitoring, atmospheric concentrations of carbon dioxide exceeded 410 ppm, averaged across an entire month (Mooney, 2018). This alarming news arrived just as the Trump administration canceled a crucial carbon monitoring system at NASA, a move that climate activists denounced as an effort by the administration to continue its "war on science" (Corbett, 2018a). As the Keeling Curve continues to soar, some scientists fear that carbon dioxide concentrations appear to be headed for 550 or 650 ppm by the end of this century. Hansen (2009) has argued that the only safe level—that is, one that would not risk further global warming—is 350 ppm (hence the name of Bill McKibben's organization, 350.org). We are a long way from safe.

Global atmospheric carbon dioxide concentrations were first measured over the critical 400 ppm threshold at the Mauna Loa observation site in 2013, and some climate scientists now predict that we have entered an era in which global concentrations will permanently exceed this mark (Kahn, 2016b; Betts, Jones, Knight, et al., 2016). In another alarming development, NOAA announced in June 2016 that the South Pole Observatory, in the remote reaches of Antarctica, measured carbon dioxide levels at 400 ppm for the first time in four million years (Kahn, 2016a). And, since greenhouse gases stay in the atmosphere for such a very long period of time, even if we stopped all carbon emissions today, great damage has already been done and will continue.

A massive amount of scientific research—about which there is over-whelming consensus—provides compelling evidence that these rising levels of greenhouse gas emissions (through the burning of fossil fuels), which result in global warming, are causing significant ecological and social damage (for overviews of this research, see; Bennett, 2016; Dow and Downing, 2011; Griffin, 2015; Hansen, 2009; Henson, 2019; IPCC, 2013, 2018; Jamail, 2019; M. Mann and Kump, 2015; McKibben, 2010b, 2019b; Nesbitt, 2018; Richardson, Steffan, and Liverman, 2011; Romm, 2019; Royal Society and NAS, 2014; U.S. Global Change Research Program, 2017, 2018; Wallace-Wells, 2019). As criminologists Paul B. Stretesky, Michael A. Long, and Michael J. Lynch (2014: 4) point out, for green criminology, such scientific research "plays an important role in objectively identifying dangerous eco-logical harms," harms, they add, "the law (often) fails to recognize." Indeed, this large body of climate research plays a critical role in my analysis of climate crimes.

And the research is clear. The environmental and human impacts of extracting and burning fossil fuels are increasingly dire. Rising sea levels are one major concern, and climate scientists warn that significant sea-level rise could produce "transformative catastrophes" on a global scale that could lead to the "remaking of the civilized world" (Goodell, 2017b). More than 90 percent of the excess heat trapped by greenhouse gases in the atmosphere is absorbed into the oceans of the world, raising their temperatures (Shank-man and Horn, 2017). A 2018 study published in *Nature* found that the oceans retain 60 percent more heat than previously thought (Resplandy, Keeling, et al., 2018). This warming of ocean waters has doubled since 1992 and, among other dangers (stronger storms, coral bleaching), causes the thermal expansion implicated in a measurable rise in sea level. An overall rise of 8 inches has already been measured since 1880, and ongoing thermal expansion could cause a rise of another 12 inches by 2100 (Bennett, 2016).

In addition to thermal expansion, the melting of glacial ice, particularly in Greenland and Antarctica, is contributing to rising sea levels. Rising seas will eventually flood coastal areas, deltas, and low-lying islands, causing enormous destruction and disruption. Many cities will be left below sea level, and hundreds of millions of urban dwellers will be devastated by the rising water. According to the IPCC, overall sea level is currently projected to rise between 0.5 and 1.2 meters (1.6 and 3.9 feet) by 2100 (M. Mann and Kump, 2015: 111). But a study published in the journal *Atmospheric Chemistry and Physics* by Hansen and eighteen other scientists (Hansen, Sato, et al., 2016) argues that sea levels could rise much higher and more rapidly than previously estimated, in the worst case resulting in a sea-level rise of several feet over the next five decades and 9 feet by 2100. As American journalist Jeff Goodell (2017b: 69) points out: "The difference between three feet and six feet is the difference between a manageable coastal crisis and a decades-long refugee disaster. For many Pacific island nations, it is the difference between survival and extinction." Given these stakes, it is quite ominous that a paper published in the *Proceedings of the National Academy of Sciences of the United States of America* (Kopp. Kemp, et. al., 2016) documents that a stunning acceleration of sea-level rise has already started to occur.

Climate scientists also expect that both the frequency and intensity of extreme weather events will increase. Based on the IPCC reports, Michael Mann and fellow climate scientist Lee R. Kump (2015: 112) point out that in the future, "New trends are expected to emerge for various types of extreme weather including heat waves, heavy downpours, and frosts." There will be fewer frost days, and heat waves will become more frequent, more intense, and longer lasting. A warmer atmosphere holds greater amounts of moisture, and thus scientists predict more intense precipitation events that will produce massive flooding. Severe storms such as hurricanes, cyclones, and tornadoes are likely to increase and produce more deluges and greater storm surge (M. Mann and Kump, 2015). In late summer 2017, major hurricanes Harvey, Irma, and Maria struck Texas, Florida, and Puerto Rico, respectively, causing severe damage and joining Katrina and Sandy among the most devastating storms to strike the United States. Then in the fall of 2018, two more intense hurricanes, Florence and Michael, hit the United States, causing historic flooding in the Carolinas and massive destruction in the Florida Panhandle. As the cryosphere (the frozen portion of the earth) melts due to planetary heating, Arctic warming and disappearing polar ice can disrupt the jet stream, causing it to "get loopy," which results in extreme weather events such as heat waves that get "stuck" in place during the summer, hurricanes that linger dumping more rainfall (as with Florence in 2018 and Dorian in 2019), and the movement of the frigid polar vortex south from the North Pole over the Midwest and northeastern regions of the

United States during the winter (Berwyn, 2018a; Coumou, Di Capua, et al., 2018; Shankman and Berwyn, 2017). Although not all extreme weather events can be directly attributed to climate change, there is an emerging scientific consensus that global warming does have an impact on these events, and the debate now is simply over the magnitude and extent of that influence (J. Schwartz, 2015).

While precipitation intensity and flooding increases in some parts of the globe, other regions will experience longer dry spells and extreme droughts. David Ray Griffin (2015: 40) points out that "drought is the climate effect that has thus far been the most harmful to people." Hot weather droughts cause food shortages, desertification, and more intense wildfires. In an alarming study published in early 2018, a group of climate scientists warned that 25 percent of the earth could face a permanent state of drought by 2050 if the planet warms by 2 degrees Celsius (Park, Jeong, et al., 2018). Since 1999 the western region of North America has been gripped by "the most persistent and severe drought on record" (M. Mann and Kump, 2015: 52), a drought more widespread than the great "Dust Bowl" of the 1930s and "symptomatic of a longer-term, global pattern of increasing drought" (M. Mann and Kump, 2015: 55). Because of extreme drought conditions, the state of California has experienced massive wildfires in recent years, and in 2017 and 2018 these catastrophic events extended even into the late fall, when such fires have been rare. The Camp Fire that engulfed the city of Paradise in Northern California in November 2018 was the deadliest and most destructive wildfire in California history, and the Union of Concerned Scientists (2018) asserts that global warming is increasing the risk of such devastating fires. In 2017, wildfires, hurricanes, and other climate-related disasters created cumulative damage that cost the United States $306 billion, a new record (NOAA, 2018).

Global warming also has a destructive impact on a variety of valuable ecosystems and may lead to a mass extinction of animals and other living organisms. Michael Mann and Lee Kump (2015: 124) define an ecosystem as "an interdependent community of plants, animals and microscopic organisms and their physical environment." They point out that these different elements interact with each other, form complex wholes, produce unique properties, and provide important material provisions, environmental functions, and cultural benefits to human societies. Coral reefs are one of the world's most valuable and diverse ecosystems threatened by climate change (Nesbit, 2018). Warmer ocean waters and increasing acidification (carbon dioxide dissolved in water transforms into carbonic acid) have a destructive impact on coral reefs, causing them to expel the symbiotic algae that live inside. This process turns the coral white (bleaching) and erodes its structure, producing significant ecological and economic harm. A June 2016 report from NOAA's Coral Reef Watch indicated that the world's oceans are enduring the longest-lasting and most widespread coral bleaching event ever recorded (Berwyn and Hirji, 2016). According to NOAA, Australia's Great Barrier Reef and other important coral reefs are becoming "ghostly underwater graveyards. They are perhaps the starkest reminders—like the melting Arctic—that a thickening blanket of greenhouse gases is irrevocably changing the face of the Earth" (Berwyn and Hirrji, 2016: 1).

Increasing deforestation, particularly in the Amazon rain forest, and the expansion of the tropics that increases the desertification of dry subtropics are still another effect of global warming on important ecosystems. A recent study concluded that the world's tropical forests are becoming so degraded that they have become a source rather than a sink of carbon emissions (Watts, 2017a). In 2018, Global Forest Watch reported that 2017 "was the second worst on record for tropical tree cover loss" (Weisse and Goldman, 2018). As these studies make clear, climate goals and biodiversity are threatened as the tropics continue to lose trees at such alarming rates. In October 2018, a group of scientists warned that the international political community must focus on ending deforestation as well as cutting carbon emissions to address the climate crisis (Conley, 2018).

Climate destabilization in general causes habitat changes that result in phenological shifts (species migrations and redistribution) and accelerate plant and animal extinctions due to ecosystem disruption (Nesbit, 2018; Zukowski, 2017). There have been five major mass extinction events on Earth since life forms first emerged. Scientists now warn that the sixth great extinction, predicted to be the most devastating since the demise of the dinosaurs, is already under way and that the major force behind this extinction event is anthropogenic global warming and the resulting climate disruption (Kolbert, 2014). The 2013 IPCC report expressed high confidence that 20 to 30 percent of plants and animals will be subject to the risk of extinction if global temperatures rise 2 degrees (Celsius) above the pre-industrial average, and 40 to 70 percent will be at increased risk of extinction if temperatures rise by 4 degrees (M. Mann and Kump, 2015). A major report produced by the World Wildlife Fund (WWF) in October 2018 documented that "humanity has wiped out 60% of animal populations since 1970" (Carrington, 2018d: 1). And a sweeping new United Nations assessment in 2019 concluded that "humans are transforming Earth's natural landscapes so dramatically that as many as one million plant and animal species are now at risk of extinction, posing a dire threat to ecosystems that people all over the world depend on for their survival" (Plumer, 2019: 1).

Climate scientists also raise another alarming possibility. As Hansen (2009) points out, if global warming continues at the current pace, positive feedback effects could occur that would in turn trigger runaway heating that is essentially uncontrollable and irreversible for thousands of years. Positive feedback tipping points that will dramatically accelerate the heating of the planet include the loss of the Arctic albedo effect (ice reflects solar radiation; thus the loss of reflective ice leads to darker open water in the Arctic, which absorbs more solar radiation); the release of huge quantities of methane (a greenhouse gas more powerful than carbon dioxide) from the melting of permafrost and hydrates in the Antarctic ice sheet; and the dieback of the Amazon rain forest. Two recent studies show that the melting ice sheets in Antarctica (Silvano, 2018) and melting permafrost (Gasser, Kechlar, et al., 2018), respectively, could both trigger feedback loops much sooner than expected and put the climate goals of the 2015 Paris Agreement at risk. The loss of the rain forest will also further increase the concentration of carbon dioxide in the global atmosphere anywhere from 20 to 200 ppm by the end of the century, with devastating and potentially lethal impacts on many forms of life (Richardson et al., 2011: 86). In 2012, a group of scientists warned that we are fast approaching a "state shift" in Earth's biosphere, a planetary-scale "critical transition" because of human activity including energy production and consumption (Barnosky, Hadly, et. al., 2012). A key impediment to an effective response to these problems is that the consequences of greenhouse gas emissions involve complex causal chains. As Australian philosopher Clive Hamilton (2010: 25) points out, "The lag between emissions and their effects on climate and the irreversibility of those effects makes global warming a uniquely dangerous and intractable problem for humanity." The lag between cause and effect also makes it uniquely difficult to mobilize the political will necessary to address the problem.

The environmental damage caused by global warming and climate disruption will in turn result in a wide range of social, economic, and political harms to human communities and the social systems on which they depend (Crank and Jacoby, 2015; Dyer, 2010; Griffin, 2015; M. Mann and Kump, 2015; Nesbit, 2018; Parenti, 2011; Romm, 2019; Wallace-Wells, 2019). As Robin Leichenko and Karen O'Brien (2019: 154, 157) point out, "Climate change presents profound yet unevenly distributed threats to human security," impacts that will be "felt most acutely by vulnerable populations that are already facing other forms of social and economic disadvantage." These social impacts also need to be taken into account.

The rise in sea levels, extreme heat, and chronic droughts will lead to drastic shortages in freshwater and the food supply, increasing famine and mass migrations. The large movement of people across borders seeking freshwater, food, and an escape from the environmental consequences of increased temperatures will cause many social and political problems. In the view of many criminologists (Agnew, 2011a; Crank and Jacoby, 2015; R. White, 2018), the presence of these "climate refugees" may fuel violent conflicts, genocides, and other crimes. A study published in the *Proceedings of the National Academy of Science of the United States of America* (Kelly, Shahrzad, Cane, et al., 2015) argues that severe drought has played a critical role in the Syrian civil war. While some sociologists raise critical questions concerning the causal link between climate and the incidence of violence (Bonds, 2016), there is no doubt that migrations related to climate disruptions have already led to the militarization and securitization of borders in the Global North as neoliberal economic policies intersect with climate change impacts like heat waves and drought to produce declines in agricultural and pastoral economies in the Global South. This in turn leads to food insecurity and health problems and increases the pressure on people to become irregular migrants seeking a minimum of food, medical care, and physical security (Dunn, 1995; Miller, 2017, 2019; Parenti, 2011). As McKibben (2018b: 49) puts it, as we lose parts of the habitable earth, we are experiencing "life on a shrinking planet."

The term *ecocide* is increasingly being used to characterize the severe ecological and social upheaval caused by climate disruption. According to the UK attorney Polly Higgins (2012: 3), ecocide can be defined as "the extensive damage, destruction to or loss of ecosystems of a given territory, whether by human agency or by other causes, to such an extent that peaceful enjoyment by the inhabitants of that territory has been severely diminished." The massive social upheavals, class conflict, disease, and negative health effects caused by climate change, for example, will stress existing social institutions, create ideological turmoil, and generate political crises. Examining these social impacts led Rob White and me (2015b) to use the term *climate change ecocide* and White (2018) to later analyze more extensively global warming as a form of ecocide.

Galtung's (1969) concept of *structural violence* has also been applied to the preventable suffering and premature deaths caused by social conditions related to global warming and climate disruption (Bonds, 2016). Poverty and other structural inequalities created and/or aggravated by climatic changes will predictably generate massive suffering and otherwise preventable deaths, particularly in the Global South. The World Health Organization estimates that factors such as heat waves, famines, and epidemics related to climate disruption will result in 250,000 excess deaths between 2030 and 2050 (Bonds, 2016). A recent report of the *Lancet Countdown* on health and climate change (Watts, Amann, et al., 2018) suggests that climate disruption could eventually force as many as a billion people to become climate refugees in the future, potentially triggering a global health crisis. Using a shorter time frame, the World Bank estimates that more than 140 million people in just three regions of the developing world are likely to migrate within their own countries between 2020 and 2050, pouring into urban "hotspots" and swelling already overcrowded slums (Harvey, 2018a).

The ecological and social harms of global warming are dispersed across time and space; they often occur gradually and out of sight. Thus, South African writer and environmentalist Rob Nixon's concept of *slow violence* can also be applied to many of the impacts of climate disruption. According to Nixon (2011: 2), slow violence is "delayed destruction . . . an attritional violence that is not typically viewed as violence at all." Slow violence "might well include forms of structural violence," he writes, "but has a wider descriptive range in calling attention, not simply to questions of agency [who caused the harm], but to broader, more complex descriptive categories of violence enacted slowly over time," ecological harm that "is decoupled from its original causes by the workings of time" (Nixon, 2011: 11).

As governments struggle to deal with climate change ecocide, the number of failed and failing states will increase as their incapacity to adapt to climate disruption increases poverty and violence around the globe, particularly in those parts of the Global South that American sociologist and journalist Christian Parenti (2011) terms "the tropic of chaos." Resource wars and other forms of international conflict will increase and, perhaps in the most extreme case, even provoke the use of nuclear weapons (Dyer, 2010; Klare, 2012). Increased warfare could also sabotage the very planetary cooperation needed to reduce further global warming (R. White, 2011). From a moral standpoint, one of the most disturbing elements of climate change ecocide is that it most harms those living in geographic areas and countries that have contributed the least to the problem (Lynch and Stretesky, 2010; Leichenko and O'Brien, 2019). The northern industrial nations have, in the words of American sociologist John Bellamy Foster (2009: 243), accumulated a huge "ecological debt" toward developing countries in the south due to resource plundering and the infliction of environmental harms. But it is a debt that the Global North has demonstrated little interest in paying. Speaking of equity issues, women are much more likely to be differentially affected by climate disruptions (Nagel, 2016; Wonders and Danner, 2015), a necessary consideration for future climate justice policies (M. Robinson, 2018). In general, we can conclude that climate disruptions have enormously destructive social impacts. Global warming is a threat multiplier, and its harmful effects on humans will also be beyond catastrophic.

Who Is Responsible for the Climate Crisis?

Given this mountain of scientific evidence documenting a hotter planet, climate devastation, and the resulting social disruption and ecocide, I argue that the continuing extraction, marketing, and burning of fossil fuels—which leads to greenhouse gas emissions that produce such ecological and social disorganization and destruction—are morally blameworthy harms that deserve to be labeled as climate crimes, whether or not the public has yet to condemn them or states have legally sanctioned them (although there has been significant movement in both of these arenas in recent years). A key question is: Who is responsible for these criminal harms? Who are the carbon criminals? As Peter Frumhoff, Richard Heede, and Naomi Oreskes (2015: 157) point out, "Responsibility for climate change lies at the heart of societal debate over actions to address it."

Concerning responsibility for the climate crisis, there is consensus within the IPCC and the larger scientific community that the global warming that has occurred since the dawn of the industrial age is "anthropogenic," that is, human-caused. It results from the extraction and burning of fossil fuels and the resulting emission of greenhouse gases, not natural planetary variations. Some notable Earth System scientists like Nobel Prize-winning chemist Paul J. Crutzen and biologist Eugene F. Stoermer have advanced the striking notion that we have left the Holocene, a period of relative climate stability that started about 11,700 years ago after the last ice age, and entered a new geological epoch in Earth's history: the Anthropocene, the epoch of humanity. The Anthropocene is considered "a new and dangerous stage in planetary evolution" (Angus, 2016: 19) and "a turning point in the history of humanity, the history of life, and the history of the Earth itself" (Lewis and Maslin, 2018: 5). The concept suggests that humans are now a new geological force, a "geological superpower" in the view of British scientists Simon Lewis and Mark Maslin, transforming the planet through their activities that put increasing pressure on the earth system. The continual search for new ways to satisfy human wants and desires, the evolution of industrial society, the development of consumerism and consumption fostered by capitalist marketing, and other human activities usually organized around fossil fuel energy collectively constitute that transforming force. The release of greenhouse gases, and other human activities that disrupt critical "planetary boundaries," has caused an abrupt and qualitative change in the state of the earth system (Angus, 2016).

Will Steffen, Paul Crutzen, and John R. McNeill (2007) posit that the Anthropocene developed in two distinct stages. The first stage was the Industrial Era (early 1800s to 1945) when atmospheric carbon dioxide levels began to exceed the upper limits of the Holocene. They call the second—and most critical—stage, from 1945 to the present, the Great Acceleration (a deliberate homage to Karl Polanyi's influential 1944 book *The Great Transformation*). After World War II, "a remarkable explosion" of human activity occurred, resulting in global impacts on the earth system and its functioning.

It was during the Great Acceleration "when the most rapid and pervasive shift in the human-environment relationship began" (Steffen et al., 2007: 617). It was during this critical time that greenhouse gas emissions began to soar and climate disruption began to accelerate.

Some scholars, like Jason W. Moore (2015) and Andreas Malm (2016), are critical of the Anthropocene concept, protesting that the idea that all of humanity is to blame for global warming is, in the final analysis, an indefensible abstraction. If everyone is to blame, they argue, then no one is, and the broader structural and cultural forces and the specific classes, organizations, and institutions that predominantly drive climate disruption are shielded from view. Canadian scholar Ian Angus (2016: 227) has responded to these criticisms by pointing out the specific ways in which "the scientists in the forefront of the Anthropocene project have repeatedly and explicitly rejected any 'all humans are to blame' narrative." Angus (2016: 232) goes on to argue that "Anthropocene does not refer to all humans, but to an epoch of global change that would not have occurred in the absence of human activity," activity that involves "the decisive issues of class and power." British writer Jeremy Davies (2016: 7) concurs, pointing out that "To say that the earth is undergoing an epoch-level physical transition, in which the activities of sundry groups of humans are playing key roles, does not imply in the least that all human beings have thus far acted in unison, or that they are all collectively responsible for the new state of affairs." The question is: What specific groups of humans and what specific activities are responsible for the climate crisis?

Writing on global climate change, sociologists have attempted to explain the anthropogenic driving forces, that is, "the human activities and societal characteristics . . . principally responsible for increasing the concentration of [greenhouse gases] and changing land surfaces" (Rosa, Rudel, York, et al., 2015: 32). While this literature examines broad social processes—such as population size and growth, affluence and consumption, and technological developments—as drivers of climate change, "most work is based on comparing political units such as nation-states across recent decades" (Rosa et al., 2015: 34). But even though the nation-state is the primary unit of analysis in many sociological investigations of the driving forces behind global warming, a cursory analysis reveals that states are not equally responsible. Current national emissions "are highly uneven across countries and world regions" (Leichenko and O'Brien, 2019: 80), with the most populous countries and largest economies having higher levels of greenhouse gas emissions. An examination of carbon emissions per capita also reveals dramatic differences between nations, showing, for example, that although China has now passed the United States in terms of total emissions per year, Americans still emit far more greenhouse gases per person than China (Henson, 2019).

When thinking about national distributions of greenhouse gases it is also important to "consider historical patterns and cumulative total emissions" (Leichenko and O'Brien, 2019: 81). When we do, the United States, the United Kingdom, and other early industrializing nations rank high in total emissions on a cumulative basis while China, India, and other countries in the Global South are low. This important fact has had a huge impact on international climate negotiations. The principle of "common but differentiated responsibilities" among nations, for example, was central to the United Nations Framework Convention on Climate Change (UNFCCC). Formalized at the Earth Summit in Rio de Janeiro in 1992, it focused attention on the greater responsibility of the industrialized nations that have historically produced the larger share of greenhouse gas emissions and should therefore as a matter of fairness take the lead in mitigating these emissions.

While the important role that nation-states play in causing (and responding to) global warming must be analyzed, focus must also be given to private industrial organizations, particularly the "distinctive responsibilities of the investor-owned fossil fuel producers" (Frumhoff, Heede, and Oreskes, 2015: 158). Concerning the question of responsibility, then, I argue that corporations in the fossil fuel industry, certain nation-states that rank high in per capita greenhouse gas emissions and total cumulative emissions, and specific government agencies (particularly in the United States)— all located within and shaped by historically specific forms of national and global social structures (particularly the political economy of global capitalism)—create what British sociologist John Urry (2011) calls a "high carbon system" that is primarily responsible for the crime of heating the planet and causing climate change ecocide.

What are the specific components of this "high carbon" social system? As American sociologists Charles Perrow and Simone Pulver (2015: 61) point out, "Through the direct emission of greenhouse gases (GHGs), or through the indirect encouragement of behaviors that result in GHG emissions, *organizations* [emphasis added] are responsible for most of the world's carbon pollution." A recent quantitative analysis conducted by Richard Heede at the Climate Accountability Institute in Colorado and published in the journal *Climatic Change* helps to identify the specific organizations responsible for this form of climate crime. Heede (2014) documented that 63 percent of cumulative global emissions of carbon dioxide and methane between 1751 and 2010 could be attributed to just ninety "carbon major" entities (organizations), including fifty leading investor-owned, thirty-one state-owned, and nine nation-state producers of oil, natural gas, coal, and cement. A follow-up study quantified the historical (1880–2010) and recent

(1980–2010) emissions traced to these ninety carbon major producers, demonstrating that they have significantly contributed to the rise in atmospheric carbon dioxide, global temperatures, and sea level (Ekwurzel, Boneham, Dalton, et al., 2017).

In a review of sociological research on the major institutional sources of carbon emissions, Perrow and Pulver (2015) point to the disproportionate contributions of private-sector, for-profit organizations (what they call "market organizations") to environmental degradation, particularly carbon pollution. They note that coal accounts for 37 percent of carbon dioxide emissions in the United States, while 44 percent comes from the use of petroleum products. Both the coal and oil industries are dominated by large profitable corporations such as the coal company Peabody Energy and the Big Five independent oil companies: BP, Chevron, ConocoPhillips, ExxonMobil, and Royal Dutch Shell.

Other social scientists concur that corporations in the fossil fuel industry are primarily responsible for the greenhouse gas emissions that cause global warming. In *Climate Change, Capitalism, and Corporations: Processes of Creative Self-Destruction*, Christopher Wright and Daniel Nyberg (2015: 15) observe that "The first of the roles business corporations play in relation to climate change, then, is that of a *producer* of GHG emissions." Eric Bonds (2016: 14) also argues that the "major oil, gas, and coal corporations have played and continue to play, a critical role in the unfolding crisis of climate change by both extracting and selling the fossil fuels that are driving global warming, but also by successfully working to defeat or water down public efforts to reduce fossil fuel dependency." He goes on to add that "because this behavior will contribute to the deaths, displacement, and untold suffering climate change will ultimately cause, it can be seen as a form of *structural violence* [emphasis added]." And in *Inequality, Democracy, and the Environment*, American sociologist Liam Downey (2015) argues that the goals pursued by "undemocratic" and "elite controlled" organizations, institutions, and networks—and the mechanisms they use to pursue these goals—are the key drivers of environmental degradation in general and climate disruption specifically.

Frumhoff, Heede, and Oreskes (2015) argue that the major investor-owned producers of fossil fuels should be the focus of critical examination for several reasons. First, they have contributed the largest proportion of the total historic emissions that cause climate disruptions. Second, these major corporations have a high level of internal scientific expertise that allows them to be aware of and understand the findings of climate scientists. And, from a blameworthy harms perspective, Frumhoff, Heede, and Oreskes's third reason for highlighting these producers may be the most important. As they point out, "An alternative was available to them: they could have adjusted their business models to anticipate policies motivating a transition to low-carbon energy by substantially investing in low-carbon energy technologies, constructively engaging in policy design, and taking other steps to reduce the adverse impact of their products. But they did not. Even today, they continue to explore for new and increasingly more carbon-polluting sources of fossil fuels, encouraging the expanded use of the products that they know to be responsible for disruptive climate change" (Frumhoff et al., 2015: 159). Given that alternatives were available, the continued extraction of fossil fuels and the carbon emissions that result from the marketing and burning of these fuels can be considered blameworthy harms.

In addition to the fact that alternatives were available, the continuation of these organizational actions well past the time that the scientific evidence demonstrated the harms that resulted from greenhouse gas emissions is another major factor in establishing the blameworthiness of fossil fuel producers. As Heede's (2014) analysis shows, more than half of the greenhouse gas emissions from carbon major organizations have occurred since 1986, the year that Hansen first testified to Congress about the greenhouse effect and two years before his more famous Senate testimony brought the issue to greater public awareness. And, as a recent report from the Energy Information Administration (EIA) of the U.S. Department of Energy documents, the extraction and consumption of oil, coal, and natural gas continues to grow, fueling even more greenhouse gas emissions. According to the 2016 EIA report *International Energy Outlook*, even though renewable forms of energy are expanding more quickly than expected, fossil fuels are still projected to control close to 80 percent of the world energy market in 2040 and greenhouse gas emissions are expected to rise by an estimated 34 percent between 2012 and 2040. As American energy expert Michael Klare (2016b: 3) points out, "If such projections prove accurate, global temperatures will rise, possibly significantly above the 2-degree mark, with the destructive effects of climate change we are already witnessing today—the fires, heat waves, floods, droughts, storms, and sea level rise—only intensifying."

In his commentary on the EIA report, Klare (2016b) goes on to explore the roots of what he calls the world's "addiction" to fossil fuels, even in the face of what we know about their role in global warming. Part of the problem, he points out, is the result of the built-in momentum of the extractive industrial system that emerged historically around fossil fuel–powered energy systems. But Klare argues that the continuation of such high carbon energy systems, long after we have overwhelming scientific proof of the environmental destructiveness of the use of fossil fuels, can be blamed on the "vested interests" and actions (or lack thereof) of corporations and governments. He observes:

Energy is the largest and most lucrative business in the world, and the giant fossil fuel companies have long enjoyed a privileged and highly profitable status. Oil corporations like Chevron and ExxonMobil, along with their state-owned counterparts like Gazprom of Russia and Saudi Aramco, are consistently ranked among the world's most valuable enterprises. These companies—and the governments they're associated with—are not inclined to surrender the massive profits they generate year after year for the future wellbeing of the planet. As a result, it's a guarantee that they will employ any means at their disposal (including well-established, well-funded ties to friendly politicians and political parties) to slow the transition to renewables. (Klare, 2016b: 4)

Slowing the transition to renewables, seeking to charge "transmission" fees for dispersing solar energy, and in general continuing with "business as usual" means that these corporations and state agencies, despite their prior knowledge of the harmful effects and despite the existence of alternative courses of action, have made deliberate organizational decisions to continue to extract, market, and burn fossil fuels with the harmful result that greenhouse gas emissions continue to enter the atmosphere and heat the planet. It is important to reiterate that more than half of all greenhouse gas emissions produced by the carbon majors since the start of the Industrial Era have occurred within the last thirty years, "well past the date when governments and corporations became aware that rising greenhouse gas emissions from the burning of coal and oil were causing dangerous climate change" (Goldenberg, 2013b: 1).

The organizational actors who engage in these climate crimes can be examined at three distinct but interrelated levels of analysis. First, corporations and governments can be viewed as independent agents who can act on and shape their environment. As Perrow and Pulver (2015: 62) point out, market organizations can be the "independent variable," choosing to act in specific ways that have certain consequences for society. They can manufacture and advertise ecologically destructive products, choose harmful techniques of energy production, fund climate change denial campaigns, and lobby against environmental regulations. In Downey's (2015: 52) terms, these "undemocratic and elite-controlled organizations, institutions, and networks" possess "mechanisms" and "tools" that provide elites with "the means to define their interests, make decisions, and achieve their goals in the face of resistance from others." Specifically, within the framework of his inequality, democracy, and environment theory (IDE), these mechanisms "provide elites with the means to monopolize decision making power, shift environmental costs onto others, divert the public's attention away from what they are doing, frame what is and is not considered to be good for the environment, and shape individuals' knowledge, attitudes, values, beliefs, and behavior" (Downey, 2015: 54). Such organizational choices may emerge from the internal dynamics of the organization itself, such as its structure, goals, reward systems, internal culture, and past experiences (Perrow and Pulver, 2015).

Perrow and Pulver (2015) also point out, however, that market organizations exist within a larger "organizational field" and can be analyzed as a dependent variable, subjects of larger social forces in their institutional environments. Downey (2015), too, highlights the broader macro structural arrangements within which undemocratic and elite-controlled organizations, institutions, and networks exist. These sociologists argue that various external economic factors and political efforts can influence organizational decision making. Importantly, since business organizations generally operate within an industry, an additional structural influence can be examined. American criminologist Elizabeth Bradshaw (2015b: 379) observes that "an industry is comprised of numerous organizational units in competition for the same resources and customers and is differentially affected by a multitude of historical and political-economic factors." She has identified and analyzed various historical and institutional factors that have created a "criminogenic industry structure" among the companies engaged in offshore oil drilling specifically and within the larger fossil fuel industry more generally. In an important contribution to the state–corporate crime literature, Bradshaw (2015a) focuses on the historical role of the federal government in structuring the U.S. offshore oil industry and creating such a criminogenic environment.

Fossil fuel corporations and the industry as a whole also operate within a historically specific global political economy that shapes their actions. As Perrow and Pulver (2015: 64) note, "Organizations operate in market systems that embody power relationships and that are embedded in institutions that guide their functions." They are critical of the IPCC and environmental policymakers who speak of the "human drivers" of climate change without mentioning, let alone examining, the larger social structures and specific organizations that are responsible for the problem. Perrow and Pulver (2015: 64) emphasize "the need to focus on the *core driver* of [greenhouse gas] emissions, namely market organizations that are locked into and promote the *carbon economy* [emphasis added]." Thus, greenhouse gas emissions and the global environmental crisis must also be examined within the larger social, political, and economic structure of global capitalism and the process of capital accumulation.

These different levels of analysis can provide important insights into climate crimes. The organizational perspectives advanced by Perrow and Pulver (2015) and by Downey (2015) overlap to a considerable extent with the integrated theoretical model of state–corporate crime that Ray Michalowski and I (2006) have developed. Our original theoretical schema links three levels of analysis—macro (structural), meso (organizational), and micro (interactional)—with three catalysts for action: motivation (goals), opportunity (means), and formal social control (sanctions). Notably, Bradshaw (2015b) proposes industry as a fourth level of analysis in the model between structural and organizational.

Our objective in developing this schema was to inventory and highlight the key factors that may contribute to or restrain organizational crime at each intersection of a catalyst for action and a level of analysis. We viewed the organization (corporation or state agency) as the key unit of analysis, nested within an institutional and cultural environment and engaged in social action (sometimes separately and sometimes together) through the decisions of individual actors who occupied key positions within the structure of the organization. According to this model, organizational crime is most likely to occur when pressures for organizational goal attainment intersect with attractive and available illegitimate organizational means in the absence or neutralization of effective formal social controls. This integrated and dialectical approach views both organizations and the individuals who occupy positions within them as independent agents capable of shaping their environment as well as dependent variables who are also the subjects of larger structural and cultural forces.

A sociological explanation of climate crimes must identify the specific organizational actors involved, such as corporations like ExxonMobil or government agencies like the U.S. Department of Defense, as well as the individuals who occupy key decision-making (executive) positions within these organizations and the specific political economic structures or institutional environments that these organizations are located within: the fossil fuel industry, the carbon economy of global capitalism, and state institutions that produce "regimes of permission" to facilitate corporate harms (Whyte, 2014). The vast majority of Heede's carbon majors are companies in the fossil fuel industry. Adding military institutions such as the Pentagon into the mix, Urry (2011: 156) identifies what he calls a "carbon military-industrial complex," a set of "interlocking carbon interests" that "seeks to develop and extend major carbon based systems" around the globe. And there is a growing body of research that allows environmental sociologists to draw some conclusions about the characteristics of these undemocratic and elite-controlled organizational actors, to describe their goals and the mechanisms they use to achieve these goals, to analyze what they did to cause global warming, and to explain why they did what they did.

There is also a growing body of theory and research on the larger political economy within which these fossil fuel market organizations and their industry are located, that is, the institutional environment they operate within. This macro-structural context is variously referred to as global capitalism, corporate capitalism, the process of capital accumulation, the neoliberal market economy, the treadmill of production, fossil capital, state regimes of permission, or the carbon economy. In addition to this academic focus on the relationship between political and economic institutions and the climate crisis, there is a growing public awareness of these connections in popular media. Naomi Klein's (2014: 64) *This Changes Everything: Capitalism vs. the Climate* points out how the capitalist economy, neoliberal state policies (particularly concerning free trade), and free market fundamentalism have "helped overheat the planet." In an opinion piece in the *New York Times* titled "The Climate Crisis? It's Capitalism, Stupid," American scholar Benjamin Y. Fong (2017: 2) argues that "the real culprit of the climate crisis is not any particular form of consumption, production or regulation but rather the very *way* in which we globally produce, which is for profit rather than sustainability. . . . It should be stated plainly: It's *capitalism* that is at fault."

THE POLITICAL ECONOMY OF FOSSIL CAPITALISM

Climate scientists have documented that global greenhouse gas emissions began to increase during the era of the Industrial Revolution with the growing use of fossil fuels in connection with industrial production. The full emergence of the capitalist mode of production, the expansion of the market economy, and the subsequent process of industrialization were central elements of what Polanyi (1944) called the "Great Transformation" that emerged in Europe and later spread to other parts of the world primarily through the force of colonization. The discipline of sociology originated in the attempt to understand this transformation, which culminated in the rise of the "modern" world, the triumph of capitalism, the rise of the infrastructural power of the capitalist state, and the use of markets as the dominant mode for structuring economic relations, along with significant intellectual, cultural, and political revolutions such as the Enlightenment and the French Revolution. Another transformational dimension of the emergence of the modern social world, unintended and unrecognized at the time, was that it set humanity on the path to the current climate crisis. Yet, despite the clear linkage between the development of capitalism, the process of industrialization, and the rise of markets, Perrow and Pulver (2015: 63) argue that "action by capitalist, industrial organizations in markets has been only a peripheral focus in analyzing the problem of global warming."

There is a growing effort in macro-structural environmental sociology and green criminology to move this focus on the political economy of the climate crisis to the center of disciplinary concerns. One important contribution to this effort is Malm's *Fossil Capital: The Rise of Steam Power and the Roots of Global Warming*. Malm (2016: 11) analyzes the historical origins of what he calls the fossil economy, "an economy of self-sustaining growth predicated on the growing consumption of fossil fuels, and therefore generating a sustained growth in the emissions of carbon dioxide." He argues that the fossil economy emerged in the late eighteenth and early nineteenth centuries during the Industrial Revolution when "the fossil fuel of coal was coupled to the machine through the rise of stationary steam power in the mills of Britain" (Malm, 2016: 16). And he contends that British manufacturers turned from traditional sources of power such as wind and water mills (*the flow*) to an engine fired by coal (*the stock*), not because coal was a cheaper or more abundant form of energy but because steam power gave capitalists greater control over both labor and the energy source. Production could be concentrated at urban sites that were more profitable, during hours that were more convenient. The steam engine was a technology that had greater value for capital. Steam won because it augmented the power of the men who owned the means of production in the class struggle of the early Industrial Era. According to Malm (2016: 277), "capitalism gave birth to the fossil economy. The rise of steam power—the critical moment in the delivery—occurred in a country with a distinctly capitalist mode of production: two centuries later, the context remains the same, the umbilical cord as strong as ever." His comment that, centuries later, the context remains the same is important. As we examine rising greenhouse gas emissions today, it is critical to understand the history and continuing influence of fossil capitalism.

Fossil fuel–based technologies became locked into the capitalist mode of production during the Industrial Revolution and laid the foundation for current levels of carbon emissions and global warming, although, as Angus (2016) points out, they increased dramatically after World War II during the Great Acceleration. While Malm acknowledges that a fossil economy does not necessarily have to be capitalist (the extractive, industrial, state socialist system in the Soviet Union and its satellite states were also responsible for a high level of carbon pollution and environmental destruction), only fossil capital remains today. At this point in history, Malm (2016: 278) argues, "the fossil economy is coextensive with the capitalist mode of production—only now on a global scale." The climate crimes of today, therefore, must be analyzed in the context of this global capitalism. As Malm (2016: 273–274) observes, for the steam engine, and "all subsequent technologies of fossil fuel combustion in class societies: their stunning power to change the climate of the earth has followed from *their value for their owners* as distinct from non-owners."

To analyze the dialectical relationship between nature and the economy, to develop a "world ecology of capitalism," Jason Moore argues that we need to go much further back in history than the Industrial Revolution and the rise of steam power built on the stock of coal. In *Capitalism in the Web of Life: Ecology and the Accumulation of Capital*, Moore (2015: 173) examines the early modern origins of capitalism and "its extraordinary reshaping of global natures long before the steam engine." He argues that during the "long sixteenth century" and the early "epochal transition" to capitalist social relations, a "new pattern of environment-making" began in the Atlantic world. This new pattern was so powerful and consequential that he raises the question of whether we are really living in the Anthropocene or in an era we should refer to as the *Capitalocene*. As Moore (2015: 172) sums up,

> To locate the origins of the modern world with the steam engine and the coal pit is to prioritize shutting down the steam engines and the coal pits (and their twenty-first century incarnations). To locate the origins of the modern world with the rise of capitalist civilization after 1450, with its audacious strategies of global conquest, endless commodification, and relentless rationalization, is to prioritize the relations of power, capital, and nature that rendered fossil capitalism so deadly in the first place. Shut down a coal plant, and you can slow global warming for a day; shut down the relations that made the coal plant, and you can stop it for good.

Informed by these historical perspectives on the early modern origins of capitalism and the rise of the fossil economy during the Industrial Revolution, the field of environmental sociology contains many theories that analyze how the current deep-seated structural forces of global capitalism produce environmental harms and ecological disorganization. Treadmill of production theory, world system theory, ecological rift theory, regulation theory, and Marxian-rooted globalization theories all suggest that the climate crisis and other forms of ecological destruction that are occurring today are rooted intrinsically in capitalism per se (Antonio and Clark, 2015). These theories focus on the system imperative of economic growth and the enduring conflict between the capitalist mode of production and protection of the environment. They argue that the drive for endless economic growth and ever greater profits leads to increasing levels of environmental degradation. As sociologists Robert J. Antonio and Brett Clark (2015: 353) point out, these theories argue that under capitalism, "The highly stratified global interstate system imposes a division of nature that replicates the flow of economic surplus, while it also creates the structural conditions that accelerate climate change."

Out of this group of sociological theories that link global capitalism to environmental harm, the one that has had the most impact within environmental sociology and green criminology is treadmill of production (TOP) theory. Initially developed by environmental sociologist Allan Schnaiberg in his 1980 book *The Environment: From Surplus to Scarcity* and developed further by Gould, Pellow, and Schnaiberg (2008), TOP theory argues that there is an inherent and deadly contradiction between the political economy of global capitalism and ecological health. Human economic systems—particularly capitalism with its expansionary logic—extract natural resources and transform them through a production process that interferes with the organization of ecological systems and results in ecosystem disruption and ecological disorganization. The routine operation of the capitalist system with its built-in demand for endless economic growth produces both commodities and environmental devastation hand in hand, a suicidal strategy that Wright and Nyberg (2015: 32) term "creative self-destruction."

Market organizations such as transnational corporations in the fossil fuel industry, state-owned oil and gas producers, military institutions, and other state agencies "can be interpreted as connected and embedded in the treadmill and are hence influenced by treadmill factors such as the drive to expand profits and production" (Stretesky et al., 2014: 13). In addition to these corporate and state actors, labor organizations also often have short-term interests in supporting increases in production (Schnaiberg, 1980). According to TOP theorists, these organizational actors are driven by the logic of industrial capitalism to expand production by extracting natural resources (raw materials and fossil fuels), transforming them into products to be sold in the marketplace, and discharging various waste by-products such as carbon emissions and other forms of pollution into the ecosystem, causing environmental catastrophe. TOP theory defines the extraction of natural resources necessary for production as *ecological withdrawals* and the pollutants that are emitted during the production process as *ecological additions*—impacts that disorganize the ecology. According to Stretesky and colleagues (2014: 22), the concept of *ecological disorganization* occupies an important place in TOP theory, noting that "Ecological disorganization occurs because capitalist production negatively impacts the relationships between organisms and their environment and impacts the self-sufficiency of the earth's biological system." Climate disruption represents a major form of ecological disorganization.

Environmental sociologists who have examined global warming and the climate crisis through the lens of TOP theory generally find it to be empirically supported and theoretically convincing. It has been lauded as one of the "most influential research programs for investigating these socio-environmental dynamics" (D. White, Rudy, and Gareau, 2016: 93). Perrow and Pulver (2015: 71) point out that "global and cross-national comparative studies of carbon emissions trajectories suggest that the treadmill logic predominates." Following their identification of TOP theory as one of

the dominant theoretical orientations in the field, Eugene Rosa and his colleagues (2015: 52) observe: "One of the most important overall lessons of
sociological research on the anthropogenic forces driving global climate
change is that it is necessary to look beyond technical fixes and consider the
social, political, and economic structures that condition human behavior
and resource exploitation." Contrasting the treadmill argument against
other theoretical approaches, Wright and Nyberg (2015: 33) state that it
"provides a much more convincing explanation for the climate crisis we
now face."

Within green criminology, TOP theory has been most extensively
developed and applied by Stretesky, Long, and Lynch (2014) in their important
book *The Treadmill of Crime: Political Economy and Green Criminology*. These
criminologists were attracted to the treadmill approach because of their
concern with exploring the structural forces that lead to environmental
harm, specifically the political economy of green crimes. As they point out,
"Schnaiberg's theory not only provides a comprehensive, radical framework
that can aid in developing political economic explanations of green crime,
it specifically focuses attention on the forms of ecological disorganization
capitalism produces" (Stretesky et al., 2014: 4). They also argue that when
TOP theory is used as an organizing framework within green criminology,
ecological withdrawals and ecological additions come to be seen as green
crimes, acts that cause environmental harm and ecological disorganization.
Stretesky, Long, and Lynch (2014: 147) propose that "green crime should be
organized in this fashion because withdrawals and additions present a language that focuses attention on the relationship between the economy and
the ecology."

In the concluding section of *The Treadmill of Crime*, the authors draw an
explicit connection between TOP theory and the concept and theory of
state–corporate crime that I apply throughout this book. They note that
"the relationship between the state and corporations that is examined in
TOP theory is readily connected to the notion of state-corporate crime as
developed by Michalowski and Kramer" (Stretesky et al., 2014: 149). More
specifically, they point out that our concept of state-facilitated crime is
"highly compatible" with the notion of green crime and TOP theory because
it can be applied to the analysis of what I am calling *crimes of omission*, the
failure of state laws and regulatory systems to address environmental harms
like global warming and climate change. Stretesky and colleagues (2014:
149) conclude that "Connections that are identified in the state-corporate
crime literature, then, are central in better establishing a political economy
of green crime." In other words, TOP theory examines the macro-structural
context within which specific organizational actors, such as states and corporations, set goals, make decisions about which means to utilize in pursuing

these goals, and seek to influence or neutralize the controls and constraints within their environment. State–corporate crime theory complements TOP theory by examining how macro-structural forces shape the work-related thoughts and actions of those who occupy positions within organizational structures and networks. Similarly, Downey (2015) views his IDE model as a complement to TOP theory because, like state–corporate crime theory, IDE identifies the specific organizations, institutions, and networks that elites use to attain goals such as capital accumulation through specific mechanisms that result in environmental harms.

This theoretical approach is important not only to understand why the climate crimes described in this book are occurring but also to provide some guidance for how we can resist these crimes and struggle to achieve climate justice. A sociological focus on the institutional and organizational forces that drive climate change ecocide makes clear the necessity for some form of structural change, difficult as it may be to achieve. It illustrates the simple yet critical fact that, as Fong (2017: 3) puts it, "the work of saving the planet is political, not technical." As I argue in the final chapter, achieving climate justice will require the building of a powerful mass social movement that can challenge, effectively regulate, and ultimately restructure the political economy of capitalism.

CHAPTER 3

"When Did They Know?"

CLIMATE CRIMES OF CONTINUED
EXTRACTION AND RISING EMISSIONS

Big Oil evidently has no qualms about making its next set of profits directly off melting the planet. Its top executives continue to plan their futures, and so ours, knowing that their extremely profitable acts are destroying the very habitat, the very temperature range that for so long made life comfortable for humanity. Their prior knowledge of the damage they are doing is what should make this a criminal activity.
—Tom Engelhardt, *Shadow Government: Surveillance, Secret Wars, and a Global Security State in a Single Superpower World* (2014: 142)

DURING THE SUMMER of 1973, my brother Bob and I became Watergate junkies. We keenly followed all the news reports that summer related to the political scandal that was engulfing Richard Nixon's White House concerning the 1972 break-in at the headquarters of the Democratic National Committee within the Watergate Building in Washington, DC, and the subsequent cover-up that would eventually lead to Nixon's resignation in August 1974. On May 18, 1973, the Senate Select Committee on Presidential Campaign Activities, better known as the Watergate Committee, began holding televised public hearings on the growing scandal. Home on summer break from college, my brother and I religiously watched the hearings from Capitol Hill. We both had factory jobs that summer but worked different shifts. Bob watched the hearings during the morning and early afternoon while I was at work, and I took over the television-viewing duties after my shift ended as he rushed off to his job. Later, we would compare notes and catch each other up on what we had missed.

Sam Ervin, a folksy Democratic senator from North Carolina, chaired the Watergate Committee and became a media star. But it was the ranking Republican member of the committee, Senator Howard Baker of Tennessee, that most impressed me. Throughout that summer, as I watched the senators carry out their investigation into one of the most serious political crimes and cover-ups in American history, my admiration for the calm and serious Baker grew. Baker was a moderate conservative in a political party moving toward a more extreme ideological position throughout that decade. His work in the Senate had earned him the respect of his colleagues, Republican and Democratic alike. He would go on to become the White House chief of staff during the second term of President Ronald Reagan and the United States ambassador to Japan from 2001 to 2005. But Baker is best known for the questions he posed to former White House counsel John Dean on July 29, 1973, during the Watergate hearings. In retrospect, Baker was probably trying to help his party's president by pushing the blame for the scandal onto others within the administration, but his questions came to define the key issue concerning Watergate and eventually served to indict Nixon in the public mind and in the political arena. Eyeglasses in hand, gently gesturing toward Dean, Baker asked: "What did the president know, and when did he know it?" Whether intended or not, implied in these momentous questions was another one: If Nixon did know something, what did he do, or not do, with this knowledge? Ultimately it was shown that Nixon did know about the break-in early on and was involved in a criminal cover-up, which led to his resignation.

Baker's famous Watergate questions provide a useful framework for examining the climate crimes of continued extraction and rising emissions. Research shows that major investor-owned fossil fuel corporations are among the ninety largest industrial carbon producers, the products of which are responsible for nearly two-thirds of all known industrial greenhouse gas emissions since 1751 (Heede, 2014). Echoing Baker, the critical questions are: What did these carbon majors know about the relationship between greenhouse gas emissions resulting from the burning of fossil fuels and the problem of global warming, and when did they first know it? Substantial evidence shows that companies in the energy industry have known about the relationship between carbon dioxide and harmful global warming since at least the early 1960s. And once these corporations knew, what actions did they take or fail to take? Documents and insider testimony indicate that they used this knowledge to plan for future explorations of oil and to climate-proof their facilities, all the while doing nothing to reduce the resulting carbon pollution and climate disruption.

Some scholars argue that "an alternative was possible" for the industry, a path away from coal, oil, and gas and toward renewable sources of energy (Frumhoff, et al., 2015: 157). But instead of pursuing a socially responsible alternative to their destructive business practices, these corporations not only engaged in a concerted effort to deny that global warming was occurring at all but also lobbied effectively to prevent government regulation of carbon emissions and to block the development of renewable energy. Despite what they knew about the greenhouse effect, fossil fuel companies continued with business as usual, allowing greenhouse gas emissions to rise to what they were warned by their own scientists would be dangerous levels. I argue that these organizational actions and omissions are morally blameworthy harms that can be defined as climate crimes.

CONTINUED EXTRACTION AND RISING EMISSIONS AS BLAMEWORTHY HARMS

The first and most central "core characteristic" of crime in Robert Agnew's (2011b) integrated definition of crime—the definition that I am using to identify climate crimes—is that there must be a "blameworthy harm." While the other two core characteristics, public condemnation and state sanction, are important and will also be examined, the central focus of this chapter is the responsibility for blameworthy harms related to fossil fuel extraction, rising greenhouse gas emissions, and climate change. Condemnation of specific harms by the public and legal sanctioning of responsible actors by the state may emerge more slowly (if at all) once the harms are revealed (as they eventually have for the continued extraction of fossil fuels and rising carbon emissions). Due to this time lag, Agnew also identifies unrecognized blameworthy harms, a range of crimes that are often neglected by the public and the state (and, it should be noted, by most criminologists as well). Much state and corporate harm in general, and climate harms specifically, fall into this category. Agnew (2011b: 43) argues that it should be "a major mission" of criminologists to make "the public and state aware of unrecognized blameworthy harms," particularly those committed by corporate and government actors.

While the evidence of environmental and social harm caused by climate change is overwhelming, can we make the determination that market organizations like the corporations in the fossil fuel industry, that are primarily responsible for the greenhouse gas emissions that result in climate disruption, are blameworthy actors? That is, can we, at least for criminological purposes, classify their actions as *crime*? Agnew states that blameworthiness involves voluntary and intentional behavior, broadly defined as purposeful, knowledgeable, reckless, or negligent. These acts range along a continuum from those that are purposely committed to harm another to those committed not to harm but with the knowledge that harm is a highly likely outcome, a risk a "reasonable person" should have been aware of. Agnew adds that the harmful act must also be "unjustified" and "inexcusable," since most legal jurisdictions provide an array of justifications and excuses for criminal behavior such as insanity, duress, necessity, and self-defense. In conclusion, he states that "blameworthy harms are those for which individuals or groups bear some responsibility, are unjustified, and are inexcusable," adding that "researchers who employ these criteria for blameworthiness argue that many harmful acts which are *not* defined as crimes in the United States, *should be* defined as crimes" [emphasis in original] (Agnew, 2011b: 25).

The organizational actions of extracting and marketing fossil fuels, are voluntary and intentional actions, committed without justification or excuse, and I argue that they should be defined as crimes. As Klein (2014: 169) observes, the mentality of *extractivism*, "a nonreciprocal, dominance-based relationship with the earth, one of purely taking," developed under capitalism. And governments across the ideological spectrum have, for a very long time, embraced "this resource-depleting model as a road to development." Yet, this long-standing logic has been profoundly called into question by the accumulation of scientific knowledge about climate change in the twentieth century. Although in the beginning corporate executives in the fossil fuel industry did not purposely intend to harm others through their extractivist organizational decisions, at some point they did have knowledge that ecological and social harm was a highly likely outcome of their actions, or they had knowledge that there was a grave risk of harm in their decisions, or at the very least as "reasonable" social actors they should have been aware of the environmental risks their behavior imposed.

The capitalists who decided to utilize steam power during the early days of the Industrial Revolution in Britain did intend, as Andreas Malm (2016) documents, to advance their economic interests in the class struggle against the working class. These men, however, had no way of knowing that their fateful decision to shift from *the flow* (water and wind power) to *the stock* (fossil fuels) and create a *fossil economy* would one day heat the planet and cause massive environmental harm. Their twentieth-century counterparts, especially the men and women who occupy the top positions within corporations in the fossil fuel industry, cannot claim such innocence. Organizational decision makers in fossil fuel companies have had knowledge that ecological and social harm was a highly likely outcome of their actions since at least 1988, when the issue of global warming became a major public and political issue (and often much earlier). I argue that the continuing extraction and marketing of fossil fuels, and the subsequent emission of greenhouse gases, after 1988 (and on into the future) are blameworthy harms of great magnitude and consequence and can be defined as climate crimes, even though public condemnation of and legal sanctions for these acts have been slow to emerge. Frumhoff, Heede, and Oreskes (2015: 166) point out that "These companies plan for a future in which the world will continue to rely on fossil fuels at levels that will lead to highly disruptive climate impacts." And, as Engelhardt (2014: 142) argues in the epigraph that opens this chapter, concerning oil executives, "Their *prior knowledge* of the damage they are doing is what should make this a *criminal activity* [emphasis added]."

This prior knowledge predates the momentous events of 1988. Evidence has recently emerged that corporations in the oil industry knew of the relationship between the emission of carbon dioxide from the burning of fossil fuels and global warming as early as the 1950s. Enough was known by 1959, for instance, that the Hungarian American nuclear weapons physicist Edward Teller could warn more than 300 oil industry executives, government officials, and scientists gathered at the Energy and Man symposium, organized by the American Petroleum Institute (a prominent oil industry association), that due to rising concentrations of carbon dioxide in the atmosphere, the planet will get warmer, the ice caps will start melting, and the level of the oceans will begin to rise, all before the end of the twentieth century (Franta, 2018). More important, hundreds of internal documents related to the oil industry's understanding of the risks of carbon dioxide have been unearthed from various archives by a Washington, DC, environmental law organization, the Center for International Environmental Law (CIEL). The documents assembled by CIEL show that in the 1950s and 1960s oil companies monitored the published academic research into rising carbon dioxide levels, started to carry out their own internal research on this issue, and were directly informed by the industry's leading pollution control consultants about the environmental harms that could result.

In 1968, Elmer Robinson and R. C. Robbins, consultants from the Stanford Research Institute (SRI), advised in a paper to the American Petroleum Institute that carbon dioxide was "the only air pollution which has been proven to be of global importance to man's environment on the basis of a long period of scientific investigation" and that "if CO_2 levels continue to rise at present rates, it is likely that noticeable increases in temperatures could occur" (Banerjee, Cushman, et al., 2016: 2). The SRI consultants went on to warn the fossil fuel industry that "changes in temperature on the world-wide scale could cause major changes in the earth's atmosphere over the next several hundred years including change in the polar ice caps" (Banerjee et al., 2016: 2). The Robinson and Robbins paper and other documents discovered by CIEL show that corporations in the oil industry, and their major trade association, were aware of the problem of rising carbon dioxide levels and the serious risks of climate change by the 1960s. Yet, despite carrying out some internal research on these issues, fossil fuel companies would engage in decades of deceit and denial to stave off regulations that could reduce carbon emissions and slow dangerous global warming. Despite the intrusion of climate science into the industry, the pressure for profits and growth and the dominance of the logic of extractivism ensured that the level of greenhouse gases in the atmosphere would continue to rise.

THE FOSSIL FUEL INDUSTRY AND GLOBAL WARMING'S TERRIFYING NEW MATH

In the summer of 2012, climate activist Bill McKibben published an article in *Rolling Stone* magazine that graphically demonstrated how, despite what is widely known about the connection between carbon dioxide and climate disruption since at least the 1960s, the fossil fuel industry continues to extract and market oil, gas, and coal. McKibben illustrated this climate crime with what he called "global warming's terrifying new math." Based on research from a London-based think tank, the Carbon Tracker Initiative, McKibben presented three simple numbers—2, 565, and 2,795—that he argued "add up to global catastrophe—and that make clear who the real enemy is" (McKibben, 2012a).

The first number is 2 degrees Celsius (3.6 degrees Fahrenheit), which is the upper limit that has generally been accepted within the international political community that the temperature of the planet can rise over the pre-industrial average without catastrophic results. And we are halfway there already: the average temperature of the planet has already increased by 1 degree Celsius (1.8 degrees Fahrenheit). The 2-degrees Celsius target has been discussed as a political objective since the mid-1990s and was set as the acceptable upper limit as part of the accord at the end of the Copenhagen climate conference in 2009. Although that conference is widely regarded as a failure, the world community did formally recognize the scientific perspective that the increase in global temperature should be held below 2 degrees Celsius. Some scientists, however, such as James Hansen, regard this target as entirely too lax and a prescription for disaster. At the UN Climate Change Conference in Paris in 2015, there was a strong push by several nations to lower the target to 1.5 degrees Celsius, a number that gained recognition as an "aspirational" goal in the formal agreement and has since been endorsed by the IPCC in a 2018 report. But in this 2012 article McKibben used the more established 2-degree figure.

In global warming's terrifying new math, the second number was 565 gigatons. A gigaton is one billion metric tons, and 565 was the scientific estimate of how many more gigatons of carbon dioxide could be safely released into the atmosphere by midcentury without exceeding the 2-degree Celsius target for global temperature increase. McKibben pointed out that the idea of a global "carbon budget" emerged early in the twenty-first century as climate scientists began to calculate how much more oil, gas, and coal could still be safely burned without dangerously overheating the planet. The 565-gigaton figure was derived from a sophisticated computer simulation model and has been widely confirmed by climate scientists around the world. Yet, humans continue to pour record amounts of carbon into the atmosphere, and study after study predicts that these emissions will continue to grow at roughly 3 percent per year. At that rate, McKibben (2012a: 4) observed, "we'll blow through our 565-gigaton allowance in 16 years, around the time that today's [2012] preschoolers will be graduating from high school."

The third number, 2,795 gigatons, was the most terrifying number in the new math. As McKibben (2012a: 4) pointed out, "The number describes the amount of carbon already contained in the proven coal, oil and gas reserves of the fossil-fuel companies, and the countries (think Venezuela or Kuwait) that act like fossil-fuel companies." This is the coal, oil, and gas that, according to the Carbon Tracker Initiative, the industry is currently planning to extract, sell, and burn in the future—an amount more than five times higher than the 565 gigatons we can safely burn and remain below the 2-degree target. As McKibben (2012a: 5) noted, "We have five times as much oil and coal and gas on the books as climate scientists think is safe to burn. We'd have to keep 80 percent of those reserves locked away under-ground to avoid that fate. Before we knew those numbers, our fate had been likely. Now, barring some massive intervention, it seems certain." The climate math in 2012 was indeed terrifying and foretold a very bleak future unless the world quickly mobilized to "keep fossil fuels in the ground" (Cohan, 2014). But the story gets worse.

New numbers contained in a 2016 report by Oil Change International (OCI), a Washington, DC-based think tank, using data purchased from Norwegian energy consulting firm Rystad, are even scarier than McKibben's original analysis. In this report, climate scientists newly estimated that to have even a two-thirds chance of staying below the 2-degree Celsius benchmark, only 800 more gigatons of carbon dioxide can be released into the atmosphere. But according to OCI, the Rystad data indicate that the currently operating coal mines and gas wells worldwide contain 942 giga-tons worth of carbon dioxide. And these 942 gigatons are what is on the table right now, the amount of carbon in the coal, gas, and oil that fossil fuel companies are currently extracting and burning; this figure does not include the 2,795 gigatons of proven reserves that they have on their books and are planning to extract and market in the future. And the math gets even worse if we take the safer (and now officially endorsed) 1.5-degree Celsius warming mark into consideration. To have just a fifty-fifty chance to meet that goal, only 352 more gigatons of carbon dioxide can be released. The political and economic implications of this *new* new math are stark: the only realistic approach to limiting global warming and climate disruption, to avoid wreaking havoc on the planet, is to keep those fossil fuels in the ground. As McKibben (2016d: 2) argues, "To have just a break-even chance of meeting that 1.5-degree goal we solemnly set in Paris in 2015, we'll need to close all of the coal mines and some of the oil and gas fields we're currently operating long before they're exhausted." To live up to the Paris Agreement, then, we must, in the words of OCI's executive director Stephen Kretzmann, start a "managed decline" in the fossil fuel industry immediately (quoted in McKibben, 2016d: 2).

Despite what we have known about the connection between greenhouse gas emissions and global warming since at least 1988 (a generous date), and despite the new climate math, keeping coal, oil, and gas in the ground and managing the decline of the fossil fuel industry are not on the agenda of any of the large extractivist corporations that dominate this industry. And again, as Frumhoff, Heede, and Oreskes (2015: 159) point out, "an alternative was possible to them: they could have adjusted their business models to anticipate policies motivating a transition to low-carbon energy by substantially investing in low-carbon energy technologies, constructively engaging in policy design, and taking other steps to reduce the adverse impact of their products." But they did not do any of this. Alternative courses of action receive little serious consideration by the undemocratic and elite-controlled fossil fuel companies dominated by an extractivist mentality.

An obvious reason is money (profit), as the financial problem for the industry is clear. If the currently operating coal mines or oil and gas fields are closed or restricted, and if planned carbon reserves are also kept in the ground, fossil fuel corporations would be writing off more than ten trillion dollars in assets (C. Hayes, 2014). This is not something that these market organizations are willing to do within the economic constraints of the global capitalist system. Continuing to extract and market fossil fuels, with full knowledge of the harms and risks resulting from the rising greenhouse gas emissions that heat the planet, is the normal, extremely profitable business plan of this industry. As Klein (quoted in McKibben, 2012a: 7) puts it, "Lots of companies do rotten things in the course of their business . . . but these numbers make clear that with the fossil-fuel industry, wrecking the planet is their business model. It's what they do." McKibben (2012a: 7) uses even harsher language: "But what all these climate numbers make painfully, usefully clear is that the planet does indeed have an enemy—one far more committed to action than governments or individuals. Given this hard math, we need to view the fossil-fuel industry in a new light. It has become a rogue industry, reckless like no other force on Earth. It is Public Enemy Number One to the survival of our planetary civilization. . . . The numbers are simply staggering—this industry, and this industry alone, holds the power to change the physics and chemistry of our planet, and they are planning to use it." McKibben's terrifying new math (and OCI's ominous *new* new math) makes clear that fossil fuel corporations can indeed be considered carbon criminals.

WHAT DID EXXON KNOW?

To further explore the climate crimes of ecological withdrawal (extraction) and ecological addition (emissions), I turn to recent journalistic investigations of the leading corporation in the fossil fuel industry. ExxonMobil is the largest, most profitable, and most powerful private corporation in the world. Pulitzer Prize–winning American journalist Steve Coll titled his penetrating examination of the mighty oil giant, what McKibben (2012b) calls "The Ultimate Corporation," *Private Empire: ExxonMobil and American Power.* Coll (2012) meticulously documents that this powerful "private empire" sustains an extreme culture of resistance to government regulation that goes back to the early history of the company when it was split off from John D. Rockefeller's Standard Oil monopoly in 1911 and finds expression more recently in ExxonMobil's campaign of "disinformation" concerning climate science and its "opposition" to regulations on carbon emissions. First operating as Jersey Standard (Standard Oil of New Jersey), the company became Exxon in 1972, a new, unified brand name for all its former outlets, Enco, Esso, and Humble. In 1999 Exxon merged with another fossil fuel giant, Mobil (formerly known as Socony, or Standard Oil of New York), to become today's ExxonMobil (essentially undoing the anti-monopoly regulation under the Sherman Act that was imposed on Standard Oil in 1911).

The story about what Exxon knew about the relationship between carbon dioxide and global warming and what it did and did not do with that important scientific information is well documented. A trove of internal Exxon documents from the 1970s and 1980s, discovered recently by investigative journalists, demonstrate that the company's own researchers had confirmed the emerging consensus among outside scientists concerning the role that fossil fuels played in anthropogenic global warming and the threat of climate disruption. In response to this information, Exxon's Research and Engineering department launched an innovative ocean research program and began to develop computer models to predict global temperature increase. On the basis of this internal scientific information concerning global warming, Exxon also began to climate-proof its facilities and infrastructure against rising seas and melting permafrost and to plan for future oil exploration in the Arctic. At first, Exxon shared its scientific findings and participated in workshops and policy deliberations. But in the late 1980s, when global warming emerged as a major public policy issue, the company did an about-face and became a significant force in the global warming denial countermovement, in a concerted effort to reject climate science and obstruct government policies to mitigate global warming.

Exxon's attempts to corrupt the debate on global warming through the funding of proxy groups that engage in denial and deception concerning climate science, through newspaper ads and political lobbying activities that involve the dissemination of disinformation, and through campaign contributions to climate change–denying politicians have been documented by environmental groups like Greenpeace (2011, 2013a), investigative journalists like Coll (2012), and various sociologists studying the climate change denial countermovement (Brulle, 2013; Farrell, 2015; McCright and Dunlap, 2000, 2003). The internal company documents discovered in the various investigations show that despite the knowledge that its current operations and future plans were jeopardizing the planet, Exxon not only engaged in global warming denial and deception; it also continued to extract and market oil, which ensured that carbon wastes would continue to pour into the atmosphere. Even though alternatives were available, Exxon continued with "business as usual," which resulted in the crimes of ecological withdrawal and ecological additions.

Evidence concerning these blameworthy harms has emerged from three separate and independent investigations. InsideClimate News (ICN), a Pulitzer Prize–winning, nonprofit, nonpartisan news organization dedicated to covering climate change, energy, and the environment, conducted an eight-month investigation into what Exxon knew about climate change science and what the corporation did with this knowledge. The ICN investigation was based on interviews with former Exxon employees, scientists, and federal officials and hundreds of pages of internal Exxon documents, as well as documents from archives at the ExxonMobil Historical Collection at the University of Texas at Austin's Briscoe Center for American History, the Massachusetts Institute of Technology, and the American Association for the Advancement of Science. This research resulted in the publication of nine articles on its website in the fall of 2015 that were then gathered together with additional materials and published as an e-book titled *Exxon: The Road Not Taken* (Banerjee, Cushman, et al., 2015). In collaboration with ICN, a short Public Broadcasting System (PBS) *Frontline* video was broadcast on this topic on September 16, 2015.

The *Los Angeles Times*, in collaboration with the Energy and Environmental Reporting Project at Columbia University's Graduate School of Journalism (CSJ), conducted a similar but independent investigation. Based on research published in scientific journals, interviews with dozens of experts, including former Exxon employees, and documents housed in the Glenbow Museum in Calgary, Alberta, and at the ExxonMobil Historical

Collection at the University of Texas at Austin's Briscoe Center for American History, the investigation resulted in a series of articles in the *Los Angeles Times* in late 2015. Finally, the climate activist website DeSmogBlog (desmogblog.com) found additional Exxon corporate documents in an Imperial Oil (Exxon's Canadian subsidiary) archival collection at the Glenbow Museum and analyzed them for the public in a post titled "'There Is No Doubt': Exxon Knew CO_2 Pollution Was a Global Threat by Late 1970s" (April 26, 2016). Many of the archival documents related to Exxon's climate history are available at http://climatefiles.com.

The ICN investigation, carried out by veteran journalists Neela Banerjee, John H. Cushman, Jr., David Hasemyer, and Lisa Song, revealed that top executives at Exxon were warned in no uncertain terms in the late 1970s that the burning of fossil fuels would heat the planet and cause climate disruptions that could eventually endanger humanity. In July 1977, at a meeting in corporate headquarters, a senior company scientist named James F. Black delivered a sobering assessment to Exxon's Management Committee: "There is general scientific agreement that the most likely manner in which mankind is influencing the global climate is through carbon dioxide release from the burning of fossil fuels" (Banerjee, Song, and Hasemyer, 2015b: 1). A year later, Black, a top technical expert in Exxon's Research and Engineering division, delivered an updated version of his report to a broader audience of Exxon scientists and managers. According to Banerjee, Song, and Hasemyer (2015b: 1), he warned that "independent researchers estimated a doubling of the carbon dioxide (CO_2) concentration in the atmosphere would increase average global temperatures by 2 to 3 degrees Celsius (4 to 5 degrees Fahrenheit), and as much as 10 degrees Celsius (18 degrees Fahrenheit) at the poles. Rainfall might get heavier in some regions, and other places might turn to desert." While Black correctly noted some of the uncertainties about the details of global warming in the scientific literature at that time, he nonetheless urged in the 1978 summary that quick action was necessary: "Present thinking holds that man has a time window of five to ten years before the need for hard decisions regarding changes in energy strategies might become critical" (Banerjee, Song, and Hasemyer, 2015b: 2). Such information concerning global warming was also well known at Exxon's Canadian subsidiary, Imperial Oil. DeSmogBlog uncovered a 1980 report by scientists at the company titled "Review of Environmental Protection Activities for 1978–1979" that stated unequivocally: "It is assumed that the major contributors of CO_2 are the burning of fossil fuels. . . . There is no doubt that increases in fossil fuel usage . . . are aggravating the potential problem of increased CO_2 in the atmosphere" (Demelle and Grandia, 2016: 1).

Exxon responded quickly to Black's reports and other scientific information concerning the greenhouse effect by launching its own ambitious research program concerning the environmental impact of the release of carbon dioxide from fossil fuels. In 1979, the company outfitted its largest supertanker, the *Esso Atlantic*, with custom-made instruments to measure concentrations of carbon dioxide in the air over the oceans and in the water. In 1980, Exxon funded elaborate computer models to investigate important questions about the climate's sensitivity to the build-up of carbon dioxide in the atmosphere. An official from the U.S. Department of Energy lauded these corporate research contributions. And, on the basis of their interviews, Banerjee, Song, and Hasemyer (2015b: 2) note that "Working with university scientists and the U.S. Department of Energy, Exxon strove to be on the cutting edge of inquiry into what was then called the greenhouse effect. Exxon's early determination to understand rising carbon dioxide levels grew out of a corporate culture of farsightedness, former employees said. They described a company that continuously examined risks to its bottom line, including environmental factors. In the 1970s, Exxon modeled its research division after Bell Labs, staffing it with highly accomplished scientists and engineers." It seems apparent that, at least at this time, some Exxon managers and scientists wanted to create a project of excellence concerning the greenhouse effect that could be "aimed at benefitting mankind" (Banerjee, Song, and Hasemyer, 2015b: 3).

One key result of this internal research program is that Exxon possessed clear and convincing information on the causes and effects of global warming in 1980 and knew of the public importance of that knowledge. A memo from that year laid out an ambitious public relations plan and indicated that Exxon wanted a seat at the policy-making table when this issue was to be addressed. The PR plan aimed at "achieving national recognition of our CO_2 Greenhouse research program." The memo went on to state, with perhaps a bit of foreshadowing of a different approach in the future, that "It is significant to Exxon since future public decisions aimed at controlling the buildup of atmospheric CO_2 could impose limits on fossil fuel combustion. It is significant to all humanity since, although the CO_2 Greenhouse Effect is not today widely perceived as a threat, the popular media are giving increased attention to doom saying theories about dramatic climate changes and melting polar icecaps" (Hasemyer and Cushman, 2015: 3).

The ICN investigation clearly shows that Exxon was very engaged in public policy deliberations about global warming in the early 1980s. One Exxon scientist, Henry Shaw, was invited by Senator Gary Hart, a Democrat from Colorado, to attend an October 1980 conference of the National Commission on Air Quality to discuss "whether potential consequences of increased carbon dioxide levels warrant development of policies to mitigate adverse effects" (Hasemyer and Cushman, 2015: 3). In the spring of 1981, Exxon's top climate researcher, Brian Flannery, attended a large gathering of well-known scientists at Harpers Ferry, West Virginia, for a Department of Energy "Workshop on First Detection of Carbon Dioxide Effects." Flannery was one of only a few oil industry representatives invited to the workshop where he sat on a panel with the soon-to-be famous Dr. Hansen (Hasemyer and Cushman, 2015: 3). The workshop issued a declaration that indicated consensus among the scientists present that the build-up of carbon dioxide in the atmosphere was causing global warming and that this warming would bring about climate disruptions. Exxon's Flannery would also go on to co-author a "highly technical 50-page chapter" of a Department of Energy report, projecting "up to 6 degrees Celsius of warming by the end of the 21st century unless emissions of greenhouse gases were curtailed" (Hasemyer and Cushman, 2015: 4).

The results of Exxon's extensive internal research program on carbon dioxide and climate change were organized and presented in a 1982 corporate primer prepared by the company's environmental affairs office and marked "not to be distributed externally" (Banerjee, Song, and Hasemyer, 2015a: 4). The primer, written by an Exxon environmental affairs manager named Marvin Glaser, contained information that "has been given wide circulation to Exxon management" and concluded that despite many lingering unknowns, heading off global warming "would require major reductions in fossil fuel combustion." Unless those reductions were pursued, "there are some potentially catastrophic events that must be considered," and the document also warns, citing independent academic scientists, that "once the effects are measurable, they might not be reversible" (Banerjee, Song, and Hasemyer, 2015a: 4). The primer and other documents uncovered by ICN and DeSmogBlog make it very clear that by the early 1980s, before global warming became a major public and political issue, Exxon knew that the continued extraction and marketing of oil—business as usual—would result in dangerous environmental harms. And it was known that avoiding these harms would require reductions in fossil fuel combustion, changes in national energy strategies, and the pursuit of alternative business paths for the company.

So, what did Exxon do with this scientific knowledge? First and foremost, the company "used its knowledge of climate change to plan its own future" (McKibben, 2016b: 2). As Columbia University's Energy and Environmental Reporting Project and the *Los Angeles Times* have documented, between 1986 and 1992, in the far northern regions of Canada's Arctic frontier, "researchers and engineers at Exxon and Imperial Oil were quietly incorporating climate change projections into the company's planning and closely studying how to adapt the company's Arctic operations to a warming planet" (Jerving, Jennings, et al., 2015: 1). During these years, Ken Croasdale, a senior ice researcher for Exxon's Canadian subsidiary Imperial Oil, led a Calgary-based team of researchers and engineers on a project to, in his words, "assess the impacts of potential global warming," both positive and negative, on Exxon's Arctic oil operations (Jerving et al., 2015: 1). Croasdale, like many Exxon scientists at the time, was very much aware of the science linking the dangerous rise of greenhouse gases in the atmosphere to Exxon's primary product. In a later interview, he acknowledged that "The issue of CO_2 emissions was certainly well known at that time in the late 1980s," and to an audience of engineers at a conference in 1991, Croasdale stated that greenhouse gases are rising "due to the burning of fossil fuels. Nobody disputes this fact" (Jerving et al., 2015: 3–4).

According to the *Los Angeles Times*–CSJ investigation, Croasdale's team eventually reported back that "potential global warming can only help lower exploration and development costs," good news for a company that had leased large tracts of the Arctic for oil exploration efforts that had turned out to be difficult and expensive due to the extreme conditions (Jerving et al., 2015: 1). A warming planet, with the effects of that warming being most pronounced in the polar regions, might make Arctic oil exploration and production easier and cheaper in the long run. While more drilling opportunities might result from global warming, internal research also suggested that a warmer and wetter Arctic with rising seas might pose risks to the industry's facilities and infrastructure. Thus, throughout the region, Exxon and other oil companies began "raising the decks of offshore platforms, protecting pipelines from increasing coastal erosion, and designing helipads, pipelines and roads in a warmer and buckling Arctic" (Lieberman and Rust, 2015: 3). "In other words," writes McKibben (2016b: 3), "the company started climate-proofing its facilities to head off a future its own scientists knew was inevitable."

At about the same time that Exxon was using its extensive scientific knowledge of the greenhouse effect to explore for more Arctic oil and climate-proof its infrastructure in that region, the company also began to change its public stance on the issue of global warming. Throughout the 1980s, Exxon sought and earned a public reputation for its pioneering research on the relationship between carbon dioxide emissions and rising global temperatures. But as the issue of global warming emerged as a major public policy concern following Hansen's Senate testimony and the creation of the IPCC in 1988, the company saw a major threat to its profits and changed course. As climate scientists and government officials began to call for political and regulatory action to mitigate global warming, Exxon shifted "from embracing the science of climate change to publicly questioning it" (Jennings, Grandoni, and Rust, 2015: 2). According to the *Los Angeles Times*–CSJ investigation, Duane LeVine, Exxon's manager of science and strategy development, made a presentation to the company's board of directors in 1989 that, while confirming the scientific facts concerning the connection between the burning of fossil fuels and the greenhouse effect, also laid out the "Exxon Position" on this threatening information. As Jennings, Grandoni, and Rust (2015: 4) explain, the Exxon Position was to manufacture public doubt about some of the uncertainties in climate science: "In order to stop the momentum behind the issue, LeVine said Exxon should emphasize that doubt. Tell the public that more science is needed before regulatory action is taken, he argued, and emphasize the 'costs and economics' of restricting carbon dioxide emissions." In following this "extend-the-science" strategy to protect profits, Exxon was taking a page out of the tobacco industry playbook of blocking regulatory action on a dangerous product by becoming "merchants of doubt" concerning the underlying scientific knowledge (Oreskes and Conway, 2010).

In 1992 Exxon joined the Global Climate Coalition, an association of corporations within the fossil fuel industry that had banded together to deny global warming and fight against regulations on the emission of greenhouse gases—and which sought to, following the Exxon Position, emphasize "scientific uncertainty and underscoring the negative economic impact of such laws on consumers" (Jennings et al., 2015: 5). Later in the 1990s, Exxon began to fund conservative think tanks that were also engaged in a concerted public policy campaign "that was transparently designed to raise public skepticism about the science that identified fossil fuels as a cause of global warming" (Coll, 2012: 184). According to Greenpeace (2013b), Exxon has spent more than thirty million dollars in this effort. In 1997, Lee Raymond, Exxon's chairman and chief executive officer, "publicly rejected" climate science and the IPCC's latest assessment report in a speech delivered to the Fifteenth World Petroleum Congress in Beijing (Coll, 2012: 81). In its campaign to disparage climate science and block any "burdensome" government regulations on fossil fuels, Exxon also took out newspaper ads, employed lobbyists, and made lavish campaign contributions to political candidates who would deny global warming. Commenting on Exxon's efforts to deny climate science and obstruct political action to mitigate carbon emissions, Coll (2012: 184) argues that "What distinguished the corporation's activity during the late 1990s and the first Bush term was the way it crossed into disinformation."

Here I want to emphasize again the company's involvement in the climate crimes of continued extraction and rising emissions. The evidence presents a clear and convincing case that Exxon knew in the 1970s and 1980s that its normal business of extracting and marketing oil resulted in dangerous emissions of carbon dioxide that were heating the planet and leading the world to climate catastrophe. On the basis of its own strong internal research program and the warnings of its own scientists, not to mention the dire predictions of outside climate scientists, Exxon knew that for the sake of humanity it should pursue alternative energy strategies and business plans. As American science historian Naomi Oreskes (2015: A21) put it, "Exxon had a choice. As one of the most profitable companies in the world, Exxon could have acted as a corporate leader, helping to explain to political leaders, to shareholders and institutional investors, and to the public what it knew about climate change. It could have begun to shift its business model, investing in renewables and biofuels or introducing a major research and development initiative in carbon capture. It could have endorsed sensible policies to foster a profitable transition to a 21st-century energy economy." But Exxon did not do any of these things. And climate activists charge that, by failing to act on what it knew, failing to take these steps, the company engaged in a climate crime of omission.

Again, despite its knowledge about the relationship between carbon emissions and climate disruption, one of the richest and most powerful corporations in the world did not change course. Instead it used the scientific information it had acquired to explore for additional sources of oil in the Arctic while climate-proofing its infrastructure against the problems the company knew that a warming world would bring. It chose a path of disinformation, denial, and delay concerning the emerging public issue of global warming. And, as McKibben (2016b: 6) points out, outside the Arctic, ExxonMobil continues to spend money to bring other new fossil fuel projects online ($63 million a day) and to search for new sources of hydrocarbons ($1.5 billion a year, $4 million a day). Despite its knowledge of the climate problem, and its more recent acknowledgment that global warming is indeed occurring, the "ultimate corporation" continues with its normal business plan.

To be clear, Exxon was not the only oil company that knew about climate dangers in the 1970s, and the fossil fuel industry was not the only industry that had such knowledge and tried to cover it up (Sullivan, 2019; Wang, 2018). As ICN's Neela Banerjee (2015) has reported, the American Petroleum Institute together with many of the largest oil companies created a task force to monitor and share climate research between 1979 and 1983. Initially called the CO_2 and Climate Task Force, the group changed its name in 1980 to the Climate and Energy Task Force. The composition of the group indicates that it was not just Exxon alone that had knowledge about the impact their business practices were having on the world's climate. In early 2018, a trove of documents was discovered by investigative journalists that revealed that oil giant Shell also knew in the 1980s that fossil fuels created climate change risks and that company scientists had urged management to heed the warnings and work to reduce greenhouse gas emissions or face future lawsuits (Cushman, 2018). In an internal document from 1988, the company acknowledged the need to "defend against a sea level rise" and other impacts of climate change (Savage, 2018b: 1). Shell even produced a farsighted film in 1991, titled *Climate of Concern*, that warned that the continued extraction and consumption of oil could lead to extreme weather, famines, and other harms. The company noted that the dangers of climate change were "endorsed by a uniquely broad consensus of scientists" (Carrington and Mommers, 2017: 1). But ultimately, Shell too failed to act on what it knew about global warming and then also participated in political efforts to deny climate change and block regulations to reduce carbon emissions.

Other industries also had knowledge about global warming and the potential for climate disruption but took no actions to change business-as-usual practices and instead participated in the climate change denial campaign. For example, in July 2017 the Energy and Policy Institute issued a report that documented the extensive knowledge that electric utilities also had about climate change in the 1960s and 1970s. What did these companies do with this knowledge? Taking a page out of the oil industry playbook, they engaged in vigorous efforts to spread disinformation about climate science and block regulations on carbon emissions and have continued their harmful and profitable current operations into the present (D. Anderson, Kasper, and Pomerantz, 2017).

Business-as-usual practices by ExxonMobil and Shell, other fossil fuel corporations, and electric utilities—often encouraged and subsidized by the United States and other governments to the tune of $5 trillion (Coady, Parry, et al., 2016)—will blow through the "carbon budget" of the earth (the estimate of how much more carbon we can burn without overheating the planet), leading to climate catastrophe. These organizational actions of continued extraction and emissions are not only wildly irresponsible; they are also in my view *criminal*—that is, they constitute voluntary and intentional courses of action undertaken with full knowledge that grievous harms are a likely outcome. These actions are unjustified and inexcusable. And they are increasingly condemned by the public and subject to legal sanctions.

SOCIAL, POLITICAL, AND LEGAL RESPONSES TO THE EXXON CASE

The evidence revealed by the ICN reports, the *Los Angeles Times*–CSJ articles, and other investigations concerning what Exxon knew about climate change and when it knew it generated a firestorm of social, political, and legal responses critical of the company. While ExxonMobil had its defenders, these critical reactions taken together constitute a form of public condemnation of Exxon's harmful actions and omissions and even threaten the company with legal sanctions. Recall that in Agnew's definition of crime, the core characteristics of crime are blameworthy harms that are condemned by the public and sanctioned by the state. The magnitude and severity of these responses to what Exxon knew serve to further strengthen the argument that the company has participated in climate crimes and can be considered a carbon criminal.

The documentation of what Exxon knew about global warming and when it knew it was revealed in the late summer and early fall of 2015. Reaction from climate activists was swift and critical. As he has so often concerning the public issue of climate change, McKibben, cofounder of the climate activist organization 350.org, took the lead. On September 18, after the first installment of the ICN series was published, McKibben (2015c) wrote a short piece for the *New Yorker* titled "What Exxon Knew about Climate Change." On October 14, after the first *Los Angeles Times* report had also appeared, he published an article in the *Guardian* (UK) stating that "No corporation has ever done anything this big and this bad" (McKibben, 2015a: 1). The next day McKibben was arrested for protesting at an Exxon-Mobil gas station in Burlington, Vermont, to bring attention to Exxon's "unparalleled evil." During the protest, he held a sign that read "This Pump Temporarily Closed Because ExxonMobil Lied about (#Exxon Knew) Climate." The sign not only included the Twitter hashtag #ExxonKnew but also featured a link to a Tumblr web page that provided additional information about why McKibben had staged his protest. The Tumblr page also reported a new political development: Congressmen Ted Lieu and Mark DeSaulnier (Democrats from California) had sent a letter to U.S. attorney general Loretta Lynch on October 14 calling for a Department of Justice (DOJ) investigation into possible illegal actions by ExxonMobil, given the information revealed by ICN and the *Los Angeles Times*. Following suit, 350.org organized a public petition also calling on the DOJ to "Investigate Exxon" for possible crimes.

Public condemnation of Exxon in the form of calls for a criminal investigation of the company continued to build. On October 18, an article by American writer and activist David Atkins in *Washington Monthly* magazine called for ExxonMobil to be prosecuted for covering up its knowledge of fossil fuel–induced climate change. Atkins (2015: 2) argued: "A fossil fuel company intentionally and knowingly obfuscating research into climate change constitutes criminal negligence and malicious intent at best, and a crime against humanity at worst. The Department of Justice has a moral obligation to prosecute Exxon and its co-conspirators accordingly." Senator Bernie Sanders from Vermont, who had recently launched his bid for the Democratic Party nomination for president in 2016, also joined in the call for a DOJ investigation of ExxonMobil, arguing that that the oil company's actions may "ultimately qualify as a violation of federal law" (Prupis, 2015a: 1).

Following his request, a broad coalition of social and political groups—along with prominent leaders from the major environmental, civil rights, and Indigenous people's movements in the United States—also sent a joint letter to the DOJ on October 30 demanding an investigation into what Exxon knew about climate change, when it knew it, and what the company did or did not do in response to this knowledge. An article reporting on the coalition's request carried the headline "Seething with Anger, Probe Demanded into Exxon's Unparalleled *Climate Crime* [emphasis added]" (Common Dreams, 2015). Thus, as the documentation of what Exxon knew about climate change was revealed to the public in the fall of 2015, the broad-based climate justice movement and other environmental groups, joined by numerous journalists and politicians (including presidential candidates Sanders and Hillary Clinton), publicly and forcefully stated their condemnation of the corporation and issued calls for legal actions against the company. Whether these calls and actions affected public opinion about ExxonMobil or the fossil fuel industry at the time is not clear. However, over time the American public has become more receptive to regulating the coal, gas, and oil companies and making stronger efforts to reduce carbon emissions (Freedman, 2019).

In addition to these public condemnations, a variety of legal challenges have emerged or intensified since the evidence against Exxon has been revealed. As the DOJ began to consider the numerous requests calling for an investigation into Exxon's knowledge and actions concerning climate change in the past, the company faced a new legal threat. In November 2015, it was revealed that New York State attorney general Eric Schneiderman had opened an investigation of ExxonMobil to determine whether the company had engaged in fraud by lying to the public and its investors concerning the risks of climate change and how such risks could negatively affect the oil business in the future. Catalyzed by the ICN and *Los Angeles Times* reports, Schneiderman announced on November 5 that he had issued an eighteen-page subpoena demanding extensive financial records, e-mails, and other documents (Gillis and Krauss, 2015). According to ICN, Schneiderman was seeking information from Exxon "related to its past research into the causes and effects of climate change, to the integration of climate change findings into business decisions, to communications with the board of directors and to marketing and advertising materials on climate change" (Simison, 2015: 2). The *New York Times* pointed out that Schneiderman's probe "may well open a new legal front in the climate change battle" (Gillis and Krauss, 2015: 2). By refusing to act on what it knew about climate science and the clear risks of continuing with business as usual, and by failing to inform the public or its investors about these risks, ExxonMobil (and other fossil fuel companies) may have also committed the crime of fraud—on top of the climate crimes of withdrawal (extraction of fossil fuels) and addition (the release of carbon emissions)—under New York's Martin Act, a securities law that protects investors.

As the election year of 2016 dawned, ExxonMobil continued to face mounting legal challenges. In January, California State attorney general Kamala D. Harris (now a Democratic U.S. senator) also launched a probe into whether the company had lied to the public and its shareholders about the risk climate change could pose to its business, actions that could likewise constitute securities fraud as well as violations of state environmental laws (Penn, 2016). Then on March 2, the DOJ forwarded Lieu and DeSaulnier's letter to the Federal Bureau of Investigation (FBI) and requested an evaluation of whether the company's behavior concerning the climate change issue might warrant a Racketeering Influenced and Corrupt Organization (RICO) legal action (a law aimed at illegal conspiracies). "There is substantial evidence to support a full investigation and prosecution" said DeSaulnier at the time, adding: "There is going to be a moment of judgment both politically and legally . . . a moment of judgment [that] will be quite critical of the fossil fuel industry in terms of obfuscating the scientific facts and for not adhering to their moral and legal responsibility to the public" (Hasemyer, 2016b: 2).

In late March 2016, two new and stunning developments shook the fossil fuel industry. First, the Rockefeller Family Fund (RFF) announced that it would gradually divest its holdings in fossil fuel companies because of its great concern about the threat of climate change. That was shocking enough, but in a letter posted on its website the family-led charitable organization singled out ExxonMobil for immediate divestment because of its "morally reprehensible conduct." The foundation stated that it would eliminate its ExxonMobil holdings "effective immediately" because the company associated with the family's fortune had misled the public about the risks of climate change as revealed in the ICN and *Los Angeles Times* investigations. How ironic that the descendants of John D. Rockefeller, the founder of Standard Oil, the company that would become ExxonMobil, were now stating that "There is no sane rationale for companies to continue to explore for new sources of hydrocarbons" (Loki, 2016). Given the family's historical connection to the firm, the RFF announcement was one of the most significant forms of public condemnation that Exxon could have suffered. It was a stunning blow to ExxonMobil's public image and reputation, one that the company did not take lying down. Later in the year, ExxonMobil struck back by making the outlandish accusation that the Rockefeller family was masterminding a "conspiracy" against the company and denying them their First Amendment right to criticize climate science, since RFF had provided partial funding for the ICN and CSJ investigations (J. Schwartz,

2016). The family responded with a scathing two-part essay in the *New York Review of Books* (co-authored by David Kaiser, a fifth-generation Rockefeller). The essay defended the foundation's decision to divest from ExxonMobil, asserted the family's own First Amendment right "to criticize ExxonMobil on moral grounds," provided a broad overview of the ICN, *Los Angeles Times*, and Union of Concerned Scientists investigative reports, and reviewed Coll's *Private Empire* and Oreskes and Conway's *Merchants of Doubt*, two important books that provide a critical examination of Exxon's climate denial activities (Kaiser and Wasserman, 2016).

On March 29, a second bombshell dropped on the fossil fuel industry as a historic coalition of top state law enforcement officials held a press conference to announce their intention to investigate and prosecute oil companies that deceived investors and consumers about the dangers of climate change (Hasemyer and Shankman, 2016). Led by Schneiderman, seventeen state attorneys general (all Democrats) vowed to hold ExxonMobil and the other fossil fuel corporations accountable for their conduct related to global warming. At the press conference, Massachusetts State attorney general Maura Healey disclosed that she too had opened an investigation into Exxon's actions concerning climate change. "Fossil fuel companies," Healey stated, "must be held accountable," noting that there is a "Troubling disconnect between what Exxon knew, what industry folks knew, and what the company and industry chose to share with investors and the American public" (McCauley, 2016: 1). As a result, the legal scrutiny of ExxonMobil, and the fossil fuel industry in general, greatly intensified.

The coalition of attorneys general held their historic press conference under the banner "AGs United for Clean Power," which made clear that their agenda also extended to supporting President Obama's Clean Power Plan (Hasemyer and Shankman, 2016). This important EPA plan to reduce carbon emissions from power plants was the centerpiece of Obama's climate change policy (analyzed in chapter 5) and was at that time under severe attack from the fossil fuel industry and many Republican congressmen and state attorneys general. It was even put on hold by the U.S. Supreme Court pending a lower court appeal. The Democratic attorneys general were joined at the press conference by former vice president and climate champion Al Gore, who predicted that the formation of the legal alliance would be a turning point in the fight against climate change. "We cannot continue to allow the fossil fuel industry or any industry to treat our atmosphere like an open sewer or mislead the public about the impact they have on the health of our people and the health of our planet," Gore stated. He continued, "Those who are using unfair and illegal means to try and prevent the change are likely now, finally at long last, to be held accountable" (Hasemyer and Shankman, 2016: 4). ExxonMobil and the other big oil companies were being put on notice that their climate fraud and suppression of key climate science were being targeted for legal sanction by many states, if not yet by the federal government.

Later in 2016, several groups of investors and an important federal regulatory agency added to ExxonMobil's legal woes over the revelations concerning the company's knowledge of the risks of climate change. In May, Green Century Funds, a mutual funds advisory company founded, managed, and owned by a partnership of nonprofit environmental organizations, organized a coalition of investors to also call on the DOJ to investigate whether Exxon intentionally misled the public and its shareholders about climate change. Then the U.S. Securities and Exchange Commission (SEC) announced in September that it was launching an investigation into, as the *Wall Street Journal* put it, "how Exxon Mobil Corporation values its assets in a world of increasing climate-change regulations, a probe that could have far-reaching consequences for the oil and gas industry" (Olson and Viswanatha, 2016: 1). The progressive website Common Dreams (2016) called the SEC probe a "possible moment of reckoning for Exxon's *climate crimes* [emphasis added]." In November, ExxonMobil shareholders themselves became involved and filed a class-action lawsuit against the company, "alleging it misled its investors and the public by failing to disclose the risks posed to its business by climate change" (Hasemyer, 2016a: 2). Key to these public condemnations and specific legal complaints is the idea that Exxon continued to falsely assert that the corporation was on solid financial ground even though it possessed scientific knowledge that climate disruptions and climate policies would almost certainly reduce the future value of the company's stock in a significant way.

ExxonMobil responded forcefully to the ICN and *Los Angeles Times* reports by claiming that they were inaccurate and misleading. The company has not, however, challenged any of the specific findings of the investigations or presented any evidence to dispute the key facts of the reports. As Coll, now the dean of CSJ, noted in response to ExxonMobil's complaints about the *Los Angeles Times* articles his school had contributed research to: "What your letter advocates really is that the factual information accurately reported in the article, and unchallenged by you, be interpreted differently" (Somaiya, 2015: B3). In its defense, ExxonMobil now argues that it does acknowledge the risk of climate change and that the risk warrants action. Today, the company says, it recognizes that increasing carbon emissions are having a warming effect and has even expressed support for a carbon tax. While it is true that the company no longer openly denies the reality of global warming, it still has not changed its business-as-usual practices and continues to fight the regulation of greenhouse gases. As the Union of Concerned Scientists points out, "The company is still engaged in a behind the scenes effort to block sensible climate policy" (Pinko, 2018: 1). And ExxonMobil has also assured its shareholders that it is "confident that none of our hydrocarbon resources are now or will become stranded" (McKibben, 2014: 2). While the company might argue that it has a fiduciary responsibility to maximize shareholder value in the present, this myopic focus on short-term earnings ignores the larger, long-term dangers to shareholders if in fact oil does become a stranded asset in the future (an entirely likely scenario). By ignoring climate science and only advocating for short-term shareholder value, ExxonMobil may indeed be guilty of investor fraud. By continuing to extract and market the fossil fuels it has on reserve, ensuring the ongoing emission of greenhouse gases, the oil giant will remain guilty of climate crimes.

ExxonMobil's argument that it has changed its position and behavior when it comes to the public issue of climate change was also undercut by two disclosures from Schneiderman as part of his office's sweeping investigation into whether the corporation had committed fraud by failing to disclose what it knew about climate change to the public and its investors. First, Schneiderman revealed that former CEO Rex Tillerson (Trump's first secretary of state) had used a shadow e-mail account within the company when discussing global warming and the risks it posed to the company's business and then concealed the existence of the secret account from law enforcement officials (Hasemyer, 2017a). In a March 2017 letter to New York State judge Barry Ostrager, the attorney general disclosed that Tillerson had used the alias "Wayne Tracker" e-mail account for eight years within the company to hide his discussions concerning climate change issues. Schneiderman also charged that Exxon had concealed the shadow e-mails despite a 2015 subpoena for Tillerson's communications issued as part of the investigation. To make matters worse, in June the attorney general's office discovered that ExxonMobil had erased seven years of Tillerson's "Wayne Tracker" e-mails (Hasemyer, 2017b), raising questions about what ExxonMobil was attempting to hide.

Those questions intensified in June 2017 when Schneiderman also charged that the company had been using two sets of numbers in its greenhouse gas accounting. The attorney general accused ExxonMobil of "sham" accounting, arguing that the company used one set of numbers about the likely financial risks posed by climate change for investors while using a different, more accurate, and secret set of numbers for making internal business decisions (Kusnetz, 2017a). If such evidence does exist, it goes to the heart of the fraud case that Schneiderman first brought against ExxonMobil back in 2015. The "Wayne Tracker" e-mail and sham accounting charges have further damaged ExxonMobil's public image. As the ultimate corporation tries to convince the public that it has changed its position and is now attempting to be a good citizen regarding the climate issue, these charges reinforce that Exxon has long known of the dangers of carbon emissions and yet refuses to substantially change its business practices to seriously address the problem. Although the New York investigation was complicated by Schneiderman's sudden resignation in May 2018, when allegations surfaced that he had physically assaulted four women, the new attorney general, Barbara Underwood, did sue ExxonMobil in October of that year claiming that the oil giant had, in fact, defrauded investors over climate change (Hasemyer, 2018b).

This perception, that ExxonMobil is guilty of the climate crimes of continued extraction and continued emissions as well as investor fraud, was reinforced by the company's ferocious attempts to fight back against the climate fraud investigations in New York and Massachusetts and other attempts to hold it legally and socially responsible for blameworthy climate harms. The company has pursued a "bare knuckles and big dollars" approach, "a legal strategy of massive resistance to any effort to hold Exxon accountable for global warming or harm to the environment" (Hasemyer, 2018a: 2). This defiant stance, use of big scare tactics, and tenacious legal bullying serves ExxonMobil well in the short run. Thus far, the company has been able to fight the climate probes and charges to a standstill. But the countersuits, the charges of conspiracy, the use of Republican lawmakers soaked in oil industry money to harass their accusers through congressional hearings, and the absurd assertion that its First Amendment rights to free speech have been violated have only intensified the public condemnations of Exxon by the climate justice movement and other concerned citizens. The public condemnations will continue, as will the threat of state sanction. Both state and federal appellate courts have rejected ExxonMobil's attempt to shut down the climate fraud investigations (Cushman and Hasemyer, 2018; Hasemyer, 2018a); and the oil giant's legal challenges are expected to reach "critical mass" in fall 2019 "as a wave of climate liability lawsuits come to a crescendo" (Drugmand, 2019: 1).

I offer one final, and I think most critical point about Exxon's strategy of massive and intransigent resistance to any of the efforts to hold the company socially or legally accountable for what it knew about climate change and for what it did or did not do with this knowledge. By mounting this political and legal resistance, ExxonMobil (and the fossil fuel industry in general) can continue its operations unabated. As Christine Todd Whitman, former administrator of the EPA under George W. Bush, points out: "Fossil fuel corporations doggedly battle lawsuits, regulations and investigations, because it allows them to continue pumping oil, mining coal and fracturing for natural gas—and banking the profits—even if sometime later there will be a reckoning" (Hasemyer, 2017c: 3). As these legal and political battles rage, energy corporations continue to extract and market fossil fuels with the result that more carbon emissions are released into the atmosphere, making catastrophic climate disruption ever more likely. And in early 2019, a research report found that the five largest publicly traded oil and gas majors (ExxonMobil, Shell, Chevron, BP, and Total) have invested over a billion dollars of shareholder funds since the Paris Agreement "on misleading climate-related branding and lobbying" to block climate change policies (InfluenceMap, 2019: 1). In response to the report, British writer and environmental activist George Monbiot tweeted, "In a just world, it would be treated as a crime against humanity—and against the rest of the living planet" (Common Dreams, 2019: 2).

EXTREME EXTRACTION METHODS
AS CLIMATE CRIMES

As a final note on the climate crime of continued extraction, it is important to consider the expanding efforts to access unconventional forms of dirty energy through extreme extraction methods. At a time when climate scientists have warned that vast quantities of fossil fuels must stay in the ground to avoid climate catastrophe, fossil fuel companies continue to plan massive new projects. Greenpeace (2013b: 5) examined fourteen new, big, dirty energy projects that, taken together, "would produce as much new carbon dioxide emissions in 2020 as the entire U.S., and delay action on climate change for more than a decade." A more recent study released by Oil Change International and seventeen partner organizations (Trout, 2019: 5) states that "At precisely the time in which the world must begin rapidly decarbonizing to avoid runaway climate disaster, the United States is moving further and faster than any other country to expand oil and gas extraction." Both reports point out that many of these destructive new projects involve not only searching for, extracting, and marketing conventional fossil fuels, such as plans to expand coal mining and oil drilling, but also developing more risky, unconventional sources of dirty energy that require more extreme methods of extraction, such as hydraulic fracturing of shale rock to release natural gas (fracking).

By accessing unconventional fossil fuels and developing more extreme extraction procedures, corporations in the oil and gas industry have upped the ante on their climate crimes. These extreme extraction methods (crimes of ecological withdrawal) include offshore deep-water drilling for oil (Bradshaw, 2014, 2015a, 2015b), the mining of Canadian tar sands to extract heavy bitumen oil (T. Clarke, 2008; Lynch, Stretesky, and Long, 2016; Smandych and Kueneman, 2010), and fracking, which often leads to the release of methane, a major greenhouse gas that can also make a significant contribution to global warming (Doyon and Bradshaw, 2015; Hauter, 2016;

McKibben, 2016c; Oreskes, 2014). All three developments result in some form of environmental destruction, additional pollution, and increased greenhouse gas emissions (crimes of ecological addition). To understand these more extreme forms of extraction as state–corporate environmental crimes, it is helpful to know the broader historical context within which the fossil fuel industry and some governments turned to such unconventional fuels.

As Klare (2012: 8) points out in *The Race for What's Left: The Global Scramble for the World's Last Resources*, until recently "the energy industry has been able to tap into giant, easily exploited oil and gas reservoirs in relatively accessible locations, providing the world with cheap and abundant power." The availability of such vast amounts of affordable energy fueled the post–World War II industrial boom and drove the Great Acceleration. But according to Klare (2012: 8), "the era of readily accessible oil and gas has come to an end." From now on, he declares, "vital energy supplies will have to be drawn from remote and forbidden locations, at a cost far exceeding anything experienced in the past. The world is entering an era of pervasive, unprecedented resource scarcity." This new era not only has enormous geopolitical implications, setting up the potential for resource wars between states (Klare, 2001), but also creates the conditions for these new forms of climate crimes.

In this era of resource scarcity, the energy industry has made a fateful decision to continue to search for, extract, and market fossil fuels, no matter the production costs to corporations within the industry, the environmental destruction that results from the extraction process, or the continued release of greenhouse gases that will accelerate the threat of climate disruption. Rather than taking the socially responsible course of leaving fossil fuels in the ground and developing new sources of clean and renewable energy, these companies have moved to produce unconventional forms of dirty energy with these more extreme methods of extraction. As Klare (2012: 31–32) notes, "The accelerating depletion of existing oil fields, along with doubts about how many new deposits of the oil companies will find in frontier regions, has led to growing reliance on 'unconventional' oil, Canadian tar sands, shale oil, extra-heavy crude, and other materials obtained from nonstandard petroleum deposits." He goes on to observe that even this increase in more extreme methods of global production will not be enough to compensate for the decline in conventional production, "so pressure to develop new fields in problematic areas—such as the Arctic, Siberia, and the deep oceans—will only continue to grow." Although a more in-depth analysis of extreme extraction methods will not be attempted here, I would argue that these newer fossil fuel industry actions, often facilitated by government agencies, can also be defined as state–corporate climate crimes.

CHAPTER 4

"The Politics of Predatory Delay"

CLIMATE CRIMES OF POLITICAL OMISSION AND SOCIALLY ORGANIZED DENIAL

Predatory delay is the blocking or slowing of needed change, in order to make money off unsustainable, unjust systems in the meantime.
—Alex Steffen on Twitter, August 28, 2017

Failure to act with all deliberative speed . . . functionally becomes a decision to eliminate the option of preserving a habitable climate system.
—Dr. James Hansen (quoted in Blumm and Wood, 2017: 106)

But any consideration of the moral dimension of the climate crisis must begin with the villains—those who have tried to bewitch an unassuming public with uncertainty, lies, and the gratuitous fantasies of denialism. The morality of these tactics can only be described as sociopathic.
—Nathaniel Rich, *Losing Earth: A Recent History* (2019: 194)

IN AUGUST 2015, a lawsuit was filed in the federal district court of Eugene, Oregon, that may turn out to be one of the most significant legal cases of the twenty-first century. The ongoing lawsuit, *Juliana, et al. v. United States*, was brought on behalf of twenty-one children and young adults over the United States government's affirmative actions in creating a national energy system that causes climate change, its failure to mitigate greenhouse gas emissions that are heating the planet, and its general refusal to plan for effective adaptation to the looming crisis of climate disruption. The youth plaintiffs seek a legal declaration that their constitutional and public trust rights have been violated by these state actions and omissions (nondecisions) and are asking the court to direct the U.S. government to create an action plan to significantly reduce carbon dioxide emissions and develop climate change adaptation measures.

As the presiding judge in the case, U.S. District Court judge Ann Aiken (2016: 3), has noted, "This is no ordinary lawsuit." The plaintiffs in this civil rights action are a group of young people, who, at the time the case was filed, ranged in age from eight to nineteen; the Earth Guardians, an association of young environmental activists; and famed climate scientist Dr. James Hansen, acting as guardian for future generations (Hansen's granddaughter, Sophie Kivlehan, is one of the youth plaintiffs). Our Children's Trust, a group of environmental lawyers from Eugene, Oregon, provides legal support to the young challengers. Julia Olsen, executive director of Our Children's Trust, is the lead counsel for the plaintiffs. The original defendants in the case were the United States government (as an organizational actor), the president of the United States (initially Barack Obama, later amended to name Donald Trump), and specific federal executive agencies (such as the EPA). Federal law allows for outside parties to "intervene"—that is, to choose to become part of a lawsuit (a position akin to being a co-defendant)—and in late 2015 three corporate trade associations with strong ties to the fossil fuel industry requested that the court include them as "intervenors" in the case to get the lawsuit dismissed. The intervenors were the National Association of Manufacturers (NAM), the American Petroleum Institute (API), and the American Fuel and Petrochemical Manufacturers (AFPM).

In their legal action, the youth plaintiffs allege that for more than fifty years the defendants (U.S. government agents and agencies) have known that carbon dioxide produced by the burning of fossil fuels was destabilizing the climate system to such an extent that it would "significantly endanger plaintiffs, with the damage persisting for millennia." Despite knowledge of the severe harm being inflicted, federal officials, "by their exercise of sovereign authority over our country's atmosphere and fossil fuel reserves . . . permitted, encouraged, and otherwise enabled continued exploitation, production, and combustion of fossil fuels . . . deliberately allowing atmospheric CO_2 concentrations to escalate to levels unprecedented in human history." The plaintiffs claim that the U.S. government bears "a higher degree of responsibility than any other individual, entity, or country" for exposing them to the dangerous harms of disruptive climate change. The children argue that both the government's affirmative actions and political omissions concerning fossil fuels and carbon emissions "violate their substantive due process rights to life, liberty, and property, and that defendants have violated their obligation to hold natural resources in trust for the people and for future generations." The lawsuit seeks a declaration that the plaintiffs' constitutional and public trust rights have been violated and a direct remedy: a court order that the defendants develop a comprehensive national plan to reduce greenhouse gas emissions (Aiken, 2016: 2).

The plaintiffs' action in *Juliana, et al. v. United States* draws in part on "an ancient and enduring legal principle known as the public trust doctrine" (Wood, 2014: 14). Under this doctrine, well established in American case law, governments serve as trustees of natural resources held in common such as water, wildlife, and air on behalf of future generations. As American law professor Mary Christina Wood (2014: 14) points out in *Nature's Trust: Environmental Law for a New Ecological Age*, "the doctrine rests on a civic and judicial understanding that some natural resources remain so vital to public welfare and human survival that they should not fall exclusively to private property ownership and control." The public has a lasting ownership interest in these resources, which should be preserved for both present and future generations. Wood (2014: 14) adds that "Public trust law demands that government act as a trustee in controlling and managing crucial natural assets. Held to strict fiduciary obligations, government must promote the interests of the citizen beneficiaries and ensure the sustained resource abundance necessary for society's endurance." The heart of public trust law rests on this fundamental fiduciary responsibility of government trustees to protect trust resources from damage, and this case raises the question of whether the atmosphere is in fact such a crucial natural asset that it must also be protected for the benefit of present and future generations.

If the earth's atmosphere is a crucial natural asset that is being dangerously altered by rising greenhouse gas emissions, and human-caused global warming is scientifically demonstrated to be a severe threat to other natural resources held in common by the public, then the applicability of this legal duty to the issue of climate change is clear. Wood (2014: 337) declares that "government's failure to protect the planet's climate system on which all life depends amounts to the most dangerous perversion of this fiduciary responsibility." This is one of the core legal arguments animating *Juliana, et al. v. United States*.

The lawsuit passed a major hurdle on November 10, 2016, when Aiken denied the defendants' motion to dismiss the case. In her historic opinion, Aiken first rejected the government's claim that the case raises a nonjusticiable political question (hence, not a matter for the courts to consider) and then granted the youth plaintiffs standing to sue, ruling that the plaintiffs had demonstrated that they had suffered an injury, that the injury is fairly traceable to the defendants' challenged conduct, and that the injury is likely to be redressed by a favorable court decision. Concerning the due process claims, Aiken stated, "Plaintiffs have alleged that defendants played a significant role in creating the current climate crisis, that defendants acted with full knowledge of the consequences of their actions, and that defendants have failed to correct or mitigate the harms they helped create in deliberate indifference to the injuries caused by climate change." She then ruled that the plaintiffs "may proceed with their substantive due process challenge to defendants' failure to adequately regulate CO_2 emissions" (Aiken, 2016: 36).

Aiken then turned to the public trust claims of the lawsuit. Rejecting the four core arguments advanced by the defendants to counter these claims, Aiken ruled that (1) the atmosphere is indeed a public trust asset, (2) the federal government (not just individual states) has public trust obligations, (3) common-law public trust claims have not been displaced by more recent federal statutes, and (4) plaintiffs do have a right of action to enforce a federal public trust. Aiken (2016: 51–52) concluded her opinion and order to deny the motions to dismiss the case by returning to the uniqueness of the lawsuit:

> Throughout their objections, defendants and intervenors attempt to subject a lawsuit alleging constitutional injuries to case law governing statutory and common-law environmental claims. They are correct that plaintiffs likely could not obtain the relief they seek through citizen suits brought under the Clean Air Act, the Clean Water Act, or other environmental laws. But that argument misses the point. This action is of a different order than the typical environmental case. It alleges that defendants' actions and inactions—whether or not they violate any specific statutory duty—have so profoundly damaged our home planet that they threaten plaintiffs' fundamental constitutional rights to life and liberty.

With her ruling, Aiken asserted that the young plaintiffs have raised an important constitutional issue, that they have legal standing to sue, and that the government will have to stand trial for its actions and omissions concerning climate change.

After Aiken denied the motion to dismiss, *Juliana, et al. v. United States* entered the discovery phase, a period when each side seeks documents, deposes witnesses, and gathers other evidence relevant to their legal arguments. The plaintiffs requested a significant number of specific documents and materials related to fossil fuel production and climate change policy from the U.S. government and the intervenors. An important development occurred in the case on January 18, 2017, two days before Trump's inauguration, when the Obama Department of Justice formally stipulated (agreed to) ninety-eight facts pertaining to climate change, the responsibility of human activity in creating it, and the destructive impact of climate disruption. After the inauguration, the lawsuit was amended to name Trump as a defendant in the case. In May 2017, after the motion to dismiss had been denied by Aiken and the discovery process threatened to expose internal documents from the fossil fuel industry, the three intervenors—NAM, API, and AFPM—all asked for permission to withdraw from the suit. In late June 2017, Judge Thomas Coffin (the federal magistrate judge who is overseeing the pretrial and discovery portion of the case) approved their request, leaving only U.S. government agents and agencies (which at that point included Trump and his administration) as the defendants.

Seeking to delay the discovery period and avoid a trial, the Trump DOJ sought to bring an interlocutory appeal (a special request to an appellate court before the issue has been decided at trial) to the Ninth Circuit Court of Appeals. Both the federal magistrate, Coffin, and the presiding judge, Aiken, denied the defendants' motion to certify the early appeal. The Trump administration responded on June 9, 2017, by filing a motion for a *writ of mandamus* (a rarely used legal procedure that requests that a superior jurisdiction order an inferior court to take some action, in this case to dismiss the lawsuit) with the Ninth Circuit. Such writs are usually reserved for the most extraordinary and compelling circumstances in which the ordinary rules of appellate procedure must be overridden to prevent some grave injustice, a situation that did not seem to apply in this case. Various groups and legal scholars submitted *amicus curiae* briefs on behalf of the plaintiffs to the Ninth Circuit, urging the court to deny the motion for the writ of mandamus and allow the trial to begin (Blumm and Wood, 2017). On March 7, 2018, that is exactly what the court did. Ninth Circuit chief judge Sidney R. Thomas, writing for a unanimous three-judge panel, rejected the Trump administration's "drastic and extraordinary" petition for the writ of mandamus and ruled that the *Juliana* case can proceed. The affirmative actions and political omissions of the U.S. government concerning climate change can proceed to trial. A trial date of October 29, 2018, was set, but the start of the trial was delayed when the Trump DOJ filed several more petitions for a writ of mandamus with both the Ninth Circuit and the U.S. Supreme Court. As of October 2019, one of these petitions was still under consideration at the Ninth Circuit, and a new date for what some called the "trial of the century" had not yet been set. While the long-term prospects for the lawsuit are uncertain (especially if a verdict reaches the Supreme Court), the importance of the case was emphasized by Blumm and Wood (2017: 108), who argue that, given the general failure of the political system to respond to the climate crisis, the Our Children's Trust lawsuit may be "the only legal mechanism" that might work to stave off catastrophe.

POWER AND STATE CRIMES OF POLITICAL OMISSION

The young plaintiffs in *Juliana* are not alone in arguing that states in general, and the U.S. government in particular, have a political and legal obligation to reduce greenhouse gas emissions. In *The Politics of Climate Change*, British sociologist Anthony Giddens (2011: 94) argues that nation-states "must take the lead in addressing climate change. . . . The state has to be the prime actor." In his view, mitigating greenhouse gas emissions and adapting to climate disruption is more of a political challenge than a scientific or technical problem. Furthermore, state responsibility for dealing with global warming is not only an important political issue, it is also a moral and legal imperative, according to a group of experts in international law, human rights, and environmental law who came together to promulgate the Oslo Principles on Global Climate Change Obligations. The preamble to the Oslo Principles (Expert Group on Global Climate Obligations, 2015: 2) states that "While all people, individually and through all the varieties of associations that they form, share the moral duty to avert climate change, the *primary legal responsibility* rests with States and enterprises" [emphasis added]. The International Bar Association (IBA, 2014) has also asserted that under international law, states have a legal obligation to reduce greenhouse gas emissions and achieve climate change justice. Furthermore, the IBA argues that states should be held accountable under international law for their failures to fulfill these legal responsibilities.

And failed they have in the judgment of many informed commentators. American philosopher Dale Jamieson, for instance, echoes this sentiment in the subtitle of his 2014 book *Reason in a Dark Time: Why the Struggle against Climate Change Failed—And What It Means for Our Future*. On the thirtieth anniversary of his influential 1988 Senate testimony, James Hansen observed that the world is "failing miserably" to address the dangers of climate change (Milman, 2018: 1). And, reacting to a claim that the response to global warming was the greatest "market failure" of all time, George Monbiot (2010: 1) insisted instead that "The response to climate change . . . is the greatest *political failure* the world has ever seen [emphasis added]," a political failure that violates international law and the legal obligation that states must protect the commons.

On September 23, 2014, a symbolic legal event was held in New York City in an effort to advance the political argument that states must be held accountable for the failure to use their power to mitigate carbon emissions and protect the atmospheric commons. As an important United Nations Climate Summit was taking place across the street, people's organizations from around the world convened a Climate Justice Tribunal. The tribunal was modeled after the International War Crimes Tribunal organized by philosophers Bertrand Russell and Jean-Paul Sartre in 1967 to protest U.S. war crimes in Vietnam. After listening to testimony concerning the causes and consequences of climate change and the failure of the international political community to mitigate greenhouse gas emissions, a "People's Judicial Panel" concluded: "The governments of the world have a duty to protect the atmosphere that belongs in common to the world's people. Based on the evidence we have heard here today, the nations of our world are in violation of their most fundamental legal and constitutional obligations" (Brecher, 2015: 109). In the view of the climate justice movement, by failing to act decisively on the global warming crisis, governments around the world have broken international law and violated a public trust.

Given that nation-states have the greatest political capacity to mitigate climate change, and the primary legal obligation under international law to accomplish this critical goal, the failure by individual states and the international political community to take necessary and effective actions to reduce carbon emissions constitutes a crime, a climate crime. As Brecher (2015: 110) observes, "The Climate Justice Tribunal established a credible case that the governments and corporations of the world are systematically violating human rights, international laws, and their duty to protect the public trust by allowing the greenhouse gas emissions that are destroying the earth's climate." Given the devastating consequences of global warming and the fierce urgency of responding to the climate crisis, the U.S. government and the international political community have a moral and legal responsibility to take effective action. The failure of the U.S. government and other states to fulfill this responsibility and respond effectively to the climate crisis is a specific form of climate crime, a state crime of political omission, which cannot be fully understood without considering the powerful influence of the interrelated phenomenon of climate change denial. A well-funded countermovement emerged in the late 1980s and has systematically engaged in what I call the socially organized denial of climate change. Denialists have used a variety of funding sources, tactics, and strategies to successfully block effective political action on the issue, what I call climate crimes of denial.

Several criminologists have previously conceptualized the failure of states to prevent or control blameworthy harms that they have a moral and legal responsibility to address as a form of state crime. Friedrichs (2010: 140) uses the term *negligent state criminality* and argues that "The most serious form of negligent state criminality involves the unnecessary and premature loss of life that occurs when the government and its agents fail to act affirmatively in certain situations." It is well documented that unnecessary and premature loss of life will be one of the catastrophic social consequences of climate disruption. Other criminologists prefer the concept *state crimes of omission* (Barak, 1991; Kauzlarich, Mullins, and Matthews, 2003; R. Watts, 2016). According to Rob Watts (2016: 10), "State crimes of omission occur when states and their officials fail to act, or neglect to notice bad things happening, or even enter into a state of denial. This leads to significant, even life-threatening occasions for harm to particular groups of people simply because state officials have not acted. The starting premise here is that the state employs officials who are charged with acting in the public or national interest. Serious harms arise when official people who ought to both know and do better, do nothing."

I use the term *state* crimes of *political* omission, to emphasize the fact that state decisions not to act on climate change are part of what C. Wright Mills (1963: 23) called the "basic problem of power." In *The Power Elite*, Mills (1956: 4) argued that the failure to make decisions is also an exercise of political power by elites: "Their failure to act, their failure to make decisions, is itself an act that is often of greater consequence than the decisions they do make." Other sociologists have also focused attention on "non–decision making" and the setting of public agendas that exclude critical issues, what British political and social theorist Steven Lukes (2005) calls the "second dimension of power." While economic and political elites, and the organizations they control, do make explicit and consequential decisions—that is, they choose to take actions that protect their material and ideological interests in direct conflict situations (the first dimension of power)—it is important to be alert to other situations where a "mobilization of bias" operates to limit or exclude issues from public consideration and action in a way that benefits certain groups at the expense of others (Bachrach and Baratz, 1970; Molotch, 1970; Schattschneider, 1960). As political scientists Peter Bachrach and Morton S. Baratz (1970: 44) note, "The primary method for sustaining a given mobilization of bias is nondecision-making," which they define as "a decision that results in suppression or thwarting of a latent or manifest challenge to the values or interests of the decision-maker." They go on to argue that non–decision making "is a means by which demands for change in the existing allocation of benefits and privileges in the community can be suffocated before they are even voiced; or kept covert; or killed before they gain access to the relevant decision-making arena; or, failing all these things, maimed or destroyed in the decision-implementing stage of the policy process." The political failure of the U.S. government to act on climate change often involves this second dimension of power, non–decision making. Various demands to reduce greenhouse gas emissions and effectively address climate disruptions have been suppressed, thwarted, or suffocated before they gained access to the policy arena or killed and maimed before they could be implemented. These nondecisions constitute a significant portion of the state climate crimes of political omission.

In his classic article, "Oil in Santa Barbara and Power in America," American sociologist Harvey Molotch (1970) made a significant contribution to scholarship on this second dimension of power. He argued that the political response, or, more important, the lack of response, to technological "accidents" like the infamous 1969 Santa Barbara oil spill can provide critical insights into the structure of power in society that often remains hidden. Drawing on the work of Molotch and others, environmental sociologists Aaron M. McCright and Riley E. Dunlap (2000, 2003, 2010) argue that strategic tactics such as the manipulation of information, diversionary reframing, and political attacks on science were part of a mobilization of bias that successfully kept many environmental issues off the public agenda and from becoming widely defined as serious social problems. In the specific case of climate change, they analyzed how the American conservative movement more generally and the climate change denial countermovement specifically were able to use these tactics to challenge the environmental movement's definition of global warming as a legitimate social problem and block the passage of significant climate change policy.

In a more recent work on international climate policy that also advances our understanding of the different dimensions of power at play, sociologists David Ciplet, J. Timmons Roberts, and Mizan Khan (2015: 28) take a "strategic view" of power, seeing it "as a process that involves intense political struggles by competing coalitions over what constitutes legitimate leadership, as well as assertions of dominance through wielding ideological, economic and military might." Following Italian social theorist Antonio Gramsci's concept of "hegemony," Ciplet, Roberts, and Khan (2015: 27) make a distinction between "domination, or rule by coercion" and "intellectual or moral leadership, or rule by hegemony." They argue that structural, institutional, and instrumental forms of political domination involve "coercive mechanisms of influence" that rely on material force (the first dimension of power). At the international level, this can include military intervention and threats to cut foreign aid, while at the domestic level, it most often involves large campaign contributions, control over leadership positions, political lobbying, unequal access to information, and a continuous circulation of corporate and political elites between high-level organizational positions. While these structural and institutional perspectives on unequal power relations within and between states are important, in climate politics it is also "the creation and use of noncoercive legitimate power" (hegemony) that provides a key to understanding "how and why some courses of action have been embraced by state representatives, civil society groups, and industry as politically feasible, while other options have been viewed, often pre-emptively, as off the table" (Ciplet et al., 2015: 29). These theoretical insights concerning the different dimensions of power, in particular the use of the subtler form of power called non–decision making, guide my analysis of what Steffen (2016) calls the "politics of predatory delay."

In her *Juliana* opinion, Judge Aiken (2016: 25) rightly points out that the lawsuit raises the critical question of U.S. government agencies' "failure to act in areas where they have authority to do so," a failure (nondecision) to fulfill their legal responsibility (under the Nature's Trust doctrine) that allows a severe harm to occur—the essence of the climate crime that I am calling state crimes of political omission. To fully understand the "politics of predatory delay" on climate action in the United States, we must examine the relationships that the Ronald Reagan, George Herbert Walker Bush, Bill Clinton, and George W. Bush presidential administrations have had with the Congress, political parties, and administrative agencies within the government. What we call "the state" is not a unitary entity but an ensemble of governmental institutions capable of wielding various forms of political power. This ensemble is embedded in a set of relational processes with the corporate economy and civil society. As Michalowski (2010: 26) suggests, "While we can, from an organizational perspective, identify institutions of government, of economy, and of civil society, it is probably theoretically wise to remain aware that these elements of contemporary society are so tightly woven that it is risky to consider one without examining their *intersection* with the others [emphasis added]." This relational view of the state also guides the analysis of political omission and organized denial concerning climate change.

THE SOCIALLY ORGANIZED DENIAL OF CLIMATE CHANGE

One of the key intersections that requires investigation is the critical relationship between governmental institutions and those corporate- and civil-sector organizations engaged in the socially organized denial of global warming. In *States of Denial: Knowing about Atrocities and Suffering*, South African sociologist Stanley Cohen (2001) demonstrated how individuals, organizations, publics, political cultures, and governments—whether victims, perpetrators, or observers—frequently incorporate statements of denial into their social definitions, beliefs, knowledge, and practices in such a way that atrocities and suffering related to state crimes are not acknowledged or acted upon. According to Cohen (2001: 51), *denial* "refers to the maintenance of social worlds in which an undesirable situation (event, condition, phenomenon) is unrecognized, ignored or made to seem normal." He identifies three categories of denial: literal, interpretive, and implicatory, all of which can be applied to climate change. A *literal* denial is "the assertion that something did not happen or is not true"—for example, the claim that the earth is not warming. With an *interpretive* denial, the basic facts are not denied; however, "they are given a different meaning from what seems apparent to others" (Cohen, 2001: 7). Here, the event or the harm is socially and morally framed or reframed in such a way as to deny the corporate state's responsibility or culpability—for instance, when denialists admit that the earth is warming but argue that the warming is due not to human activities like burning fossil fuels but rather to natural factors such as higher energy output from the sun. Finally, the notion of *implicatory* denial "covers the multitude of vocabularies—justifications, rationalizations, evasions—that we use to deal with our awareness of so many images of unmitigated suffering." Here, "knowledge itself is not an issue. The genuine challenge is doing the 'right' thing with this knowledge" (Cohen, 2001: 7–9). American sociologist Kari Marie Norgaard (2011), for example, has analyzed how many people who know about global warming and its catastrophic consequences do little to incorporate that knowledge in their everyday life or political actions.

Starting in the late 1980s, shortly after Hansen put the issue on the public agenda with his forceful Senate testimony, a climate change denial countermovement emerged as a powerful actor in the politics of global warming. This conservative countermovement consisted of well-financed and ideologically cohesive organizations that used a variety of strategies and tactics in an effort to counter Hansen's warning, create public doubt about the science of global warming, delegitimize climate change as a serious social problem requiring governmental action, and block political efforts to mitigate greenhouse gases and adapt to climate disruption. Climate change denial efforts have been analyzed by many social scientists, climate scientists, and journalists (Antonio and Brulle, 2011; Banerjee, 2017; Beder, 1999; Brulle, 2014; Ceccarelli, 2011; Diethelm and McKee, 2009; Farrell, 2015, 2016; Gelbspan, 1998, 2004; Griffin, 2015; Hamilton, 2010; Hess, 2014; Hoffman, 2015; Hoggan, 2009; Jacques, Dunlap, and Freeman, 2008; Klein, 2014; M. Mann and Toles, 2016; McCright and Dunlap, 2000, 2003, 2010, 2011a, 2011b; Oreskes and Conway, 2010; Otto, 2016; Powell, 2011; Washington and Cook, 2011). After reviewing much of this research, Dunlap and McCright (2015: 320) concluded that "by constantly challenging the reality and significance of climate change, the denial countermovement represents a powerful obstacle to mobilizing societal action aimed at reducing [greenhouse gas] emissions." A key factor in the Republican Party's current political obstructionism on the issue of global warming, and in the general failure of the U.S. government to produce any substantive domestic or international policy to mitigate greenhouse gas emissions, is the increasing politicization and polarization of the issue of climate change within the American political system (Dunlap, McCright, and Yarosh, 2016). This politicization and the resulting policy stalemate are primarily the result of the activities of the conservative global warming denial countermovement.

Like fossil capitalism and global warming, climate change denial has deep roots in the history, culture, and politics of modern industrial societies. These social systems emerged out of Polanyi's "Great Transformation," massive economic and political changes that wrought the "modern" capitalist world. Dunlap and McCright (2015: 302–303) point out that the industrial societies that emerged out of this transformative process came to "embody" a number of significant cultural characteristics: an "anthropocentric" and "instrumentalist" view of nature as a resource for humans to exploit through technology and science for economic and social progress; a "dominant social paradigm" that valued private property, individual rights, free markets, and limited government; the prioritization of economic growth and prosperity as a social "imperative"; and the general notion that modern industrial societies have become "exempt" from nature's constraints.

These "widely held and deeply embedded beliefs" provide a "rich cultural tool kit" to those who hold a conservative worldview, especially the economic and political elites who control the world's most dominant organizations and institutions (Dunlap and McCright, 2015: 303). Environmental concerns in general, and climate disruption in particular, generate political demands for state actions and responses that threaten this ideological framework and the vested economic and political interests of these elites. As Klein (2014: 41) puts it, "Climate change detonates the ideological scaffolding on which contemporary conservatism rests." From the perspective of these elites and those who share their cultural worldview, climate change is a political and ideological threat because it challenges the basic premise of capitalism that the unregulated search for profits will, via the "invisible hand," produce the best possible society. Thus, the problem must be denied in some fashion. Dunlap and McCright (2015: 303) argue that denialists draw on this cultural tool kit to "label efforts to reduce [greenhouse gas] emissions as threatening economic growth and prosperity, the free market system, individual rights, the American way of life, and even Western civilization—discursive resources that they readily employ." And they have employed these resources with considerable political success.

The climate change denial countermovement is also rooted in the well-organized and powerful conservative political movement that developed in the United States during the 1970s. This American conservative movement (a network of business organizations, trade associations, foundations, think tanks, alternative media outlets, and public intellectuals) arose in reaction to what activist Tom Hayden (2009) called the "long sixties," a broad array of social movements and countercultures that emerged in the post–World War II era to challenge the elite "Machiavellians" who controlled the U.S. corporate state. The civil rights, anti–Vietnam War, and environmental movements played central roles in contesting state power and demanding progressive social change in this era. Although the progressive movements of the long sixties won some critical civil rights victories, ended the imperial war against Vietnam, produced some important environmental laws, and generated some significant social and cultural changes in American society, an extremely well-funded and well-organized conservative counter-movement, anchored by an ascendant "neoliberal" ideology, fought back against progressive people's movements and protected corporate interests (Antonio and Brulle, 2011; Hacker and Pierson, 2010; Madrick, 2011). The election of Ronald Reagan in 1980 represented a significant political victory for the conservative movement and served to institutionalize the neoliberal ideology of market fundamentalism, large tax cuts for corporations and the wealthy, dramatic reductions in federal regulations, and drastic cuts to social welfare programs.

The "institutionalization of neoliberalism" not only "signified a major cultural shift"; it also "produced structural changes" that further enhanced the accumulation of capital and the corporate domination of the state (Dunlap and McCright, 2015: 304). These cultural and structural changes have continued to shape political decision making on climate change and other critical issues. Today, corporate power increasingly "holds the government hostage" (Hedges, 2009: 143), and corporate interests have the ability to subvert democracy in the United States, a phenomenon that political theorist Sheldon Wolin (2008) called "inverted totalitarianism." Anthropologist and geographer David Harvey (2010: 220) points out that "raw money power wielded by the few undermines all semblances of democratic governance." The 2010 U.S. Supreme Court decision in *Citizens United v. Federal Election Commission* only enhances this subversion of democracy. The grip of the corporate forces that dominate the state is strong, and the "liberal class" that once provided a minimal level of opposition to such private tyranny now appears to be "dead" (Hedges, 2010). The limited progress against poverty, racism, militarism, and environmental destruction achieved in the 1960s and 1970s has been derailed. This is the historical, ideological, and political context within which the climate change denial countermovement developed and began to exert great influence.

One specific target of the American conservative movement, "a keystone of neoliberal antiregulatory politics" (Antonio and Brulle, 211: 197), was environmentalism in general and environmental laws specifically. McCright and Dunlap (2010: 107) point out that conservatives "mobilized against the environmental movement largely by attacking the 'impact' science upon which the environmental movement's claims and resulting environmental policy proposals are based." Schnaiberg (1980) first made the distinction between "technological-production" science and "environmental-social impact" science. Under industrial capitalism, "production science" and associated technology dominates nature and society and provides economic producers with control over both resources and people—over the environment and workers and consumers. However, in the post–World War II era, some scientists (like Rachel Carson) began to gradually identify "the negative impacts of science and technology" and "challenged the assumption that production science inevitably led to advancement and progress for society" (McCright and Dunlap, 2010: 104). As the findings of "impact science" advanced, and its warnings of ecological catastrophe grew stronger, this scientific knowledge posed a political threat to the industrial capitalist system as the environmental movement demanded change. To defend "the treadmill of production," conservative "merchants of doubt" (Oreskes and Conway, 2010) soon emerged to attack and undermine impact science in general and the threat of climate science in particular.

Specific climate change denial efforts are largely carried out by conservative think tanks, from large organizations central to the conservative movement such as the Heritage Foundation, the Hoover Institution, the Cato Institute, Americans for Prosperity, the American Legislative Exchange Council, and the Competitive Enterprise Institute (Dunlap and McCright, 2015; Union of Concerned Scientists, 2013), to smaller ones that specialize in the denial of global warming like the Marshall Institute (Oreskes and Conway, 2010), the Manhattan Institute for Policy Research (Union of Concerned Scientists, 2013), and the Heartland Institute (Ward, 2012). These conservative think tanks were funded initially by direct money from the fossil fuel industry (Gelbspan, 2004; Greenpeace, 2011, 2013a; Jacques et al., 2008; Klein, 2014; Oreskes and Conway, 2010; Powell, 2011; Union of Concerned Scientists, 2012a, 2012b). For example, Western Fuels, a large coal cooperative, Koch Industries, and the ExxonMobil corporation have each contributed millions of dollars to these conservative think tanks and the environmental "skeptics" working within them to deny global warming (Farrell, 2015; Jacques, 2009; Klein, 2014; Mayer, 2017; McNall, 2011). InfluenceMap, a British nonprofit research organization, estimates that ExxonMobil, Royal Dutch Shell, and three oil industry–affiliated groups together spend $115 million a year on "advocacy" designed to obstruct climate change policy. This corporate money supports not only direct lobbying but also "advertising, marketing, public relations, political contributions, regulatory contacts, and trade associations" (Roston, 2016: 1). Concerning the role of ExxonMobil in particular, Oreskes and Conway (2010: 247) note that the oil giant's "support for doubt-mongering and disinformation is disturbing but hardly surprising. What is surprising is to discover how extensive, organized, and interconnected these efforts have been, and for how long."

Recent research by sociologist Justin Farrell (2015: 4) shows that the successful production and diffusion of false information about climate change does indeed have a specific institutional and corporate structure. As he points out, within the countermovement's network, "power and semantic influence is not spread evenly among organizations . . . but is concentrated within a smaller group of organizations with ties to particular actors in the private sector," specifically ExxonMobil and the Koch Family Foundations (the philanthropic arm of the fossil fuel–based Koch Industries). Farrell (2016: 92) finds that this corporate funding shapes the "thematic content" of denial efforts and has significantly contributed to an increase in "ideological polarization about climate change" in American society and that this polarization in turn produces "public uncertainty" and "policy stalemate."

The broader network of organizations that make up the climate change denial countermovement (conservative think tanks, industry trade associations, public relations firms, and right-wing media outlets) are not only funded by corporations in the fossil fuel industry; they are also increasingly financed by conservative foundations. American environmental sociologist Robert Brulle (2014: 693) argues that denialist efforts "have been bankrolled and directed by organizations that receive sustained support from foundations and funders known for their overall commitments to conservative causes." He examined ninety-one organizations that make up the climate change countermovement and found that conservative foundations provided around $64 million to support the active campaigns of these groups to manipulate and mislead the public on the nature of climate science and the threat of climate disruption. Brulle also found that as direct corporate funding from the fossil fuel industry has declined in recent years, the countermovement's funding has shifted to untraceable sources through groups like the Donors Trust/Capital foundations. The names of those who contribute to private foundations like these are not revealed; hence the term "dark money" is used to characterize these secret donations. In her highly acclaimed book *Dark Money: The Hidden History of the Billionaires behind the Rise of the Radical Right*, American journalist Jane Mayer (2017: 252–253) details the way Donors Trust and other conservative foundations use dark money to bankroll climate change denial and other right-wing causes.

With direct funding from the fossil fuel industry and conservative foundations, along with an increasing amount of untraceable dark money, the climate change denial countermovement has been able to attack climate science, manufacture uncertainty about the reality of global warming, and block political efforts to reduce carbon emissions. The Union of Concerned Scientists (2012b) points out that corporations have a long history of corrupting science at the public's expense. But the climate change denial countermovement has taken this effort to a new level in its attempt to influence the U.S. dialogue on climate science and policy (Union of Concerned Scientists, 2012a). Corporate-funded conservative think tanks "function as countermovement organizations" that provide "the connective tissue" that hold the broader countermovement together and serve as "vehicles" to extend its public reach (Dunlap and McCright, 2015: 312). The conservative think tanks recruit contrarian scientists who, because they have a background in some scientific field (although not climate science), are granted legitimacy by the public and the media. These contrarian scientists and other public intellectuals have produced a vast literature that challenges climate science, attacked and harassed legitimate climate scientists, served as "expert" witnesses in congressional hearings, organized front groups for specific campaigns, and in general tried to promote the appearance of scientific credibility for the denial countermovement as it wages its "inquisition" of climate scientists and its overall "war on science" (Dunlap and McCright, 2015; Oreskes and Conway, 2010; Otto, 2016; Powell, 2011). This is a very successful war that has won many political battles over the years, blocking climate action.

In this war on climate science, denialist activities produce a large amount of ideological propaganda built around lies and deceptions masquerading as science. This corporate-friendly propaganda can be viewed as a set of what Norgaard (2011) calls "legitimating and normalizing narratives." These cultural narratives, which legitimate and normalize fossil fuel industry practices and cast doubt on climate science, are then disseminated through the countermovement's network of conservative think tanks, industry trade associations, right-wing opinion leaders, conservative politicians, the Tea Party movement and other right-wing populist organizations, the mainstream corporate media (with their notorious concern for "balance"), Fox News and other conservative media outlets, the denial blogosphere, and by some Republican Party elected officials, now including Trump (Gelbspan, 1998, 2004; Hamilton, 2010; Hoggan, 2009; Jacques, 2009; Klein, 2017; McCright and Dunlap, 2000, 2003, 2010; Oreskes and Conway, 2010; Powell, 2011; Union of Concerned Scientists, 2012a).

The denialist arguments (narratives), disseminated through these vehicles, generally fall under the following categories: conspiracy theories, fake experts, cherry-picking of data, impossible expectations of what research can deliver, and misrepresentation and logical fallacies (Diethelm and McKee, 2009). As environmental scientists Hayden Washington and John Cook (2011: 43) conclude about the denialists, "Their goal is to convince the public and the media that there are sufficient grounds *not to take the action* [emphasis added] recommended by the consensus position of mainstream science. To achieve this, the vocal minority employs rhetorical arguments that give the appearance of legitimate debate where there is none." In the sections that follow, as I examine the political failure of the U.S. government to reduce carbon emissions, this particular form of non–decision making, the specific actions and impacts of the denial countermovement, will be analyzed.

THE UNITED STATES, THE INTERNATIONAL
COMMUNITY, AND GLOBAL
ENVIRONMENTAL DANGERS

Any examination of the climate crimes of political omission and denial must consider the critical role of the U.S. government in blocking action on global warming. After all, as Giddens (2011: 89) pointed out, long before Trump came on the political scene, "the U.S., the country with the greatest responsibility to develop a far-reaching climate change policy, has done nothing at all on a national level. It is almost alone among industrial states in this respect." Yet climate change is a global problem that demands, ultimately, a global political solution. In addition to focusing on the politics of predatory delay in the United States, it is also important to examine the role of the international political community in responding or not responding to the problem of climate change and the responsibility it too bears for climate crimes of political omission (Ciplet et al., 2015). International political negotiations concerning global warming, or "climate diplomacy" (Jamieson, 2014), also provide the larger institutional framework within which U.S. government debates and decisions must be understood.

Due to "impact science" post–World War II, a variety of environmental issues emerged that highlighted the global nature of many ecological dangers and the need for a response from the international political community. While some concern for global environmental issues existed prior to the war, two events in 1945 "played crucial roles in galvanizing movements for environmental protection at the international level" (Michalowski and Kramer, 2014: 197). These events were the atomic bombings of Hiroshima and Nagasaki and the creation of the United Nations. Radioactive fallout from the use of atomic bombs on Japan, and later the atmospheric testing of nuclear weapons during the early Cold War, generated worldwide public awareness of a specific global environmental threat and led to calls for international controls over nuclear tests. The establishment of the United Nations provided an important political forum in which impact scientists, NGOs, and states could address the threats of nuclear weapons and radioactive fallout, as well as other emerging environmental problems.

In August 1963, the United Nations approved a multilateral treaty that banned nuclear weapons tests in the atmosphere, outer space, and underwater. Although it was not the first multilateral treaty dealing with environmental issues, the test ban treaty represented a major step forward in addressing global environmental dangers and shaping the United Nations as a mechanism by which the international political community could address such threats. From this point forward, the movement for international environmental protection grew substantially. In the last quarter of the twentieth century, the scientific analysis and political discussion of global environmental problems were incorporated under the umbrella of three major issue areas: protecting biodiversity, fostering sustainable development, and mitigating global warming and climate change (Michalowski and Kramer, 2014).

In the United States, the 1962 publication of *Silent Spring* by Rachel Carson, an exemplar of Schnaiberg's (1980) "environmental-social impact science," played an important role in launching the environmental movement. This landmark book issued a grim warning about the dangers of the indiscriminate use of pesticides, but as American naturalist Ralph H. Lutts (1985: 212) argued, "People in the United States and around the world were prepared or pre-educated, to understand the basic concepts underlying Rachel Carson's *Silent Spring* by the decade-long debate over radioactive fallout preceding it." The 1969 Santa Barbara oil spill (analyzed by Molotch, 1970), the fire in the Cuyahoga River that same year, and concern over deadly air pollution events also helped raise political awareness of environmental issues. Fueled by these and other concerns, the environmental movement continued to develop during the 1960s, and on April 22, 1970, Democratic senator Gaylord Nelson from Wisconsin organized a significant event, the first "Earth Day" (co-chaired by Congressman Pete McCloskey, a Republican from California). That summer, President Richard Nixon proposed a major executive branch reorganization that would consolidate the variety of federal environmental agencies and responsibilities into one major agency. Congress approved his proposal for an Environmental Protection Agency, and the new federal regulatory agency was established in December 1970 with William Ruckelshaus serving as its first administrator. The early 1970s "witnessed a tremendous boom in federal environmental law" (Keenan, 2008: 169), including the revised Clean Air Act (1970), the Clean Water Act (1972), the Endangered Species Act (1973), the Safe Drinking Water Act (1974), the Toxic Substances Control Act (1976), and others. Klein (2014: 201) describes this as the "Golden Age of Environmental Law," adding that "This was a time when intervening directly in the market to prevent harm was still regarded as a sensible policy option." Due to the influence of impact science such as the Keeling Curve (documenting increasing levels of carbon dioxide in the atmosphere), and a rising environmental consciousness, the greenhouse effect seemed to be moving toward a place on the political agenda. As Jamieson (2014: 21) put it, "Climate change had arrived in Washington, but it was far from center stage." American environmental historian Joshua Howe (2014: 170–71) stresses an important point when he observes that climate science "was born of the Cold War, and the politics of global warming emerged against a national and international political backdrop dominated by the conflict."

Growing concerns about the greenhouse effect among some members of Congress and other federal officials resulted in the passage of the National Climate Program Act of 1978, which established a new National Climate Program Office to carry out research on climate change. This was followed by numerous congressional hearings in the 1980s and the Global Climate Protection Act of 1987, which directed the Climate Program Office to conduct even more research on the problem. But despite the increasing concern, the hearings in Congress, and the growing pace of climate research, no legislation or public policies emerged at that time to slow the rise of carbon emissions (Rich, 2018a). Even after Hansen's 1988 Senate testimony finally created a broader public awareness of the dire threat posed by the greenhouse effect, political action to deal with this serious social problem was slow to emerge. The institutionalization of the American conservative movement, with its neoliberal ideology, and the development of the climate change denial countermovement would play key roles in blocking effective climate policy in the 1990s and beyond.

As international alarm about global warming continued to build, the United Nations Environmental Programme and the World Meteorological Organization created the Intergovernmental Panel on Climate Change in 1988 to assess climate research. Earlier in that pivotal year in Toronto, the World Conference on the Changing Atmosphere: Implications for Global Security highlighted atmospheric issues such as global warming and allowed scientists to discuss emission reductions with policymakers. But the most important event organized by the international political community in response to the growing threat of disruptive climate change was the second Earth Summit held in 1992 in Rio de Janeiro, Brazil. Formally known as the United Nations Conference on Environment and Development (UNCED), this second Earth Summit resulted in the landmark UN Framework Convention on Climate Change (UNFCCC), which was based on the IPCC's First Assessment Report completed in 1990. Despite serious tensions over specific targets and timetables and the role of the developing nations in the overall process, the UNFCCC managed to create a global regime designed to negotiate international agreements that would limit the worldwide emission of greenhouse gases. Recent international developments had created a new, more encouraging environment for such negotiations. As Howe (2014: 171–172) points out, the end of the Cold War in 1991 "changed the nature of the international system within which debate about CO_2 and global warming unfolded," and the UNFCCC "reflected the optimistic spirit of international cooperation . . . sensed in the immediate post–Cold War world." A new world of "sustainable development" under global capitalism seemed possible to many.

One reason for optimism about the UNFCCC was the prior success of the Vienna Convention of 1985, which formally recognized the threat of ozone depletion and initiated a research and policy development process that resulted two years later in the Montreal Protocol on Substances that Deplete the Ozone Layer. The 1987 Montreal Protocol set deadlines for specific actions to reduce ozone depletion and represented a concrete and effective global strategy to protect the ozone layer. The development and success of the Montreal Protocol inspired hope that the UNFCCC might also lead to the negotiation of a global agreement to reduce carbon emissions and prevent climate disruption. That hope would be dashed on numerous occasions, primarily (but not only) by the U.S. government during various presidential administrations.

NEOLIBERALISM, RONALD REAGAN, AND GEORGE H. W. BUSH

If the environmental movement viewed the 1970s as the Golden Age of Environmental Law, conservative ideologues, right-wing foundations and think tanks, and polluting industries would come to see the 1980s and 1990s as the Golden Age of Environmental and Regulatory Rollback. The election of Ronald Reagan in 1980 signaled the ascendency and eventual domination of the American conservative movement and the institutionalization of a neoliberal market fundamentalism in American politics. The dominance of the neoliberal movement resulted in the deregulation of harmful corporate practices in general and the rollback of environmental protections specifically. Howe (2014: 125) refers to it as the "Reagan antienvironmental revolution," noting that it "represented the most staggering and comprehensive peacetime rollback of environmental policy in the history of the conservation movement." Reagan took down the solar panels that President Jimmy Carter had installed on the roof of the White House, froze the automobile mileage standards that had significantly cut demand for oil, appointed pro-business, anti-regulatory, anti-environmental zealots to head the EPA (Anne Gorsuch) and the Department of the Interior (James Watt), and used the federal bureaucracy to squelch ongoing research on carbon dioxide and climate change. Rich (2018a: 27) argues that "Reagan appeared determined to reverse the environmental achievements of Jimmy Carter, before undoing those of Richard Nixon, Lyndon Johnson, John F. Kennedy and, if he could get away with it, Theodore Roosevelt." The right-wing Heritage Foundation, which urged its followers to "strangle the environmental movement" that it viewed as "the single greatest threat to the American economy," became known as Reagan's "shadow government" (Kennedy, 2005: 24–25) and began pushing hard for regulatory "relief."

When it came to the issue of global warming, it became clear that "Climate policy was not going to be made on Reagan's watch, and research money was not going to be wasted on the environment when it could be spent on the military" (Jamieson, 2014: 28). While Reagan did eventually relent and sign legislation that promoted more research on climate change, his administration steadfastly opposed any concrete policies that would reduce greenhouse gas emissions. More important, Reagan cemented in place the neoliberal worldview that would lead to the greater extraction and burning of fossil fuels and a frightening rise in carbon emissions over the next thirty years (Mitchell, 2013). As McKibben (2010b: 95) observes: "[Reagan's] worldview gave us not only the Bush administrations but also the Clinton years, with their single-minded focus on economic expansion. The change was not just technological; it wasn't simply that we stopped investing in solar energy and let renewables languish. It's that we repudiated the idea of limits altogether—we laughed at the idea that there might be limits to growth. Again, not just right-wing Republicans but everyone." This illustrates what sociologists who take a strategic approach to power mean when they refer to hegemonic ideological frameworks.

Hegemonic frameworks or worldviews, like the one Reagan promoted, have the power to noncoercively manufacture consent and frame what policy options are politically feasible, what Gramsci (1971) called "a war of position." As Ciplet, Roberts, and Khan (2015: 30) detail, the hegemonic ideas that frame and constrain the politics of climate change include a focus on market-based solutions, a strong commitment to the goal of sustained economic growth, and a narrow view of the relationship between science and climate policy that excludes measuring the social impacts of global warming or proposed policies to deal with it. These hegemonic ideas concerning environmental problems, which involve "reshaping the broad interests and values evident within society" (Wright and Nyberg, 2015: 88), were given their first systematic framing by Reagan but would have their largest impact on climate politics over the course of the subsequent three presidential administrations.

Reagan's neoliberal economic ideology, anti-environmental revolution, and hawkish military policies all provoked fierce political resistance. While prior to 1988 climate change as a stand-alone issue was not yet a major public policy concern, Reagan's aggressive conservatism and anti-environmental actions "galvanized disparate groups of scientists, environmentalists, and politicians into a single community of global warming advocates." During the 1980s, the politics of global warming became "a politics of dissent" (Howe, 2014: 119, 144). Congressional resistance and public opposition to the administration's environmental policies caused conservative activists to rethink their strategy. According to McCright and Dunlap (2010: 108), they "learned it was unwise to attack environmental protection directly (exercising the first dimension of power), as Americans are in fact supportive of environmental protection and see it as a governmental responsibility." Conservative groups then changed tactics and began to challenge the seriousness of environmental problems, attack the impact science upon which these assessments were based, and raise concerns about the economic consequences of government responses to such problems. Neoliberal activists "learned to be more subtle and use the second dimension of power to prevent major decisions on environmental policy-making that might threaten their conservative interests" (McCright and Dunlap, 2010: 108). This would become the preferred strategy of the emerging climate change denial countermovement in the 1990s.

Reagan's fervent market fundamentalism and anti-environmental rollbacks caused concern even among some Republicans (Klein, 2014). Although the GOP was moving sharply to the ideological right, it had not yet become the extremist political party it would become in the twenty-first century. As the 1988 presidential election approached, the Democratic Party platform promised to address the greenhouse effect, and their candidate, former Massachusetts governor Michael Dukakis, appeared to be more likely to deal effectively with environmental issues than an anti-regulatory Reagan Republican. Thus, as George H. W. Bush, Reagan's vice president, campaigned for the presidency in his own right in that pivotal year, he felt compelled to address the issues of global warming and environmental protection.

On the campaign trail that fall, Bush stated, "I am an environmentalist" and then boldly declared, "Those who think we are powerless to do anything about the greenhouse effect forget about the 'White House effect'; as President, I intend to do something about it." Bush promised to be "the environmental president" and convene an international conference on the environment, proclaiming, "We will talk about global warming and we will act" (*New York Times*, 1989: 1). He seemed to understand the growing threat of climate change and have a genuine desire to take action to limit carbon emissions. But despite his campaign rhetoric and a pledge to pursue a "no regrets" strategy, once in office Bush accomplished very little on climate change. By the end of his term, "no regrets" would be twisted around to mean no regrets about causing economic harm to the American economy by "unnecessary" emission reduction policies. A *New York Times* editorial as early as 1989 accused the Bush administration of floundering "in confusion and timidity" on the issue. Democratic congressman James H. Scheuer (1990: 1) from New York mocked the president's promise to use the "White House effect" to counter the "greenhouse effect," charging that what he delivered instead was the "whitewash effect." While Bush was more engaged with the issue of global warming than Reagan, and the United States did

sign and ratify the UNFCCC during his tenure, the president did not propose any mandatory cuts in carbon emissions due to his fear of the adverse
economic effects such programs might have.

In early 1989, there were some encouraging signs that the new administration might address the serious problem of global warming. Bush did
create a National Security Council Policy Coordinating Committee on
Oceans, Environment and Science to coordinate climate change policy
(Pielke, 2000). Chaired by Assistant Secretary of State Frederick Bernthal,
the committee held a relatively low position within the administration and
had no formal connection to the Committee on Earth Sciences, which was
set up during the Reagan years to coordinate research on global warming.
Thus, "the science and policy of climate change were poorly linked at this
time" (Pielke, 2000: 22). Furthermore, the lower-level Bernthal Committee had to send its climate proposals up through National Security Advisor
Brent Scowcroft, who had little interest in the issue, and Bush's chief of staff
John Sununu, who, influenced by the emerging climate change denial
countermovement, doubted the science and strongly opposed any emission
reduction measures because he thought they would slow economic growth.
As a result, the committee "had little impact" on climate policy development (Pielke, 2000: 22).

In his first speech as Bush's secretary of state in early 1989, James Baker
addressed the newly created IPCC and offered another encouraging note.
Baker stated, "We can probably not afford to wait until all of the uncertainties about global climate change have been resolved. Time will not make
the problem go away." But soon after this speech, Baker received a visit
from Sununu, who told him in no uncertain terms to "stay clear of this
greenhouse-effect nonsense. You don't know what you're talking about"
(Rich, 2018a: 51). Baker never spoke about the issue again during his term
as secretary of state. In addition to Sununu's "strident opposition" to
addressing the greenhouse effect, Bush's Office of Management and Budget
(OMB) censored the congressional testimony of NASA's James Hansen for a
May 1989 hearing. The OMB went so far as to force Hansen to argue that
Congress should only pass climate legislation that immediately benefited
the economy, "independent of concerns about an increasing greenhouse
effect" (Rich, 2018a: 54). Concerned about the alteration of his scientific
conclusions, Hansen alerted Senator Al Gore to the censorship and Gore
went to the press, charging that the Bush administration was engaging in a
form of "science fraud" (Rich, 2018a: 54). A public uproar occurred, with
the Los Angeles Times calling the censorship "an outrageous assault." The
bad publicity forced the Bush administration to backtrack, and the next day
the president pledged to host a climate workshop at the White House (Rich,
2018a: 55). That conference never took place, but Bush would soon have

an opportunity to demonstrate that he was, in fact, an environmental president.

By 1989, climate change was not just a domestic political issue in the United States but a "global project" (Jamieson, 2014: 31). Members of the European Community agreed that they should begin to take steps to combat global warming. In November of that year, environmental ministers from sixty-seven nations, along with representatives from a number of international organizations, met in the Netherlands to talk about targets and timetables for stabilizing and then reducing greenhouse gas emissions. But the "White House Effect" failed to materialize at the conference. The climate skeptics led by John Sununu won out, and the Bush administration ended up expressing strong opposition to any such agreement during the negotiations. Daniel Becker of the environmental organization the Sierra Club concluded that the conference was a "failure" and called the United States and the other dissenting nations "the skunks at the garden party" (Rich, 2018a: 59). According to Jamieson (2014: 35), "Throughout the negotiating process American rejectionism was consistently on display." He goes on to observe, "Indeed, the policies of Reagan and both Bushes were remarkably consistent: do as little as possible on climate change, rationalized by casting doubt on the science and exaggerating the costs of action." This was the mobilization of bias and the power of non–decision making in action.

George H. W. Bush's first year in office marked the opening moves and growing influence of the climate change denial countermovement. Alarmed by the emergence of climate change as a global political project, the anti–environmental movement promoted by Reagan mobilized to meet the developing threat posed by the political issue of global warming. The Global Climate Coalition, an Exxon-organized fossil fuel industry front group, was formed in 1989 to question the reality of global warming and attack climate scientists in general and the newly created IPCC in particular (Oreskes and Conway, 2010; Otto, 2016). That same year, "discrediting global climate change claims began in earnest" when the George C. Marshall Institute, an influential conservative think tank, issued its first report attacking climate science (Antonio and Brulle, 2011: 197). As Oreskes and Conway (2010: 186) point out, "[The Institute's] initial strategy wasn't to deny the fact of global warming, but to blame it on the sun," a natural cause, not a human one that required reducing emissions. Following the publication of a small book in 1990 titled *Global Warming: What Does the Science Tell Us?*, the Institute's Washington staff gave a briefing to several important federal agencies "to set the record straight on global warming." According to Oreskes and Conway (2010: 186), "The briefing had a big impact, stopping the positive momentum that had been building early in the Bush administration." A year later, in a letter to the vice president of the

American Petroleum Institute, the founder of the Marshall Institute, Robert Jastrow, would brag, "It is generally considered in the scientific community that the Marshall report was responsible for the [Bush] Administration's opposition to carbon taxes and restrictions on fossil fuel consumption" (Oreskes and Conway, 2010: 190). The climate change denial countermovement was off to a successful start and the "White House Effect" was effectively blocked.

In addition to internal opposition and the influence of the climate change denial countermovement, the end of the Cold War played an important role in shaping the Bush administration's approach to the issue of global warming. The fall of the Berlin Wall in 1989 and the collapse of the Soviet Union in 1991 brought about a major change in the discourse about climate change and environmental problems more broadly. Howe (2014: 172) argues that the end of the Cold War coincided with "the rise of economics as the key rubric for environmental decision making." The much-touted concept at the time of a "new world order" really meant a new global *economic* order in which corporate capitalism and environmental protection could co-exist under the banner of *sustainable development*. This fusion of economic priorities with environmental considerations "would help define the politics of both the UNFCCC and the Kyoto Protocol" (Howe, 2014: 172). But as the Bush administration approached the 1992 Earth Summit in Rio and entered into negotiations over the UNFCCC, it became clear that economic concerns trumped environmental concerns. U.S. resistance to binding emissions targets at this and other conferences to follow was based on a commitment to the goal of economic growth and the maintenance of the profitability of U.S. corporate interests. The "no regrets" policy of Bush's 1988 campaign had been transformed into an "economic precautionary principle" where "any action or policy to curb environmental degradation was assumed to represent an economic threat unless proven otherwise" (Howe, 2014: 184). Binding emission targets and significant Global North financial commitments to sustainable development in the Global South were judged to violate this principle.

Congressional Democrats, environmental groups, and some newspapers heavily criticized Bush "for dragging his feet on environmental issues in favor of domestic corporate profits," with the *New York Times* even calling him "the Darth Vader of the Rio meeting" (Howe, 2014: 180). It was unclear if he would even join other world leaders at the conference. Eventually Bush did fly to Rio de Janeiro to sign the UNFCCC, pledging to translate the written document into "concrete action to protect the planet" (Oreskes and Conway, 2010: 197). But the document, important as it was at the time, contained no enforcement mechanisms. Binding limits on emissions would have to be determined in future annual negotiations called the

Conference of the Parties (COP), like the one in Kyoto in 1997. In the end, the George H. W. Bush "White House Effect" failed to reduce carbon emissions and slow global warming. The "environmental president" did very little to counter the greenhouse effect due to concerns about the economic costs of a more aggressive stance. An opportunity to act was lost. The politics of predatory delay won out. Would his predecessor do better, or would political omission continue?

BILL CLINTON, THE KYOTO PROTOCOL, AND THE PRIORITY OF ECONOMIC GROWTH

William Jefferson Clinton defeated the incumbent president George H. W. Bush and upstart Ross Perot in the 1992 election. After fighting fierce political battles with obstructionist Republican presidents on a number of policy fronts for the past twelve years, environmental groups were relieved and very optimistic about the new administration, especially on the issue of global warming. Clinton came into office with a strong commitment to reduce carbon emissions, and environmentalists were ecstatic that Al Gore, a long-standing and outspoken advocate for action on the problem of climate change, was his vice president. In Jamieson's (2014: 38) opinion, Gore's 1992 book *Earth in the Balance* was "certainly the most moving and knowledgeable book about the environment ever written by an American politician." On Earth Day in April 1993, the new president promised to reduce greenhouse gas emissions to their 1990 levels by the year 2000 in accordance with the new UNFCCC. Clinton and Gore would go on to propose various policies and programs over the years to achieve this ambitious goal. From the perspective of environmentalists, they made "good" appointments to key cabinet positions like the EPA and the Department of the Interior. The Clinton administration also took very seriously the international climate negotiations that were getting started within the UN Framework Convention process (which came into force in March 1994) and worked extremely hard to achieve results at the various COPs that were held in the 1990s. One significant result of these efforts was the Kyoto Protocol negotiated at COP 3 in 1997. But despite a genuine commitment to make a difference on this issue, and despite all the painstaking efforts expended, Clinton's climate change legacy, like Bush's before him, ended up as one of political failure. When Clinton, the so-called green president, left office in 2001, greenhouse gas emissions were 15 percent above the 1990 target levels (Wapner, 2001), and the United States had not ratified the Kyoto Protocol. What accounts for this failure? Can it also be defined as a climate crime of political omission?

The first thing to note is that Clinton faced serious opposition to his climate proposals both at home and within the international political community. Pro-business and anti–environmental regulation Republicans, significantly influenced by the fossil fuel industry and the climate change denial countermovement, opposed his policies at every turn. And after the Republican Party captured both the House and the Senate in the 1994 midterm elections, Republicans were able to block all of his proposals in Congress. Asked in an interview after Barack Obama had become president why it had taken so long for Democrats to gain any momentum on the issue of climate change, Clinton offered the explanation, "We didn't have the votes before" (P. Baker, 2009: 12). Fossil fuel corporations were able to directly provide large campaign contributions and conduct intense lobbying efforts, and indirectly manipulate public opinion through denialist public relations campaigns, to assist the Republican Party as it battled the administration's proposals to reduce carbon emissions. At the international level, the Clinton administration also had serious conflicts with both the European Union (EU) and the G77 (a large block of developing countries in the United Nations) during the climate change negotiating process. The EU wanted the United States to do more to reduce carbon emissions from its industrial sector, while China, India, and other developing nations resisted efforts to force them to be more involved in reducing their greenhouse gas emissions, which, under the UNFCCC rules, they were not required to do at that time (Royden, 2002).

These very real institutional constraints and unresolved political conflicts prevented Clinton from making any progress on the goal of reducing emissions. But beyond the structural level, the president's failure to achieve his ambitious agenda on global warming can also be attributed in part to his acceptance of a set of socially shared, hegemonic ideas about the nature of the global corporate economy, the imperative of economic growth, the promotion of consumerism, and what environmental policies were politically feasible within this framework. Clinton's commitment to the goal of economic prosperity within a taken-for-granted corporate capitalist system often blinded him to other issues. The most prominent meme of his 1992 presidential campaign was "It's the economy, stupid!" Clinton's embrace of what environmental politics scholar Paul Wapner (2001: 12) calls "the hegemony of economic idolatry" influenced all of his political decisions, including those concerned with climate policy. Like Reagan and Bush before him, Clinton's single-minded focus on economic expansion meant that any proposed solutions to the problem of global warming had to be "market-based" and crafted to do no harm to the American economy.

The power of this economic prosperity paradigm to shape political decision making on climate issues was illustrated at the very beginning of Clinton's first term with the controversial North American Free Trade Agreement (NAFTA). The agreement had been negotiated by the previous administration, but Bush ran out of time and had to pass the required signing and ratification of the pact on to Clinton. Many environmentalists and congressional Democrats expressed concern that NAFTA, as written, would drive down both environmental and labor standards, and they encouraged the new president to oppose the deal. During the 1992 campaign, Clinton had pledged that he would not sign the agreement unless those critical issues were addressed. One key question that should have concerned the president was: What impact would NAFTA have on his goal of reducing greenhouse gas emissions? Trade negotiations like NAFTA developed in parallel to international climate talks like the UNFCCC in the changing post–Cold War world. But as Klein (2014: 76) points out, "these parallel processes—trade on the one hand, climate on the other" were remarkable in that "they functioned as two solitudes." Each pretended that the other did not exist, and the question of how they would impact each other was ignored.

But the potential impact of free trade agreements on environmental issues like global warming was enormous. The provisions of NAFTA and other trade deals being negotiated at this time allowed private companies to sue national governments within international organizations (such as the World Trade Organization) for enacting regulatory laws that might reduce corporate profits. Such policies could result, for example, in the "criminalization" of state efforts to control corporate crimes such as the release of harmful greenhouse gases into the atmosphere. By making state regulatory efforts "trade illegal," these provisions were clearly intended to "dissuade governments from adopting tough antipollution regulations, for fear of getting sued" (Klein, 2014: 76). And while trade deals contained legal sanctions that could impose severe economic consequences on governments attempting to protect the environment, climate negotiations never considered such sanctions as a tool to enforce emissions reduction policies. As Klein (2014: 76) points out, "the commitments made in the climate negotiations all effectively functioned on the honor system, with a weak and unthreatening mechanism to penalize countries that failed to keep their promises. The commitments made under trade agreements, however, were enforced by a dispute settlement system with real teeth, and failure to comply would land governments in trade court, often facing harsh penalties." If it came down to a battle between trade (and corporate profits) on the one hand and the climate (reducing carbon emissions) on the other, trade would always win.

Such was the dilemma Clinton faced in early 1993. He and Gore wanted to do something about the climate crisis as they entered the White House, but they also recognized that the new rules contained in NAFTA would weaken environmental (and labor) standards and make dealing with the crisis more difficult. It was a tough decision, but even though Clinton had promised to reduce greenhouse gas emissions, the goals of liberalized trade and expanding the global economy were more important to him. After all, "It was the economy, stupid!" Thus, once he was in office, "the deal was left intact," and "two toothless side agreements were tacked on" for political cover (Klein, 2014: 83). NAFTA was then signed and ratified. For many reasons, some of which I will explore in the final chapter, a number of large, mainstream environmental organizations ended up backing the agreement. At the signing ceremony, Clinton thanked the supportive environmental groups and assured them, "We will seek new institutional arrangements to ensure that trade leaves the world cleaner than before" (Klein, 2014: 85). It was a hollow promise. The new trade architecture that was built in the 1990s and beyond would undercut global climate change negotiations and allow carbon emissions to soar.

Clinton's support for NAFTA and other neoliberal trade agreements like the World Trade Organization reflected his commitment to global corporate capitalism and the expansionary logic of that economic system. As treadmill of production theorists argue, there is an "enduring conflict" between the system imperative of economic growth under capitalism and the protection of the environment. Clinton was unable to challenge the widely shared and deeply entrenched logic of unlimited economic growth and unrestrained consumerism, and this would end up constraining his climate policies. Clinton, of course, did not see it that way. He attempted to portray the conflict between growth and environmental protection as a "false choice." In an October 1997 speech at the National Geographic Society, he argued that his emissions-reduction strategy, "if properly implemented, will create a wealth of new opportunities for entrepreneurs at home, uphold our leadership abroad, and harness the power of free markets to free our planet from an unacceptable risk; a strategy consistent with our commitment to reject false choices." Clinton believed that it was possible to reduce greenhouse gas emissions by using market tools and without impeding economic growth or causing economic harm. He was a champion of the idea of sustainable development. Clinton's Climate Change Action Plan (CCAP), initially proposed in October 1993, "consisted of over 50 new or expanded initiatives that the administration estimated would bring U.S. emissions back to 1990 levels by 2000" (Royden, 2002: 420). To the dismay of environmental groups, the CCAP did not contain any mandatory emissions reductions. Instead, the CCAP, and the administration's overall strategy, relied on various "tools of the market" to reduce carbon emissions such as "voluntary programs with industry, including electric utilities and the transportation and buildings sectors" and "spending and tax incentives designed to stimulate the use of energy efficient technologies in building, industrial processes, vehicles, and power generation" (Royden, 2002: 416).

The idea was that corporations could be induced to lower emissions and still make profits at the same time.

Perhaps this market-based strategy could have brought carbon emissions down somewhat at the end of the twentieth century, although treadmill of production theorists and many environmental activists continue to argue, persuasively in my view, that only more radical changes to the political economy can accomplish this goal. But it became a moot point in 1995 when both houses of Congress passed to the control of the Republican Party (for the first time since the 1950s). As a result, "the Newt Gingrich-led 'Republican Revolution' attempted to repeal existing environmental legislation, underfund the environmental science programs at government agencies, and generally cripple the functioning of environmental regulatory agencies" (McCright and Dunlap, 2010: 108). A key part of this Republican campaign was the 1995 elimination of the Office of Technology Assessment (OTA), which was Congress's only source of (impact) scientific advice. Influenced by the now very active climate change denial countermovement, the new Republican majority rejected most of Clinton's emission reduction proposals, and political gridlock set in on the climate issue. Blocked on the domestic front, the Clinton administration turned to international climate change negotiations as the first Conference of the Parties (COP 1) to be carried out within the framework of the UNFCCC was held in Berlin in March 1995. In this venue, too, Clinton encountered a number of conflicts with both other members of the international political community and a bipartisan group of opponents to a climate treaty at home.

One critical issue on both fronts concerned what role the developing countries would play in reducing carbon emissions under an international agreement. Because the developed nations (the Global North) represent the largest historical and cumulative source of greenhouse gas emissions, the UNFCCC had established different and higher standards for these industrialized nations than for developing ones (the Global South), who historically have done little to contribute to the overall problem of global warming (the principle of common but differentiated responsibilities). This only seems fair, but it would become a controversial issue in the U.S. Congress. The "Berlin Mandate" that emerged out of COP 1 specified that "developed countries go first" in reducing greenhouse gas emissions, an agreement that "would come to haunt the U.S. delegation in future negotiations because of domestic concerns about the trade implications of excluding China, India and Brazil from emissions reduction commitments" (Royden, 2002: 425). Although the Clinton team accepted this principle at Berlin, due to the domestic political controversy over the issue, they would have to revisit it repeatedly and try, to their dismay, to get the developing countries to do more to reduce their own emissions.

In late 1995, the IPCC issued its second assessment report, which confirmed the major findings of the first assessment and continued to build the scientific case for human influence on the global climate through the emission of greenhouse gases. In response to the report, political demands for taking immediate action to reduce these carbon emissions escalated, as did the climate change denial countermovement's personal attacks on climate scientists involved in writing the IPCC report (Oreskes and Conway, 2010). At the COP 2 summit held in Geneva in 1996, one of the key questions was whether the United States would support a legally binding agreement concerning the reduction of greenhouse gas emissions. During treaty negotiations for the UNFCCC under the first Bush administration, the United States was the only industrialized nation that refused to accept binding rules for the limitation of emissions, and the U.S. delegation eventually succeeded in obtaining treaty language that made compliance with greenhouse gas reduction goals voluntary. At Geneva, the Clinton administration reversed course and expressed public support for a legally binding treaty. "This is a big deal," stated then undersecretary for global affairs Timothy Wirth. "Saying we want to have a target that is binding is a clear indication that the United States is very serious about taking steps and leading the rest of the world" (quoted in Cushman, 1996: 2). But in prepared remarks addressing the conference, Wirth rejected proposals for steep and immediate cuts in emissions, calling that goal "unrealistic" given the economic disruption that would occur if energy consumption were restricted so sharply. Instead, he called for outcomes that are "real and achievable," using "market-based approaches" that provide "flexibility" to allow "all countries" to meet the new targets as best they can (quoted in Cushman, 1996: 2). No treaty resulted from this meeting, but the groundwork was laid for a potential breakthrough at COP 3 in 1997.

Clinton was re-elected president in November 1996, defeating Republican senator Bob Dole, but the House and Senate remained in the control of the GOP. The first year of Clinton's second term (1997) would prove to be a pivotal year for climate policy. Delegates from 170 nations were scheduled to meet in Kyoto, Japan, in December for COP 3. There was anticipation that this conference would finally produce a strict climate treaty, a breakthrough achievement, to fulfill the general principles agreed to in Rio de Janeiro in 1992 (the UNFCCC). As the Clinton administration painstakingly prepared for Kyoto, the fossil fuel industry and other powerful business interests began spending millions of dollars "on highly effective lobbying campaigns designed to limit the White House's options" (Cushman, 1997: 1). While private sector lobbying and socially organized denial on the issue of climate change had been building since the late 1980s when the greenhouse effect had finally emerged onto the public policy agenda, it became "more intense, prolonged and costly than usual in the realm of diplomacy" in the build-up to Kyoto (Cushman, 1997: 1). In July 1997, this intense corporate lobbying campaign bore fruit when the U.S. Senate passed a damaging resolution sponsored by Robert C. Byrd, Democrat of West Virginia (coal country), and Chuck Hagel of Nebraska, a Republican who strongly opposed a new climate agreement. The Byrd–Hagel Resolution, which passed by a vote of 95–0, directed the president not to sign any climate treaty that, first, did not require developing countries to also reduce their emissions or, second, would result in serious economic harm to the United States. Coming from the body that would have to ratify any treaty, this bipartisan resolution directly contradicted the Berlin Mandate that developing countries go first, further polarized the relationship between the Clinton administration and Congress, and undercut the president's negotiating team before they even arrived in Japan (Royden, 2002).

Assessing the impact of Byrd–Hagel, Oreskes and Conway (2010: 215) conclude that "Scientifically, global warming was an established fact. Politically, global warming was dead." One important reason that "the political calculus had changed between 1992 and 1997," according to Howe (2014: 190), was that climate denialists and Republican pro-business lobbyists "had begun to court traditionally Democratic labor organizations" (such as the auto workers and mine workers unions) and successfully persuaded them to actively oppose the Kyoto treaty over concerns about job losses. Howe concludes that "fueled by the dual rhetoric of scientific uncertainty and economic precaution (the latter cast in terms of lost American jobs), opposition to a robust climate treaty had grown to such a degree that not a single senator was willing to support a binding commitment to emissions reductions limited to industrialized nations" (Howe, 2014: 190). Despite this politically harmful resolution, the Clinton administration continued to prepare diligently for the summit in Kyoto.

Given the anger and dismay that some of its positions had aroused within the international community (concerning targets, voluntary programs, and flexible mechanisms), the Clinton team was surprisingly successful in accomplishing most of its objectives during the COP 3 negotiations in December. Gore, for example, helped negotiate what he called "a comprehensive plan . . . one with realistic targets and timetables, market mechanisms, and the meaningful participation of key developing countries" (Jamieson, 2014: 43). The result of these negotiations was the Kyoto Protocol, a multilateral agreement on legally binding targets for the reduction of greenhouse gas emissions by the developed countries. The protocol mandated a 5 percent reduction of greenhouse gases by 2012, with the United States agreeing to cut emissions by 7 percent, the European Union committed to 8 percent reductions, and Japan committed to 6 percent reductions. Achieving these goals would require the United States and other developed countries to reduce pollution and consumption in key areas by as much as 30 percent (Yamin, 1998). In recognition of the difficulty that developing nations would have in modernizing their economies while simultaneously reducing greenhouse gases, no binding targets or timetables were set for those nations (Leaf, 2001). The unanimous adoption of legally binding controls of green-house gasses by the Global North at the Kyoto conference represented a historic milestone in the development of multilateral environmental trea-ties, and emotions ran high as the chairman of the conference, Raúl Estrada-Oyuela, suggested that December 10 might come to be recognized as an international "day of the atmosphere" in commemoration of the signing (Yamin, 1998).

For the United States, however, the Kyoto Protocol would become legally binding only after ratification by the U.S. Senate. But by passing the Byrd–Hagel Resolution, the Senate had clearly indicated that it would not ratify any climate change protocol that failed to meet its two-pronged test: impose legally binding standards for reduction of greenhouse gas emissions on the developing world and not cause serious economic harm to the United States. According to its critics, the pact failed this test. Controversy over climate change "intensified in the United States after the adoption of the Kyoto Protocol . . . and the Republican-controlled Congress became even more oppositional" (Jamieson, 2014: 43). Facing certain rejection of the treaty, the Clinton administration, although it had formally signed the agreement in 1998 at COP 4 in Buenos Aires, made the decision not to submit the protocol to the Senate for ratification. At Buenos Aires, the United States had once again attempted to spur "meaningful participation" in greenhouse gas reductions by "key developing countries" (Leaf, 2001: 1219), but it was too little and too late. Clinton had been weakened politi-cally by the Monica Lewinsky sex scandal in early 1998, which led to his impeachment by the House and subsequent trial in the Senate (where he was acquitted). The scandal, combined with strong bipartisan opposition in the Congress, was too much to overcome. At the end of Clinton's two terms in office, the United States was not party to any international climate agree-ment, and the president had not been able to cut emissions as he promised. Instead, those emissions had risen as the politics of predatory delay prevailed.

Clinton did continue to speak out on the issue of climate change. In his State of the Union address in 2000, he called global warming "the greatest challenge of the new century" and insisted that cutting carbon emissions would not slow economic growth, but instead, given new technologies, these reductions could actually spur even more growth. Years later, in an interview with American journalist Joe Conason (2016: 264), Clinton acknowledged that climate change had been one of the issues that had "troubled him most" as president "because he felt the consequences for civilization were likely to be so grave." Clinton recalled: "During my second term, I spent a lot of time on it. I gave a couple of very serious speeches on it that elicited a giant yawn. And then even Democrats voted against the Kyoto Treaty, Democrats from energy producing states. I just could not get anybody to take it seriously" (quoted in Conason, 2016: 264). To the contrary, quite a few people were taking climate change seriously at that time, but a combination of Clinton's economic idolatry, his political missteps, Republican control of Congress, an increasingly active climate change denial countermovement, and strong organized corporate resistance led to the Clinton administration's political failure to mitigate carbon emissions. By failing to reduce harmful greenhouse gas emissions and failing to become a party to an international climate treaty that might have limited climate disruption, the U.S. government did engage in the state–corporate climate crime of political omission during his presidency.

George W. Bush and the Growing Extremism of the Republican Party

In December 2000, George W. Bush, son of George Herbert Walker Bush, was selected by the U.S. Supreme Court as president of the United States. The court halted the recount of votes in Florida that, if allowed to continue, might have given an Electoral College victory to then vice president Al Gore, who had won the popular vote. Gore had disappointed many environmentalists during the campaign by focusing mainly on projects such as "reinventing government" and "developing the internet" instead of climate change (Jamieson, 2014: 39), but there was no doubt about his genuine commitment to the issue. The electoral result, therefore, was not only a stinging defeat for Gore and the Democratic Party; it would turn out to impose a huge political obstacle on the movement to reduce greenhouse gases and combat climate change. After his unprecedented defeat in this most crucial election, Gore would, a few years later, bring the problem of global warming to public attention through his film and book *An Inconvenient Truth* (Gore, 2006), which would earn him (and the IPCC) the 2007 Nobel Peace Prize. But after the 2000 election he was not in a political position to make critical decisions on the issue.

During the 2000 presidential campaign, George W. Bush appeared to be at least somewhat concerned with the issue of climate change. Despite his very destructive environmental record as governor of Texas, Bush said, during his second debate with Gore, that global warming "needs to be taken very seriously" (Kennedy, 2005: 45). While he expressed doubts about the Kyoto Protocol, Bush also asserted that under his leadership the United States would still address the problem by strictly regulating greenhouse gases, and more specifically he promised to reduce emissions from coal-fired power plants (Lynch et al., 2010). Despite these campaign promises, environmentalists were not very optimistic. As international trade and finance policy expert Antonia Juhasz (2006: 324) pointed out, "Of all administrations in American history, the Bush administration may be the most beholden and interconnected to the energy sector." Bush and his vice president, Dick Cheney, were former energy company officials, and they had received huge campaign contributions from the industry. Given their strong ties to fossil fuel corporations, it was not surprising that Bush and Cheney, along with other members of their cabinet, would engage in numerous serious climate crimes (blameworthy harms) during their two terms in office. These crimes included the vigorous promotion and expansion of the extraction of fossil fuels, which would lead to the highest oil company profits in history and a sharp rise in greenhouse emissions; the killing of U.S. involvement in the Kyoto Protocol and the failure to negotiate any other binding international treaty to mitigate greenhouse gas emissions; the institutionalization of climate change denial and censorship of climate science in the White House; the failure to propose or enact any significant climate policies; and, after the September 11 terror attacks in 2001, the illegal invasions of Afghanistan and Iraq (in part, a war for access to and control over oil) and other state crimes during the so-called global war on terrorism, which also generated enormous carbon emissions from the U.S. military.

Bush's climate crimes began soon after his selection as president. He appointed Christine Todd Whitman, the former governor of New Jersey, to be the head of the EPA, and she surprised many by putting global warming at the top of her agenda. But soon after taking office, the president would undercut Whitman's agenda and renege on his campaign promises to reduce greenhouse gas emissions. According to environmentalist Robert F. Kennedy, Jr. (2005: 45), these moves "revealed the depth of industry clout at the White House." In early 2001, Whitman prepared to attend her first international meeting in Trieste, Italy, a session that was intended to prepare the major industrial nations for the official Kyoto Protocol meetings that summer. Recognizing climate change as a genuine global crisis, Whitman was prepared to outline the new administration's plan to regulate carbon dioxide as a pollutant (without necessarily committing the administration to the Kyoto Protocol). Confident that she had Bush's blessing, Whitman appeared on CNN to discuss the plan. A few days later, four right-wing Republicans with strong ties to the fossil fuel industry sent a letter to Bush complaining about Whitman's television appearance and demanding a "clarification of your administration's policy on climate change" (Kennedy, 2005: 51). Bush was also pressured by Haley Barbour, a former Republican National Committee chair and then a member of a powerful lobby group in Washington,

DC, to reverse his campaign pledge to reduce emissions from coal-fired plants (Lynch et al., 2010). The chief lobbyist from ExxonMobil also sent a letter to the White House in early 2001 requesting that the administration fire scientists and officials concerned with global warming (including one scientist who had chaired the IPCC in 1995) and replace them with climate skeptics and contrarian scientists (Kaiser and Wasserman, 2016: 6). This intense political pressure, amplified within the administration by Cheney, had the desired effect on the new president. ExxonMobil got its wishes when the climate scientists were fired, and Bush not only backed off the plan to regulate carbon emissions, he also announced his strong opposition to the Kyoto Protocol. When Bush repudiated his campaign promise to reduce greenhouse gas emissions, Whitman was reported to have said, in dismay, "We just gave away the environment" (Kaiser and Wasserman, 2016: 6). In March 2001, the new administration finally pulled the trigger and formally withdrew the United States from the Kyoto Protocol, thereby opting "out of all debate and negotiations with the rest of the world on global warming" (Kennedy, 2005: 52). A headline in the UK *Guardian* newspaper proclaimed, "Bush Kills Global Warming Treaty" (Borger, 2001). It was later revealed that the president's decision to rescind Clinton's signature of the Kyoto Protocol was partly a result of pressure he received from ExxonMobil (Vidal, 2005). The rejection was a huge slap in the face to the international political community. As international law expert Philippe Sands (2005: 70) observed, "The decision was seen as an arrogant step aimed at refashioning the global order, putting American lifestyles above foreign lives, American economic well-being above all other interests, and manifesting a refusal to be constrained by new international rules." For the next eight years under George W. Bush, there would be no possibility for an international climate treaty to be negotiated.

In addition to withdrawing the United States from the Kyoto Protocol and reneging on an election pledge to reduce carbon emissions, the Bush administration further signaled its approach to energy and the environment with a May 2001 report by the National Energy Policy Development Group (NEPDG). In its report, the NEPDG—which was chaired by Cheney and stacked with representatives from the fossil fuel industry—promoted the expansion of America's fossil fuel economy (a climate crime of continued extraction) and did not seriously address the problem of climate change (a climate crime of political omission). At the time that the report was issued, Cheney "belittled conservation and suggested that upping oil, coal, and natural gas production was our only viable option" (Klare, 2012: 58). Subsequently, the Bush administration would seek to "increase the availability" of fossil fuels, "not restrict it" (Klare, 2012: 59). Following the release of the third IPCC assessment report that same month, which further documented ominous global warming trends, the administration, now facing a storm of criticism for its fossil fuel–soaked energy policy, did introduce a series of half-hearted initiatives on climate change in June. The proposals included funding for further research on the issue and several unilateral moves to reduce greenhouse gas emissions, such as selling cleaner-burning U.S. technology to the developing world and voluntary energy-efficiency programs for U.S. consumers. These weak initiatives, however, did not include any binding targets or timetables for the reduction of U.S. emissions and were quickly abandoned as the issue of global warming became increasingly polarized in U.S. political discourse (Leaf, 2001).

The Bush administration itself played a leading role in this politicization of the climate change issue by pursuing a cooperative state–corporate strategy that "actively attempted to refute the science of global warming and install in its place economic and environmental policies that not only ignore but deny the views of the scientific community on climate change" (Lynch et al., 2010: 213). Examining the politicization of global warming under Bush as a form of state–corporate crime, green criminologists Michael J. Lynch, Ronald G. Burns, and Paul B. Stretesky (2010) analyzed the coordinated intersection of state and corporate activity and interests that resulted in collusive agreements and arrangements between the fossil fuel industry and the George W. Bush administration with regard to a number of energy and environmental policies related to global warming. They documented how corporate actors from the fossil fuel industry made their interests known to the administration and how they sought to directly influence energy and environmental policies through lobbying, behind-the-scenes meetings, and the drafting of proposed policy language. The White House under Bush in turn adopted the proposed policies, appointed industry executives and leaders to key policy-making positions within the government, attempted to censure scientific reports to cast doubt on climate science, and tried to fire or at least muzzle federal climate scientists such as NASA's James Hansen. Lynch, Burns, and Stretesky (2010: 227) conclude that the administration was able to "produce domestic policies that contributed to rather than impeded the progress of global warming." Furthermore, they note that "Because of the immense power the U.S. wields internationally, the Bush Administration was able to forestall implementation of international treaties on global warming." It was this dismal record of denial and delay, this fundamental failure of the Bush administration to meet its legal obligation to mitigate carbon emissions, that moved German physicist Hans Joachim Schellnhuber to declare, "This was a crime."

Research by McCright and Dunlap (2010: 109) has also analyzed the variety of ways that the George W. Bush White House was turned into a key component of the climate change denial countermovement, giving it "vastly more leverage to challenge environmental science and policy from *within* the state structure than it had ever enjoyed before." While Lynch, Burns, and Stretesky (2010) focused more on direct decision making by the administration concerning global warming (the first dimension of power), McCright and Dunlap (2010: 111) document how conservative activists within the Bush administration employed "four non-decision-making techniques associated with the second dimension of power to make climate change a non-issue and prevent significant progress on climate policy-making." These techniques were (1) obfuscating, misrepresenting, manipulating, and suppressing research results (attacking the evidence of impact science and presenting contrarian propaganda); (2) intimidating, harassing, censoring, or threatening individual scientists (such as Hansen and Michael E. Mann); (3) invoking existing (or creating new) rules or procedures of the political system from which the administration could disproportionately benefit (changing OMB rules for reviewing and disseminating scientific reports); and (4) invoking the existing bias of the media's balancing norm (equating "objectivity" with presenting "both sides of the story") to attack climate science and policy. McCright and Dunlap (2010: 124) conclude that the skillful employment of these non–decision making techniques resulted in the Bush administration's "successful subversion of climate policy action."

It is important to point out that the George W. Bush administration's climate crimes of political omission were aided and abetted not only by the climate change denial countermovement but also by a Republican Party that has become ever more ideologically and politically extreme. The GOP, supported by wealthy donors, conservative foundations, right-wing think tanks, and alternative media, has lurched far to the right of the ideological spectrum and become more partisan and polarizing in its politics since the Reagan era. This polarization is especially true concerning environmental issues. In the past, going back as far as Teddy Roosevelt, many Republicans supported environmental causes. The EPA itself was signed into law by a conservative Republican president, and a number of prominent Republican members of Congress held hearings and supported research on global warming in the 1980s. The current Republican Party, however, under the sway of the American conservative movement and its neoliberal ideology, not only rejects the science about global warming and obstructs any action to mitigate climate change but has also waged an unprecedented legislative assault on environmental regulations and safeguards of any type. This was true of the party in general during the Reagan, first Bush, and Clinton administrations, although it was not quite as extreme. But during the first term of George W. Bush, the president and his neoliberal party "launched over 300 major rollbacks of U.S. environmental laws, rollbacks that are weakening the protection of our country's air, water, public lands, and wildlife" (Kennedy, 2005: 3). And the party has continued to move in this direction during both the Obama and Trump presidencies. After Republicans took control of the U.S. House of Representatives in the 2010 midterm elections, they voted nearly 200 separate times to block, delay, or weaken foundational environmental laws that protect the air, water, wildlife, and lands (Deans, 2012). On the specific issue of global warming, House Republicans voted in 2011 to repeal the EPA's legal authority to regulate greenhouse gas emissions, an authority that had been confirmed in 2007 by the U.S. Supreme Court in *Massachusetts v. Environmental Protection Agency*. Not a single Republican voted against the attempted repeal, which did not survive in the Senate, where a Democratic majority still prevailed (Deans, 2012). That same year, House Republicans also voted down an amendment that simply stated that climate change is real, caused by human activities, and puts public health at risk.

The corrupting influence of money in the form of corporate campaign contributions provides a partial explanation for these political efforts by the GOP to block climate change policy and other environmental actions. American legal scholar Lawrence Lessig (2011) argues that money does indeed corrupt the legislative process and that the current campaign finance system has rendered Congress politically bankrupt. With billions of dollars of profits at stake, corporations in the fossil fuel industry have made a huge investment in lobbying efforts on Capitol Hill and used large campaign contributions to individual legislators to advance their economic interests in the Congress. According to American environmental journalist Bob Deans (2012: 88), since 1990 the oil and gas industry has contributed some $239 million in campaign contributions, and he notes that "Most has gone to Republicans who support industry goals like limiting environmental protections, blocking measures to reduce climate change, and allowing drilling in the Alaska National Wildlife Refuge." The *Citizens United v. Federal Election Commission* decision by the Supreme Court in 2010, which allows for unlimited corporate campaign contributions, has made the situation even worse.

Other scholars have highlighted the growing ideological extremism of the Republican Party and analyzed this phenomenon as a factor that is responsible for the general political failure to mitigate carbon emissions. Geoffrey Kabaservice (2012) and Nancy MacLean (2017) have independently documented that the current GOP has evolved into one of the most uniformly right-wing ideological parties in U.S. history. This shift toward ideological extremism was also fueled in the early years of Obama's first term by the rise of the "Tea Party," a right-wing populist movement. Theda Skocpol and Vanessa Williamson (2012) concluded in their analysis of this movement that grassroots activism, billionaire industrialist supporters (such as the Koch brothers, who made most of their money in hydrocarbons), and a right-wing media machine have combined to create and power the Tea Party and give it an extraordinary ability (disproportionate to its numbers) to reshape the Republican Party and the national political discourse. As a result, they argue that the Republican Party has become an obstacle to any policies that might help the United States respond to the political and economic challenges of the day, including climate change. Two well-known scholars of Congress, Thomas Mann and Norman Ornstein, concur with this assessment, concluding: "The Republican Party has become an insurgent outlier—ideologically extreme; contemptuous of the inherited social and economic policy regime; scornful of compromise; unpersuaded by conventional understanding of facts, evidence, and science; and dismissive of the legitimacy of its political opposition" (T. Mann and Ornstein, 2016: xv). They go on to add that "one of our two major political parties has become so radicalized that at critical times and on critical occasions it will not or cannot engage constructively in the governing process anticipated by our constitutional charter." Furthermore, a recent study by Norwegian political scientist Sondre Batstrand (2015) reveals that, even among ideologically conservative political parties around the world, the U.S. Republican Party is an anomaly in continuing to deny anthropogenic global warming and attempting to block climate policies.

During the second Bush administration, the Republican Party became a wholly owned subsidiary of the climate change denial countermovement. As American economist Paul Krugman (2018b: A22) has observed, "Denying climate change, no matter what the evidence, has become a core Republican principle." He compared it to the denial that smoking causes cancer, arguing that "it's depravity on a scale that makes cancer denial seem trivial. Smoking kills people . . . but climate change isn't just killing people; it may well kill civilization." Krugman (2018a: A25) also argues that climate change denial was "the crucible" for Trumpism, asserting that "Donald Trump isn't an aberration, he's the culmination of where his party has been going for years. You could say that Trumpism is just the application of the depravity of climate denial to every aspect of politics." This is the extremist, denialist political party that Barack Obama would have to contend with as he attempted to address the climate crisis after the Bush years.

I would make one final observation concerning the politics of predatory delay: Climate change is a time-sensitive problem. The world community has only so much time to stave off disaster. As Malm (2018: 7) observes, every year we fail to reduce carbon emissions, "the shadow of committed warming extends further into the future." That is why the decisions and omissions analyzed in this chapter are so important. They represent thirty years of lost opportunities. Thirty years when decisive actions by the U.S.

government and the international political community could have made a difference. Thirty years when strong and effective policies could have limited global warming and prevented climate change ecocide. And the fact that this critical delay in responding to the climate crisis was the result of deliberate decisions and omissions by fossil fuel corporations, the climate change denial countermovement, political parties, and government officials makes that delay predatory and criminal.

CHAPTER 5

"Slowing the Rise of the Oceans?"

OBAMA'S MIXED LEGACY AND
TRUMP'S CLIMATE CRIMES

The journey will be difficult. The road will be long. I face this challenge with profound humility, and knowledge of my own limitations. But I also face it with limitless faith in the capacity of the American people. Because if we are willing to work for it, and fight for it, and believe in it, then I am absolutely certain that generations from now, we will be able to look back and tell our children that this was the moment when we began to provide care for the sick and good jobs to the jobless; this was the moment when the rise of the oceans began to slow and our planet began to heal.
—Barack Obama, upon securing the nomination of the Democratic Party for president, June 3, 2008, in St. Paul, Minnesota (*New York Times*, 2008: 9)

It is not yet widely understood, though it will be, that when a government relaxes regulations on coal-fired plants or erases scientific data from a federal website, it is guilty of more than merely bowing to corporate interests; it commits crimes against humanity.
—Nathaniel Rich, *Losing Earth: A Recent History* (2019: 195)

ON THE EVENING OF July 27, 2004, I watched the keynote address at the Democratic Party National Convention on cable television. Delivered by a young state senator from Illinois, who was at that time a candidate for the U.S. Senate, the address was electrifying. The young senator's name was, of course, Barack Hussein Obama. Like many Americans, I had never heard of this eloquent young politician. But I was so impressed with him and the content of his speech, that, later that night, I wrote in my personal journal that I had just watched on television the man who would become the first African American President of the United States. Obama went on to win that U.S. Senate seat in Illinois in 2004 and just four short years later, on November 4, 2008, was indeed elected president, defeating

Republican Senator John McCain with 52.9 percent of the popular vote and 365 electoral votes. It was a remarkable story. I remember getting up early the next morning and going down to a local supermarket to buy the *New York Times* and other newspapers as keepsakes of this historic event. I have never felt more hopeful for my country than I did that day. I had great "hope" for "change" (to use two of candidate Obama's campaign memes). The Obama presidency would not always meet my early expectations, and on the critical issue of climate change the president often hedged and settled for half-measures. In the end, for reasons both beyond and under his control, he did not do enough to "slow the rise of the oceans" as he promised that night in St. Paul. Yet, after some early stumbles on the issue, he did develop a climate action plan and create some important policies to try to mitigate carbon emissions. More than any of his predecessors, he seriously addressed the problem of global warming and in his last year in office committed the United States to the historic Paris Agreement—a flawed but solid climate policy legacy. But rather than building on these positive steps, his successor, in one of the greatest climate crimes in history, would attempt instead to roll back these accomplishments and demolish any progress that had been made in responding to the climate crisis.

OBAMA'S FIRST TERM

While George W. Bush was clearly engaged in the state crime of political omission concerning climate change, it is much harder to reach a singular conclusion about the overall record of the Obama administration. Obama's election in 2008 caused some environmental activists and scholars to hope that there would be a major change in U.S. climate change policy. As Jennifer Hadden (2017: 130) points out, both strands of the environmental movement at the time, the "policy-focused movement in Washington and the more decentralized grassroots movement," perceived opportunities to act on climate change under Obama. During his 2008 campaign, Obama had promised to tackle global warming (slowing the rise of the oceans), and he appeared to have a better grasp of the issue and to be more committed to crafting a stronger policy to mitigate greenhouse gas emissions than any previous president. He was also committed to engaging constructively with the international community on climate change negotiations.

Within several months of taking office, the Obama administration proposed a cap-and-trade bill to Congress. This legislation, the American Clean Energy and Security Act of 2009 (also known as the Waxman–Markey Bill after its congressional authors), would have created a regulatory system to "cap" the carbon emissions a company can produce at a specified level but permit that organization to "trade," that is, to buy rights to produce additional emissions from a company that has not used up its own allowance. The emissions trading bill was supported by a coordinated alliance of environmental groups called the U.S. Climate Action Partnership (USCAP), which was created to promote climate change policy. This political support, however, put the U.S. groups in a difficult position with their international allies in the environmental movement who were pushing for more stringent emissions-reduction targets in the lead-up to the 2009 UN climate conference (COP 15) to be held in Copenhagen (Hadden, 2017). With the assistance of USCAP, the so-called more realistic Obama cap-and-trade bill passed in the U.S. House of Representatives in June 2009 by a narrow vote of 219–212. The legislation, however, faced an uncertain future in the U.S. Senate.

Expectations were high for COP 15 in Copenhagen during December 2009. Many hoped that this conference would mark a turning point for international climate change policy. Under Obama, the United States had already "re-engaged in U.N. climate negotiations" and breathed "fresh life into the Major Economies Forum—a grouping of the 17 largest economies originally convened under the [G. W.] Bush administration—and brought climate change within its remit" (Giddens, 2011: 190). Obama promised that the United States would "reduce emissions by 17 per cent over 2005 levels by 2020 and by more than 80 per cent by 2050" (Giddens, 2011: 190). Australia and Japan, previous resistors, made pledges to significantly reduce their emissions, and the large developing countries, China and India, also set targets for carbon reductions. Many commentators and political leaders were optimistic about the chances for success, and more than 40,000 delegates and some 5,000 journalists attended the conference.

The Copenhagen climate meeting, however, ended in a disastrous failure. The whole process was described as "chaotic and disjointed" (Giddens, 2011: 190). Obama did not arrive at the conference until the last day and then scrambled to set up meetings with other leaders in an attempt to salvage some agreement (Kerry, 2018). At the last minute, a minimalist accord was negotiated that had no emissions targets or timetables. Giddens (2011: 192) notes that "Far from being the detailed, comprehensive and binding framework originally envisaged, the Accord was essentially a short statement of intent, no more than three pages long." Some blamed China for the failure, and some blamed Obama and the United States. As climate change activist Bill McKibben (2010a: 1) defined the debacle, "The world came together and looked climate change fairly straight in the eye, and then the most powerful nations blinked." Recall that John Sauven, the executive director of Greenpeace UK, invoked the language of criminality to describe the failure: "The city of Copenhagen is a crime scene tonight, with the guilty men and women fleeing to the airport. There are no targets for carbon cuts and no agreement on a legally binding treaty" (BBC, 2009: 3). Rob White (2011: 148) contends that the political omission that occurred at the Copenhagen climate conference was a "state–corporate crime," noting that "The abject failure of the Copenhagen talks to actually do something about carbon emissions and to address climate change issues in a substantive fashion is a striking example of the fusion of state and corporate interests to the detriment of the majority." The existing global power structure held, and the young Obama administration was judged by the environmental movement to have clearly failed its first big test in the international political arena.

The devastating failure at Copenhagen was followed the next summer by the defeat of Obama's cap-and-trade legislation in the U.S. Senate. Although the American Clean Energy and Security Act of 2009 had squeaked through the House in June 2009, the effort to pass the bill collapsed in the Senate in July 2010. Facing a unified Republican Party bolstered by the emergent right-wing Tea Party movement, strong opposition from the fossil fuel industry, and resistance from some politically vulnerable Democrats in energy-producing states, Obama and Senate majority leader Harry Reid were unable to even bring the bill up for a vote (Weiss, 2010). As Mayer (2017: 245) points out, "If there was a single ultra-wealthy interest group that hoped to see Obama fail as he took office, it was the fossil fuel industry. And if there was one test of its members' concentrated financial power over the machinery of American democracy, it was this minority's ability to stave off government action on climate change as science and the rest of the world were moving in the opposite direction." Coming only a few months after the breakdown at Copenhagen, the collapse of the cap-and-trade bill was also a major political defeat for the Obama administration and the U.S. climate movement. The fossil fuel industry and its staunch ally, the climate change denial countermovement, had won a major victory.

Subsequent UN climate conferences during Obama's first term, at Cancun in 2010 (COP 16), Durbin in 2011 (COP 17), and Rio de Janeiro in 2012 (Rio+20: The Conference on Sustainable Development), also failed to produce the strong international accord that environmental activists insisted was necessary to head off climate change disaster. A last-minute deal at Durbin in December 2011 did extend the moribund Kyoto Protocol and launch negotiations on a more comprehensive and ambitious treaty regime (to take effect by 2020), but in the estimation of the Union of Concerned Scientists (2011: 1), the Durbin agreement "will do little to accelerate near-term emissions reductions." Many hoped that the Rio+20 conference, taking place in the same city twenty years after the United Nations Framework Convention on Climate Change (UNFCCC) had been negotiated, would assess why that treaty had been inadequate and take new legally binding steps to deal with the impending ecological catastrophe caused by global warming. But, as the renowned Indian environmental leader Dr. Vandana Shiva (2012: 1) pointed out, "the entire energy of the official process was focused on how to avoid any commitment. Rio+20 will be remembered for what it failed to do during a period of severe and multiple crises and not for what it achieved."

With the failures of Copenhagen, Durbin, and Rio+20, and the defeat of the cap-and-trade legislation, the high hopes some held for the Obama administration's ability to make a difference on the issue of climate change policy were clearly dashed (Hertsgaard, 2011a, 2012). Stinging criticism was directed at Obama's overall energy policy. McKibben (2017b: 2) later noted that, during his first term, "Barack Obama drove environmentalists crazy with his 'all-of-the-above' energy policy, which treated sun and wind as two items on a menu that included coal, gas, and oil." When the Obama administration gave Shell Oil the go-ahead to drill in the Arctic, McKibben (2015b: 1) charged the administration with "catastrophic climate-change denial." He pointed out that this was "not 'climate denial' of the Republican sort, where people simply pretend the science isn't real. This is climate denial of the status quo sort, where people accept the science, and indeed make long speeches about the immorality of passing on a ruined world to our children. They just deny the meaning of the science, which is that we must keep carbon in the ground." Based on this first-term record of political failure related to climate change, and the lack of White House interest in or leadership on this issue, many climate activists bemoaned more lost opportunities and judged the Obama administration too as guilty of the state crime of political omission.

The inability of the Obama administration to successfully conclude a strong international agreement on climate change during his first term, or to pass any significant legislation concerning this problem, must, however, be placed within a broader economic and political context. First, Obama took office during one of the worst economic crises in American history. The Wall Street financial crash of 2008 and the subsequent global recession, itself a monumental state–corporate crime (Barak, 2012; Scheer, 2010), produced a political situation that demanded that the new administration first respond to the economic crisis. The president believed that this crisis severely limited his options for undertaking any new responses to climate change that might further weaken the economy in the short run. Second, the Obama administration made a fateful decision to prioritize health care early in the first term and expended enormous political capital in passing the Affordable Care Act (ACA, aka Obamacare) in March 2010. Third, and even more important, Obama faced an ideologically extreme Republican Party that obstructed any effort to deal with the issue of global warming in any way (Deans, 2012; Mooney, 2005, 2012). The Obama administration was fully aware that any climate change treaty it submitted to the U.S. Senate had little chance to be ratified due to extreme and unified Republican opposition, strongly influenced by fossil fuel industry money and climate change denial propaganda.

The 2008 economic crisis, the expenditure of political capital on the ACA, and the ideological extremism and legislative obstructionism of the Republican Party provide a partial explanation of how Washington politics limited the Obama administration's climate change plans. However, as Harvard political sociologist Theda Skocpol (2013) argues in her report *Naming the Problem: What It Will Take to Counter Extremism and Engage Americans in the Fight against Global Warming*, climate change inaction during Obama's first term was also the fault of the U.S. environmental movement. She charged that environmental groups had overlooked the growing opposition to environmental regulations among more conservative American voters, underestimated the growing power of the Tea Party movement, and failed to appreciate the growing political polarization of the climate change issue in Congress due to the efforts of the powerful climate change denial movement. Skocpol (2013: 129) argued that the USCAP insider-grand-bargaining strategy on the cap-and-trade bill "was based on misplaced hopes for bipartisan bargains and a failure to grasp that support from Republicans was not going to be forthcoming." Some argue that the *Naming the Problem* report "lets Obama off the hook for the political inaction on climate change" and charges environmental groups with "political malpractice" for failing "to build the broad grassroots organizations needed to push for change" (Goldenberg, 2013a: 2). While that may be true, the major point that the report makes is that a strong social movement is necessary to implement a climate action plan, whoever is in the White House. As Skocpol (2013: 130) concludes, "The only way to counter right-wing and elite popular forces is to build a broad popular movement to tackle climate change. Ways must be found to use policy ideas as tools to knit-together inside-outside links among many organizations, including some that can draw masses of ordinary citizens into the transition to a green economy. . . . Citizens must mobilize and many organizations must work together in a sustained democratic movement."

This call for greater public mobilization and a broader movement pressing politicians for change was heard by many environmental activists such as McKibben and his 350.org group, and it would lead to a new vision for climate politics and a new wave of climate organizing (Hadden, 2017). These new targets, strategies, and tactics would help produce some important results during Obama's second term.

Obama's Second Term and the 2015 Paris Agreement

After the failure to achieve a strong and binding international accord to reduce emissions at Copenhagen in 2009 and the defeat of the cap-and-trade bill in 2010, the Obama administration did little on the climate change issue for the rest of the first term. Some critics charged that Obama mostly "ignored climate change" after the early failures (McKibben, 2015b: 1). In August 2012, however, the administration did take an important executive action to enact new Corporate Average Fuel Economy (CAFE) standards that were intended to double car and truck mileage per gallon by 2025 (Meyer, 2016b). But these new fuel-efficiency regulations were not scheduled to go into effect until 2017 and, although they could lead to significant greenhouse gas reductions in the future, appeared to not win the president any favors with environmental groups. Obama barely mentioned climate change during his successful run for re-election in 2012 against Mitt Romney until forced to do so by Hurricane Sandy late in the campaign. In fact, as he entered his second term, Obama appeared to take several giant steps backward on the issue by opening the Powder River basin to new coal mining, delaying a final ruling on the proposed Keystone XL pipeline, and then giving Shell Oil the go-ahead to drill in the Arctic. McKibben's (2015b:1) charge that Obama was guilty of a catastrophic form of "climate-change denial" by not doing all he could to "keep carbon in the ground," despite his political rhetoric concerning the issue, rang true with many climate activists.

But the Obama administration, despite that dubious general "all-of-the-above energy policy," was beginning to take some positive steps on the climate change issue. During his second term the president began to make some important decisions that would culminate in the historic Paris accord in late 2015 (Meyer, 2016a). These decisions included setting those new rules on fuel efficiency for cars and trucks, enacted in late 2012, new limits on methane and aviation emissions, a "climate action plan" announced in 2013, a landmark climate agreement with China in 2014, new regulations to reduce emissions from coal-fired power plants in 2015, and the rejection of the construction of the Keystone XL oil pipeline in 2016. Despite the huge obstacle of fierce, unrelenting, and unified opposition from the fossil fuel industry, the climate change denial countermovement, and Republicans in the Congress, Obama would make some important progress on the climate issue during his second term. As American journalist Robinson Meyer (2016a: 2) would observe in late 2016, "In the past two years, President Obama has converted climate change from a Democratic wedge issue into a major party plank." A more mobilized, contentious, and decentralized grassroots climate justice movement, which emerged after the failure of the cap-and-trade bill in 2010, was there every step of the way to pressure the president to take these important steps and to criticize him if, in their estimation, those actions did not go far enough.

At the start of his second term, Obama bluntly informed the Republicans in Congress that if they would not act on climate change through the legislative process, he was prepared to take executive action on the issue. In his State of the Union address in February 2013, Obama warned that "If Congress won't act soon to protect future generations, I will. I will direct my cabinet to come up with executive actions we can take, now and in the future, to reduce pollution, prepare our communities for the consequences of climate change, and speed the transition to more sustainable sources of energy." The president followed through on his warning in June 2013 when he unveiled his Climate Action Plan. This plan, described by the World Resources Institute as "the most comprehensive climate plan by a U.S. president to date" (Morgan and Kennedy, 2013: 1), included measures to reduce greenhouse gas emissions, accelerate renewable energy permitting on public lands, and address the question of how to protect the infrastructure of the country from devastating climate-related impacts. The centerpiece of the Climate Action Plan was the announcement that the EPA would soon be developing new rules to regulate greenhouse gas emissions from existing power plants, regulations that would eventually be called the Clean Power Plan when unveiled in 2015.

One of the most significant accomplishments during the second term of the Obama administration was the signing in 2014 of a historic agreement with China on climate change. The United States and China, often antagonists on climate policy, announced a joint plan to curb carbon emissions that they hoped would inspire other nations to also make commitments to cut greenhouse gases at the upcoming climate conferences in Lima (2014) and then Paris (2015). The U.S.–China accord was the outcome of a concerted effort that started in 2013 when U.S. secretary of state John Kerry and Chinese state councilor Yang Jiechi formed the United States–China Working Group. Kerry had reached out to the Chinese foreign minister shortly after being sworn in as secretary of state on February 1, 2013, suggesting to Yang that "If we can find constructive ways to approach this [climate change], we can set an example for the world" (Kerry, 2018: 561). Kerry traveled to China in April and found a receptive audience, and the negotiating process began in earnest. Although the negotiations were difficult at times, the two countries were able to develop "a new level of understanding between us" (Kerry, 2018: 564) and arrive at an agreement the following year. On November 11, 2014, in Beijing, Obama and Chinese president Xi Jinping jointly announced targets to reduce carbon emissions (Landler, 2014). The United States pledged to reduce net greenhouse gas emissions by 26 to 28 percent below 2005 levels by 2025. China pledged that clean energy sources (solar and wind) would account for 20 percent of its total energy production by 2030. As Kerry noted at the announcement of the new agreement, these countries are the world's two largest economies, two largest consumers of energy, and two largest emitters of greenhouse gases (accounting for about 40 to 45 percent of the world's emissions). Reflecting on the signing of the agreement, Kerry (2018: 564) wrote, "We believed this day would galvanize countries everywhere to follow suit with their own ambitious targets. We wanted to send them a message: success in Paris was possible."

The historic agreement between the United States and China did appear to galvanize the international political community. It was significant because it helped "inject new life into global climate talks" and served as a "catalyst" for the drafting of a proposed international accord to fight climate change at the COP 20 climate summit held in Lima, Peru, in December 2014 (Davenport, 2014: 2). The Lima draft global climate accord represented "a fundamental breakthrough in the impasse that has plagued the United Nations for two decades as it tried to forge a new treaty to counter global warming" (Davenport, 2014: 1). In Lima, for the first time, all the nations of the world made voluntary commitments to cut emissions. The draft agreement set the stage for a final deal to be adopted by world leaders at the climate meetings scheduled the following year in Paris.

After signing the accord with China, and endorsing the draft agreement at Lima, Obama had to take decisive action in 2015 if he hoped to achieve a final deal on climate change at the Paris conference in December. As Kerry (2018: 565) put it, "As the date for the Paris talks neared, it was imperative we ensure our own house was in order as well." Having pledged to cut emissions by 26 to 28 percent below 2005 levels by 2025, the Obama administration now had to transform those pledges into actual political outcomes that would make them a reality. Given the solid Republican opposition, there was no chance that any national legislation to reduce emissions could be passed by the Congress. This was even more evident after the midterm elections in 2014 when the Republicans re-captured the Senate and held on to the House. With both houses of Congress in Republican hands, Obama's only political tools were regulatory rulings and other executive branch actions. And, as he had warned Congressional Republicans during his 2013 State of the Union address, he was now moving to use these tools. Fortunately, the U.S. Supreme Court had opened the door to greater executive branch efforts to curb carbon emissions with *Massachusetts v. Environmental Protection Agency* in 2007. The court held that greenhouse gases are pollutants under the Clean Air Act and that the EPA has the legal authority under that act to regulate carbon pollution, if it made the determination that those emissions "endangered" public health and welfare. On December 15, 2009, the Obama EPA followed through and, based on a careful review of the scientific record, made the determination that greenhouse gas emissions do in fact endanger the public health and welfare of the American people. The "endangerment finding" was a critically "important tool" for the president (*New York Times*, 2009), a necessary precondition under the Clean Air Act for executive branch regulatory actions, like those the administration was about to unveil.

In March 2015, Obama announced his overall program to cut U.S. carbon emissions (as his 2013 Climate Action Plan had pledged to do). The emissions reduction plan was submitted to the UNFCCC as part of the planning and negotiating process leading up to COP 21 in Paris. All the participating countries were asked to submit their Intended Nationally Determined Contribution (INDC), national commitments that were to serve as the building blocks for an international accord to reduce greenhouse gas emissions and limit global warming. The Obama administration's plan included some existing regulatory policies, such as the 2012 fuel-efficiency standards, and outlined some pending measures like new rules to limit methane emissions in the oil and gas industry. But, again, the centerpiece of the Obama plan was the Clean Power Plan, a pledge to propose new rules or standards that would for the first time reduce carbon emissions from power plants.

The Clean Power Plan, initially promised in the Climate Action Plan, was formally proposed in June 2014 and finalized on August 3, 2015, when Obama and the EPA announced "a historic and important step in reducing carbon pollution from power plants that takes real action on climate change" (EPA, 2015). Prior to this, there were no rules governing the amount of carbon dioxide that power plants were allowed to dump into the atmosphere. The new regulations, which were to go into effect in 2022, were aimed primarily at coal-fired power plants. The plan established state-by-state targets for emission reductions and a flexible time frame (fifteen years) under which those reduction goals could be met. States were to be provided with a number of options to cut emissions: they could improve efficiency in burning coal, swap natural gas for coal, or replace power generated from burning fossil fuels with electricity that came from renewable resources (solar, wind, water, or geothermal heat). The EPA estimated that, when the Clean Power Plan was fully in place in 2030, carbon pollution from power plants would be 32 percent below 2005 levels.

The fossil fuel industry, other business groups, and the Republican Party immediately criticized the Clean Power Plan as illegal (under the Clean Air Act and the Constitution) and as an ideologically driven "war on coal." The coal industry, twenty-seven Republican-governed states, and the U.S. Chamber of Commerce soon filed a legal challenge to the plan in federal court (Meyer, 2016a). On the other side, environmental scholars and activists also criticized the plan, arguing that it did not go far enough. Environmental sociologist John Bellamy Foster (2017: 97) later noted that "Whatever its ambitions, Obama's climate initiative fell far short of the emissions reductions that wealthy states would need to introduce if humanity were to maintain a safe and secure relation to the climate." Environmentalists charged that with the emissions-trading provisions and allowable offsets, most of the roughly 1,000 fossil fuel–fired power plants in the United States would largely continue to operate as usual. Furthermore, even if the plan were fully implemented, U.S. emissions from generating electricity would still add more than 1.7 billion metric tons of carbon dioxide to the atmosphere in 2030 (Biello, 2015).

The Obama administration responded to its industry and Republican critics by pointing out that *Massachusetts v. EPA* and the subsequent Endangerment Finding gave the EPA the legal authority under the Clean Air Act to regulate carbon emissions. Responding to the environmental movement, the administration argued that the emissions reductions achieved under the Clean Power Plan would make it possible to achieve a global climate change accord at the upcoming Paris conference (meeting the pledged INDC). Perhaps both the climate activists and the administration could agree with American journalist David Biello's (2015: 8) assessment at the time that "This plan is likely the most the U.S. can do given current political realities and therefore is an important step, but that doesn't mean it's sufficient." Soon the Paris Agreement itself would be assessed in much the same way.

Before the U.S. delegation headed off to Paris in late 2015, the Obama administration made one more major move on the climate change front. The president took an executive action that the climate justice movement had long demanded that he take and that involved one of the most contentious issues that his administration had faced. On November 6, 2015, Obama announced that he had rejected the request from the TransCanada Corporation for a permit to build the Keystone XL oil pipeline that would be used to transport crude oil extracted from the Canadian tar sands to refineries in the United States. The controversial tar sands project in Alberta, Canada, involved the extraction of crude oil (bitumen) from these sands in various locations around the province. In addition to the local ecological devastation caused by the mining process (mostly on the land of Indigenous First Nations), the climate movement had grave concerns that the extraction, transportation, and eventual utilization of this heavy crude oil would contribute to greater global warming and climate disruption.

The proposed Keystone XL pipeline would carry 800,000 barrels of the carbon-heavy bitumen oil 1,700 miles from Alberta to the Gulf Coast to be refined and distributed each day. Without the pipeline to transport the diluted crude oil to the United States, the tar sands project would not be as profitable and could eventually be abandoned. Because the Keystone XL pipeline would cross a U.S. border, the decision to allow or deny it was an executive branch decision. No congressional action could be taken on the issue. For this reason, the climate movement made opposition to the Keystone XL a top priority after the failure of the cap-and-trade bill in 2010. A network of Indigenous First Nations groups in Canada had long opposed the entire tar sands project, and because the risky pipeline was slated to pass through the famous Sand Hills in Nebraska and over the Ogallala Aquifer (which supplies water for much of the Great Plains), an organization called Bold Nebraska organized a broad coalition to also oppose the Keystone XL pipeline (Brecher, 2015).

The defeat of the pipeline had both practical and symbolic significance to the more aggressive climate movement that emerged after the defeat of the cap-and-trade bill. The transformed movement argues that most fossil fuels, especially the dirtiest forms, must be left in the ground to maintain a viable planet. James Hansen wrote on his blog that "if the tar sands are thrown into the mix it is essentially game over" (Brecher, 2015: 44). But the Keystone XL pipeline took on a larger political significance as well. If the U.S. climate movement could convince Obama to reject the pipeline, it would represent a major victory for a movement that had little to celebrate in recent years. McKibben and 350.org made defeating the Keystone XL project a major priority and, in June 2011, organized a large civil disobedience event (Tar Sands Action) in Washington, DC, where more than a thousand people were arrested (Brecher, 2015). The "grassroots power" of the movement, McKibben (2013: 32) stated, had "gone beyond education to resistance," and that resistance appears to have helped sway Obama on the Keystone XL pipeline and the climate issue in general. At the time of the decision to reject the pipeline, one reporter noted that Obama's seven-year review of the proposal had become "a symbol of the debate over his climate policies" and "comes as he seeks to build an ambitious legacy on climate change" (Davenport, 2015a: 1). That legacy included the fuel-efficiency standards, the accord with China, the Clean Power Plan, and the rejection of the Keystone XL pipeline. Now in late 2015, another major opportunity to respond to the climate crisis was at hand for Obama and the international political community as the world headed to Paris to adopt an effective agreement on cutting greenhouse gases.

With the "dire predictions" (M. Mann and Kump, 2015) of the IPCC's Fifth Assessment Report (issued in 2013) framing their task, representatives from nearly 200 nations assembled in Paris on November 30, 2015, to conclude and adopt a landmark accord that would commit, for the first time, nearly every country in the world to lower greenhouse gas emissions and limit the most drastic effects of global warming. Organized through the

1992 UN Framework Convention on Climate Change, which at least in principle committed the world to take measures to prevent dangerous climate change, the 21st Conference of the Parties (COP 21) would end with a historic breakthrough agreement. The final text was adopted by 195 nations on December 12, 2015. The Paris pact was to "enter into force" when fifty-five parties representing 55 percent of global emissions joined the climate deal. That threshold was achieved in October 2016, and the agreement entered into force on November 4, 2016, just days before Donald Trump would be elected president (a historic electoral outcome that would later place the pact in considerable jeopardy).

The Paris Agreement is the first comprehensive climate accord to be agreed to by the international political community. UN secretary-general Ban Ki-moon stated, "This is truly a historic moment. For the first time, we have a truly universal agreement on climate change, one of the most crucial problems on earth" (Davenport, 2015b: A1). The pact is based on voluntary and nationally determined reductions in greenhouse gas emissions. There are no legally binding targets for these emission reductions. Rather, each nation came to Paris having calculated its INDC to the reduction of global carbon emissions. These national plans vary greatly in scope and ambition. The overall goal of the voluntary INDC pledges is to keep global warming below the 2-degree Celsius threshold that had been set in earlier climate negotiations. It is important to note, however, that, pressured by a number of groupings of developed and developing countries such as the Climate Vulnerable Forum, a reference to a more stringent target of 1.5 degrees Celsius was included in the final document (D. Green, 2016). This "aspirational" goal of the Paris accord would later be reinforced by the October 2018 IPCC report that global warming must be limited to that 1.5-degree Celsius target to avoid climate catastrophe. At the time, this concession, along with pledges to continue to provide some economic assistance to address "loss and damage" experienced by poor countries (climate finance through the mechanism of the Green Climate Fund), helped to ease some of the long-standing antagonisms between the Global North and Global South that had hindered previous climate negotiations. Thus, the Kyoto Protocol's division of the world into Annex 1 (North) and Non-Annex 1 (South), in which only Annex 1 nations made commitments to cut emissions, was replaced in Paris by a system in which all countries made commitments to reduce greenhouse gases.

While the INDC national emission plans are not legally binding, countries are legally required to publicly report on the progress they are making on their planned cuts. This global "stocktake" assessment program starts in 2023 and requires states to monitor, verify, and report their greenhouse gas emissions. A review process with a "ratchet mechanism," it asks countries to revisit their INDC every five years, report on their progress, and readjust their commitments upward if necessary to speed up their cuts. This hybrid legal structure—voluntary emission plans with a legally required assessment process to ratchet up the targets—was explicitly designed by the Obama administration in response to the political reality they faced at home with staunch Republican opposition. As *New York Times* reporter Coral Davenport (2015b: A1) pointed out, "A deal that would have assigned legal requirements for countries to cut emissions at specific levels would need to go before the United States Senate for ratification," and such a proposal "would have been dead on arrival in the Republican-controlled Senate, where many members question the established science of human-caused climate change, and still more wish to thwart Mr. Obama's climate change agenda." In fact, this political reality nearly derailed the final agreement as Kerry and the U.S. delegation strongly objected to the legally binding word *shall* in the pact, insisting that the less-binding diplomatic term *should* ("an ambiguous word, without legal implications for missing our target") must be used instead (Kerry, 2018: 170).

After the Copenhagen conference in 2009 ended in such disarray, an international climate deal seemed "politically impossible" (Davenport, 2015b). The Paris Agreement, therefore, represented a "remarkable turnaround" (D. Green, 2016: 171). A number of factors have been cited to account for this turnaround. Global activist Duncan Green (2016) argues that the single most important factor was the joint U.S.–China announcement in late 2014. With the two largest emitters of greenhouse gases shifting to a commitment to make significant reductions in carbon emissions, it was easier to persuade other nations to follow course. The joint agreement also helped ease some of the North–South tensions that had been an obstacle in past negotiations. A second major shift was that a significant component of the non–fossil fuel business community became more involved in pressing for emissions reductions. The greater involvement of some major corporations in the process served as a powerful countervailing force to the fossil fuel lobby's efforts to block an agreement. Duncan Green (2016: 172) points out that "Hundreds of companies and CEOs made pledges to reduce their own carbon footprint and called for governments to set more ambitious targets." Third, Davenport (2015b: A1) notes, "a crucial moment in the path to the Paris accord came . . . when Mr. Obama enacted the nation's first climate change policy—a set of stringent new Environmental Protection Agency regulations designed to slash greenhouse gas pollution from the nation's coal-fired power plants." Still other factors included the dramatic drop in the cost of renewable energy and the greater involvement of nations from the Global South in the international political process. All these elements shaped the Paris process.

Just as important, I would argue, was the greater mobilization of the climate movement worldwide. After the defeat of the cap-and-trade bill in 2010, U.S. environmental groups fighting global warming responded to the harsh criticisms directed at their "inside politics" approach by Skocpol (2013) and others by mobilizing resistance at the grassroots level, diversifying their targets, and changing their moral frame to one that focuses on climate justice (Hadden, 2017). McKibben and 350.org often led the way, but a diverse coalition of groups created a more decentralized climate movement and broadened their engagement with the public on the issue. The fossil fuel industry was labeled as a "rogue industry" (the equivalent to calling them carbon criminals). Civil disobedience in opposition to the Keystone XL pipeline emerged, a fossil fuels divestment campaign developed on university campuses and elsewhere, and an estimated 400,000 people marched through the streets of New York in September 2014 (called the Peoples Climate March) as world leaders met at the United Nations to discuss the climate accord being drafted for Paris (Hadden, 2017). An outspoken Pope Francis contributed an important papal encyclical, *Laudato Si* (On Care for Our Common Home), that also put pressure on world leaders to protect the earth from climate disruption. Finally, the climate justice movement also made its presence known in Paris, demonstrating in the streets, lobbying negotiators, and demanding a stronger accord. Without the sustained pressure from climate activists and environmental groups, the Paris Agreement would not have been possible.

How did the climate justice movement react to the final accord that was adopted at the end of the long process leading to Paris? Perhaps George Monbiot (2015: 1), writing in the *Guardian*, summed it up best: "By comparison to what it could have been, it's a miracle. By comparison to what it should have been, it's a disaster." This seemed to be the common refrain: As far as it goes, it's great, but it does not go nearly far enough. Tim Gore, head of food and climate policy at the humanitarian aid group Oxfam, observed that "The U.N.'s verdict reveals that while the world is making progress, much more needs to be done. While this round of pledges is a step in the right direction, they only take us from a 4 C [Celsius] catastrophe to a 3 C disaster" (Prupis, 2015b: 2). In a similar vein, ecopsychologist Zhiwa Woodbury (2015: 2) pointed out that the nonbinding "intended nationally determined contributions" set out in the agreement would still result in an average rise in global temperatures of 3.5 degrees Celsius: "In other words, the accord mirrors President Obama's own climate legacy of saying all the right things while doing all the wrong things." Climate insurgent Jeremy Brecher (2017: 1) offered an even harsher assessment: "The illusion that world leaders were fixing climate change was exemplified by the 2015 Paris Agreement in which 195 countries acknowledged their individual and collective duty to protect the earth's climate—and willfully refused to perform that duty." In the judgment of most climate justice activists, then, while Paris represented an important, and maybe even radical step forward (Dobson, 2018), the United States and the rest of the international political community were still guilty of the state climate crime of political omission. More needs to be done to mitigate carbon emissions and achieve even a rough form of climate justice. To that end, the Children's Trust lawsuit (*Juliana, et al. v. United States*) pressed on in the courts, and the climate justice movement more generally continued its advocacy campaigns. But after the election of 2016, the already shaky prospects for mitigating greenhouse gases and achieving climate justice would suffer a devastating blow.

"THE GREATEST CRIMINALS OF ALL TIME"?

Writing in the *New York Times* on December 12, 2015, reporter Coral Davenport (2015b: A1) made a prescient observation: "Despite the historic nature of the Paris climate accord, its success still depends heavily on two factors outside the parameters of the deal: *global peer pressure* and *the actions of future governments* [emphasis added]." It did not take long for a future U.S. government, hostile to global peer pressure, to attempt to deliver a fatal blow to the agreement. Donald Trump's surprise Electoral College victory over Hillary Clinton (who won the popular vote by over three million) in the November 2016 U.S. presidential election would change almost everything about U.S. climate change policy and present a severe challenge to the international political community's efforts to mitigate carbon emissions. While all recent U.S. presidents (including Barack Obama) have been involved to varying degrees in the politics of predatory delay and the climate crime of political omission, Trump and his administration may go down in history as, to use Tom Engelhardt's (2018: 61) powerful phrase, "the greatest criminals of all time" for their overall policies concerning fossil fuels and the environment.

As soon as it took office in 2017, the Trump administration began to pursue a "suicidal course" (Hertsgaard, 2018: 4) by engaging in all four of the morally blameworthy harms that I have identified as climate crimes. Trump not only rejects impact science and denies that climate change is real and humanly caused, he appears determined to rescind or undermine all previous U.S. government policies that attempt to deal with global warming, including American participation in the Paris Agreement. He has made it clear that he will not undertake any effort to deal with the problem in any way. As if these climate crimes of denial and political omission were not bad enough, the fossil fuel–friendly president also enthusiastically promotes all forms of carbon extraction, a set of criminal policies that could result in "drowning the world in oil" (Klare, 2016a: 2). Like the George W. Bush administration, Trump plans to create a "regime of permission" (Whyte, 2014: 237) for the fossil fuel industry by not only removing all constraints and regulations on the extraction and marketing of coal, oil, and natural gas but also energetically promoting and encouraging the extension of these destructive activities and the greater emission of greenhouse gases that will result. Klare (2017a: 2) labels Trump as "America's carbon-pusher in chief" and argues that he "will do whatever it takes to prolong the reign of fossil fuels by sabotaging efforts to curb carbon emissions and promoting the global consumption of U.S. oil, coal, and natural gas." To cap off his climate crimes, as an expression of his racist and xenophobic consciousness, Trump hopes to eliminate funding for mitigation and adaptation projects in the Global South and further militarize U.S. borders and climate adaptation policies in general. Trump has set out to aggressively pursue an unjust state policy of responding to climate disruptions by further arming the United States and policing, excluding, repressing, and forgetting the victims of global warming in the Global South (Miller, 2017, 2019).

During the 2016 election campaign, candidate Trump had several hot-button, crowd-pleasing issues he would run through during his standard stump speech to rile up his supporters (build the border wall, lock up "crooked" Hillary, bring back coal jobs). Climate change was not always on that list, although the candidate would deliver an angry rant on the topic if asked. Still, while the issue seemed peripheral to his campaign most of the time, Trump was a well-known climate change denier who had tweeted in 2012 that global warming was a "hoax" that was "created by and for the Chinese in order to make U.S. manufacturing non-competitive" (Wong, 2016: 1). In addition to this famous tweet, Trump had also tweeted about his climate change skepticism at least 115 other times (Matthews, 2017) and dismissed global warming as "very expensive . . . bullshit" (Phillips, 2016: 1). Throughout the campaign, it was clear to close observers that if elected president Trump intended to not only eliminate all of Obama's climate policies but also vigorously promote expanded extraction of fossil fuels, including "untapped energy" on federal lands the candidate valued at $50 trillion (Bastasch, 2016). The threat of a Trump presidency was alarming to many in the climate movement. It "could be game over for the climate," said Michael E. Mann, a prominent American climate scientist (I. Johnston, 2016: 2). Noam Chomsky went so far as to say that Trump's election would be "almost a death knell for the human species—not tomorrow, but the decisions we take now are going to affect things in a couple of decades, and in a couple of generations it could be catastrophic" (Benedictus, 2016: 3). Ominous charges of "crimes against humanity" and "climate crimes" came up frequently in critical commentary about the Trump campaign and his position on the topic of global warming.

Mainstream media coverage of the climate change issue in general, and Trump's specific position on it during the campaign, however, was "utterly miserable" (Negin, 2016) and did little to bring the real story of the Republican candidate's intended assault on climate regulations to public consciousness. The major television networks barely reported on the topic of global warming, and during the presidential debates between Clinton and Trump there was not a single question devoted to the issue of the changing climate. Newspapers were not much better, although some print journalists, such as columnist Eugene Robinson (2016: 1) of the *Washington Post*, did try to warn the public that "a vote for Trump is a vote to undo vital progress on climate change." But these were lonely voices. Other issues commanded greater media attention during the campaign, some salacious, as with Trump's open admission that he engaged in sexual assault, or manufactured, such as the controversy surrounding Clinton's e-mails.

As the election approached, many Democrats were complacent, convinced that Trump was a political clown who had no chance to win. Then, in a stunning turn of events, Trump won the 2016 election in the critical Electoral College, even though he lost the popular vote. A volatile transition period followed, and the Trump team confronted the country with their extreme approach to many critical issues. The issue of climate change soon emerged as a central concern during this time as the president-elect began to make appointments and draw up an anti-regulatory environmental policy agenda. Trump's climate crimes were about to begin.

The major goals of the Trump administration concerning climate change were revealed in a leaked memo from Thomas Pyle, head of the American Energy Alliance, who was selected to lead the transition team for the Department of Energy. Pyle, a former oil industry lobbyist for the Koch brothers, stated in the memo that the immediate goals of the new administration were to withdraw from the Paris climate agreement, dismantle Obama's Clean Power Plan, and expedite approval of pipeline projects such as the Keystone XL (Foster, 2017: 100). To the climate movement, this was an alarming agenda. The news that the transition process at the EPA was to be headed by Myron Ebell, a prominent climate change denier from the libertarian Competitive Enterprise Institute (CEI), particularly concerned climate activists. The CEI, which questions what it calls "global warming alarmism," was founded in 1984 and has received funding from Exxon-Mobil and the Donors Trust. The conservative think tank advocates for an extreme neoliberal political agenda and has engaged in a number of activities intended to create doubt about human-caused global warming.

Ebell, the director of CEI's Center for Energy and Environment, had been a key player in the right-wing campaign to convince George W. Bush to withdraw from the Kyoto climate accord, and he had also attacked Bush's

EPA head Christine Todd Whitman (who supported climate action) for "pushing this rubbish" (Helvarg, 2017: 22). During the Obama years, this extreme climate denier also played a critical role, along with many others in conservative think tanks, in manufacturing uncertainty about global warming, undermining the field of climate science, and attacking individual climate scientists. During the transition, Ebell told the British press that "The environmental movement is, in my view, the greatest threat to freedom and prosperity in the modern world," and he promised that the United States would soon withdraw from the global agreement to fight global warming adopted at Paris in 2015 (Carrington, 2017: 2). This is the ideological mindset of the person who advised the new president about how to run the EPA and subvert previous environmental protection policies.

In December 2016, with input from Ebell, Trump appointed Oklahoma attorney general Scott Pruitt, also an aggressive climate change denier with close ties to oil and gas corporations, to lead the EPA. Given that Pruitt was a ferocious opponent of the agency he was now to head, with a history of suing the EPA over carbon pollution regulations, it was an alarming selection. Many of Trump's other cabinet appointments were equally ominous for those committed to environmental protection and climate action. Rex Tillerson, the CEO of ExxonMobil, the powerful company with "a grim history . . . of, in short, destroying the planet in the eternal search for record profits" (Engelhardt, 2018: 62), was appointed secretary of state. Former Texas governor Rick Perry, also an advocate for the oil industry, became the head of the Department of Energy. Congressman Ryan Zinke, from coal-producing Montana, was selected as the secretary of the interior. Considering the political ideology and positions on climate change of Vice President Mike Pence and the influential White House chief strategist Steve Bannon, it was clear that the new Trump administration's top policy posts were being filled with officials who, except for Tillerson, had a public record of openly denying the established science of human-caused climate change.

The Trump administration not only institutionalized climate change denial within the US government (as had the George W. Bush administration), it intensified its role. Not only were the "climate denialists in charge" (Davenport, 2017a), but moreover, given their deep ties to fossil fuel corporations, it was clear that "the oil and gas industry is quickly amassing power in Trump's Washington" (Eilperin, Mufson, and Rucker, 2016). Concerning the new cabinet's "fossil fuel infatuation," McKibben (2017c: 3) wrote: "These guys know nothing about science, but they love coal and gas and oil—they come from big carbon states like Oklahoma and Texas, and their careers have been lubed and greased with oil money." Given a president who believed that climate change was a hoax and that the continued extraction of fossil fuels should be promoted to the max, and a cabinet filled with climate change denialists who are strong supporters of the fossil fuel industry, it was clear that Obama's climate policies, as limited as they were, would be targeted for elimination. A new group of carbon criminals were heading to Washington, DC, intending to carry out an ambitious "shock and awe" agenda of climate crimes, and they made no effort to hide their intentions.

The first specific climate crime that the new administration was to engage in is what the *New York Times* (2017b) called "Trump's war on science." One element of this war would involve censoring scientific information and inquiry that could inform public policies concerning the climate issue. As the Trump administration prepared to take office in January 2017, a group of green criminologists issued a warning to citizens and scholars concerned with ecological harms to expect to "encounter problems gaining access to important environmental data." They went on to predict that "the Trump administration will do away with important data reporting programs, producing a rupture in the historical record of important environmental data and a significant reduction in the enforcement of environmental regulations" (Lynch, Stretesky, et al., 2017: 9). This is exactly what happened. It started during the transition as the Trump team attempted what was called a "climate purge" when they demanded the names of government employees who had studied global warming or worked on climate policies under Obama (McCauley, 2016), a demand that was refused.

Once in office, the Trump administration climate change denialists cleansed the White House website of any references to climate change. Alterations were also made to many federal agency Web resources about climate change. In a 2018 report, *Changing the Digital Climate*, the EDGI (Environmental Data and Governance Initiative, 2018: 1) stated that "Although there is no evidence of any removals of climate data, we have documented overhauls and removals of documents, webpages, and entire websites, as well as significant language shifts." In addition, the Trump administration also proposed cutting the NASA program that used satellites to collect data on disappearing glaciers, rising seas, and other ecological changes that are taking place; proposed eliminating a NOAA program that provided coastal protection assistance; and planned budget cuts at the Federal Emergency Management Agency (FEMA) and the EPA. Concerning these environmental cuts, Trump's budget director (and later chief of staff), Mick Mulvaney, bluntly stated: "Regarding the question as to climate change, I think the President was fairly straightforward—we're not spending money on that anymore. We consider that to be a waste of your money to go out and do that" (Klein, 2017: 75). These actions and this statement are a fairly straightforward admission that the Trump administration intended to engage in the climate crimes of political omission and denial. And as Davenport and Landler (2019) reported in the *New York Times*, the administration hardened its attack on science in 2019, targeting in particular the National Climate Assessment.

The war on impact science in general was a central feature of the EPA under Pruitt, who Rhode Island Democratic senator Sheldon Whitehouse called "a puppet of the fossil fuel industry" (Goodell, 2017a: 49). Shortly after Pruitt was appointed to head the EPA, a large batch of e-mails were uncovered that showed that, during his tenure as attorney general of Oklahoma, he had "closely coordinated" his activities with the fossil fuel industry and political groups tied to the billionaire Koch brothers in an effort "to roll back environmental regulations" (Davenport and Lipton, 2017: 1). The *Washington Post* documented that, throughout his first year at the EPA, Pruitt's calendar was filled with meetings, fancy dinners, and speaking engagements with fossil fuel industry representatives (Dennis and Eilperin, 2017). Environmental sociologist Lindsey Dillion (2018) argues that the general pro-business direction of the EPA under Trump is enabling a form of "regulatory capture" (control of the agency by the corporations it is supposed to be regulating). In a lengthy article in *Rolling Stone* titled "Scott Pruitt's Crimes against Nature," Jeff Goodell (2017a: 44) provided convincing evidence of the many ways that Trump's EPA chief was "gutting the agency, defunding science and serving the fossil fuel industry." And a *New Yorker* article by Margaret Talbot (2018: 36) titled "Dirty Politics" detailed how Pruitt was "giving even ostentatious polluters a reprieve."

In addition to his documented deep ties to the industries he was supposed to be regulating, Pruitt was also an outspoken skeptic of the human role in causing climate change. In March 2017, he said that carbon dioxide was not "a primary contributor to the global warming that we see," a statement that the *New York Times* pointed out contradicts decades of scientific research and analysis (Davenport, 2017b: A1). Given his denialist position, it was no surprise that Pruitt brought with him to the agency "a team of experienced EPA bashers and climate change obstructionists," many of whom had previously worked for the Republican senator from Oklahoma James Inhofe, one of the "most notorious and flamboyant" climate change deniers in the Congress (Goodell, 2017a: 47). This veteran team of climate policy rejectionists came to the EPA ready to use their new organizational positions and power to create a pervasive culture of climate change denialism within the agency and instigate a war on environmental science.

During the first year of the Trump presidency, Pruitt and his staff took a series of actions that sought to undermine, restrict, or purge climate science and scientists from the EPA. First, they put a political operative (a former Trump campaign aide) with little environmental policy experience in charge of signing off on the agency's grant applications and funding awards. This political aide, John Konkus, reviewed every award given out and every grant solicitation that was issued at the EPA, and he openly "told staff that he is on the lookout for the 'double C-word'—climate change—and repeatedly instructed grant officers to eliminate references to the subject in solicitations" (Eilperin, 2017: 1). Early on, Pruitt's team also began to remove climate data and other scientific information from the agency's website. They scrubbed dozens of online resources that helped local governments address climate change, "part of an apparent effort by the agency to play down the threat of global warming" (Friedman, 2017b: 1). Then, in October 2017, the EPA prohibited three agency scientists from speaking at a professional conference where they were scheduled to discuss climate change. The *New York Times* observed that "The move highlights widespread concern that the EPA will silence government scientists from speaking publicly or conducting work on climate change" (Friedman, 2017a: 1). And later that month, Pruitt issued a directive that banned scientists and academics that receive EPA grant money from serving on the agency's federal advisory boards. Ostensibly, the move was to avoid conflicts of interest and ensure the independence of the panels that counsel the agency on scientific decisions. But in their place, the EPA head planned to appoint industry representatives without imposing "any new restrictions to prevent them from offering advice on environmental regulations that may affect their businesses" (Friedman, 2017c: A9), which would truly be a massive conflict of interest. Climate scientist Michael E. Mann called this new form of organizational denialism "malicious stupidity" (Goodell, 2017a: 49). Malicious stupidity, however, would not be the worst of Trump's and Pruitt's climate crimes.

In addition to its war on climate science and its organizational denialism, the Trump administration also took direct aim at almost every climate change policy that Obama had put into place. The clear goal was to eliminate or undermine all of these policies. Just days after the inauguration, Trump signed executive memorandums that resurrected the Keystone XL pipeline (by inviting TransCanada to resubmit its application for a presidential permit rejected by Obama) and expedited the Dakota Access pipeline in North Dakota (a project that had been strongly opposed by the Standing Rock Sioux Tribe). According to the *New York Times*, these decisions were part of this broader effort to "dismantle the legacy" and "unravel the policy structure" of Obama, "who made fighting climate change a central priority" (P. Baker and Davenport, 2017: A1). Outraged environmental activists denounced the decisions that green-lighted the pipelines and vowed to resist. Michael Brune, the executive director of the Sierra Club, stated that "Donald Trump has been in office for four days, and he's already proving to be the dangerous threat to our climate we feared he would be" (P. Baker and Davenport, 2017: A1). Still more climate crimes were to come.

In early March 2017, the Trump administration began the process to roll back the carbon pollution standards for vehicles that had been set by the EPA and the Department of Transportation under Obama. Trump's action came after a meeting with automobile corporation executives in Detroit. The auto industry had complained that the restrictions on tailpipe emissions of carbon dioxide were too burdensome and expensive. The Obama regulations would have required the manufacturers to build passenger cars that achieve an average of 54.5 miles per gallon by 2025 (the current average is 36). As the *New York Times* reported, "The tailpipe pollution regulations were among Mr. Obama's major initiatives to reduce global warming" and eventually "would have drastically reduced the nation's vehicle tailpipe pollution which accounts for about a third of the United States total greenhouse gas emissions" (Davenport, 2017c: A12). Trump's attempted regulatory rollback of fuel economy standards for vehicles, which did not require action by the Congress, would bolster U.S. auto industry profits and rescind one of Obama's significant environmental achievements. In 2019 Trump attempted to bolster this regulatory rollback effort by seeking to revoke California's special authority to set stricter emission standards, which the state had been historically granted (Davenport, 2019).

On March 28, 2017, Trump escalated his assault on the Obama environmental legacy with an expansive executive order intended to dismantle a variety of federal efforts to address climate change. Flanked by coal miners at the signing ceremony, Trump issued an executive order titled "Promoting Energy Independence and Economic Growth" that established a policy requiring all federal agencies to review existing regulations "that could potentially burden the development or use of domestically produced energy resources, with particular attention to oil, natural gas, coal and nuclear energy" and to eliminate or revise those that "unduly burden" the development of these resources. *Inside Climate News* described the executive order as "the most significant declaration of the administration's intent to retreat from action on climate change" (Lavelle, 2017: 1). In the most important directive, the president charged the EPA to start the complex and lengthy legal process of withdrawing and rewriting the Obama-era Clean Power Plan. Although the Obama plan had been put on hold by the U.S. Supreme Court in February 2016, if ever fully enforced it "would have closed hundreds of coal-fired power plants, frozen construction of new plants and replaced them with vast new wind and solar farms" (Davenport and Rubin, 2017: A1). In addition, Zinke was directed by Trump to formally lift the moratorium on new coal-mining leases on federal land, a ban the Obama administration had put in place back in 2016. The "Promoting Energy

Independence" executive order also allows for the rewriting of methane emission limits that were placed on the oil and gas industry by Obama and instructs the EPA to ignore calculations of the "social cost" of carbon pollution, a requirement also imposed by the Obama administration. Furthermore, Trump's order rescinds Obama's 2013 Climate Action Plan, eliminates the Task Force on Climate Preparedness and Resilience that had been established to advise the federal government on how best to respond to the needs of communities around the nation that are experiencing the ecological and social impacts of climate change, and reverses the Obama directive to federal agencies to address the national security implications of anthropogenic climate disruption (Jamail, 2017).

Critics of Trump's March 28 executive order charged that the new president was dismantling, nullifying, and even obliterating Obama's major climate policies. This was not only the crime of political omission (Trump's complete failure to act to mitigate greenhouse gas emissions in any way); it was also a dangerous attack intended to annihilate the few, mostly still inadequate policies that had been enacted by Obama to respond to the climate crisis. The *New York Times* (2017a: A26) editorialized that "Only 10 weeks into his presidency, and at great risk to future generations, Donald Trump has ordered the demolition of most of president Barack Obama's policies to combat climate change by reducing emissions from fossil fuels." The *New York Times* editors went on to say they were "deeply dismayed" by Trump's false complaints of job-killing environmental regulations, his repudiation of the rock-solid scientific evidence of climate change, and his administration's support for older and dirtier energy sources (*New York Times*, 2017a: A26). Trump's executive order made it clear that the new U.S. government did not intend to meet the already limited pledges of emissions reductions that the Obama administration had made at Paris in 2015, and that reckless decision threatened to undermine the global agreement. Trump's clear and present danger to any sane climate policy led American journalist Dahr Jamail (2017: 2) to conclude, even as early as March 2017, that this "means the planet, which is already well along the road of runaway abrupt climate disruption, has little chance of achieving anything like the serious mitigation steps needed to lessen the impacts, which are already inevitable." Trump and his team were well on the way to becoming "the greatest criminals of all time," but they were not done yet.

Despite the fact that Trump had filled his administration with global warming deniers, carried out a broad attack on climate science, and issued an executive order that demolished Obama's climate policies, some people still held out hope that the new president would keep the United States in the 2015 Paris Agreement. It was a vain hope. Although some climate activists questioned the overall value of the global agreement (Brecher, 2017), most still wanted to see the United States remain in the pact, as did most international political leaders. More than 200 institutional investors also lobbied for Trump to keep the United States in the accord (Kusnetz, 2017b). And inside the White House itself, there was a "war" over whether to stay in the Paris Agreement or not, with then secretary of state Tillerson, Trump's daughter Ivanka Trump, and her husband Jared Kushner arguing to remain and then EPA head Pruitt and former advisor Bannon pushing to leave (Cushman and Lavelle, 2017). In the end, President Trump did what candidate Trump had promised to do on many occasions: he formally stated his intention to abandon the global climate accord. The announcement to withdraw from the Paris Agreement was made on June 1, 2017, in the White House Rose Garden. In a fatuous and dishonest statement, Trump said, "In order to fulfill my solemn duty to protect America and its citizens, the United States will withdraw from the Paris climate accord" (O'Harrow, 2017). In addition to pulling out of the Paris pact, the president also announced that the United States would "terminate" its contributions to the Green Climate Fund. Typical for Trump, his announcement demonstrated that he did not understand or care about the details of the accord or the fund; almost all the "facts" he asserted were wrong or misleading (Kaufman, 2017; Kessler and Lee, 2017; Kotchen, 2017). He never even mentioned global warming that day.

While many Republicans and other climate change deniers on the right praised the decision to exit Paris, most of the reactions to the announcement were harshly critical. Trump's move was savaged as "stupid, reckless, reprehensible, grossly immoral, an international disgrace, one of the most ignorant and dangerous actions ever taken by a president, and an inexplicable abdication of any semblance of responsibility or leadership" (Zimet, 2017: 1). McKibben (2018a: 20) has argued that "Of all the actions that Trump has taken during his reckless and infantile months in the White House, none will do longer-lasting damage than his abandonment of the Paris climate accord." Many commentators once again used the language of crime to denounce the decision to exit the agreement. Hertsgaard (2017) called it "a crime against humanity." May Boeve, executive director of 350.org (2017), called the decision "a travesty, a crime against the future of people and the planet." And Friends of the Earth (2017) stated that leaving the Paris Agreement makes the United States "the foremost climate villain." Trump's decision to withdraw from the Paris climate agreement does indeed warrant the label of climate crime. It poses a direct threat to the ability of the global community to mitigate greenhouse gas emissions and limit the heating of the planet to 2 degrees Celsius over the preindustrial average (let alone 1.5 degrees Celsius). Not only is the Trump administration refusing to take any steps on its own to reduce carbon emissions; it is also undermining the ability of the international political community to mitigate emissions. Given what we know about the science of global warming and the ecological and social impacts of climate disruption, the decision to exit the Paris accord seems destined to be labeled as one of the greatest climate crimes in history.

Trump's crime of withdrawal from the Paris Agreement was shaped in large part by his infatuation with and close economic and political connections to the fossil fuel industry, that is, fossil capitalism. This same state–corporate nexus operated to obstruct climate change policies during the second Bush administration when individuals and corporations within this industry also played a strong role in the climate change denial countermovement and in blocking and resisting global warming policy in the political arena. In an investigative report, the *Washington Post* concluded that the decades-long "crusade" by climate change deniers in conservative think tanks and foundations had "fueled" Trump's decision to withdraw from the Paris Agreement (O'Harrow, 2017). Given the overwhelming evidence concerning the intersections of business and government with regard to climate change policy in general and this decision in particular, it seems appropriate to also call Trump's withdrawal from the Paris Agreement a form of state–corporate climate crime.

Michael Klare has presented a broader geopolitical perspective on the Trump decision to abandon Paris, a perspective that grants a more calculated strategy than the resistance often gives Trump credit for. In an examination of Trump's overall foreign policy, Klare agrees with those analysts who argue that the new administration has set out to dismantle the post–World War II liberal world order. Klare (2017b: 2), however, asserts that "Donald Trump is not only trying to obliterate the existing world order, but also attempting to lay the foundations for a new one, a world in which fossil-fuel powers will contend for supremacy with post-carbon, green-energy states." He argues that Trump's view of the world is "largely governed by energy preference," and he identifies a "grand strategic design" in the administration's international relations. The first step in this process, according to Klare (2017b: 3), "was to revitalize the historic U.S. alliance with Saudi Arabia, the world's leading oil producer," which Trump attempted to do on a trip to Riyadh in May 2017. The second step was the "enfeeblement" of the North Atlantic Treaty Organization (NATO) alliance and the European Union and the improvement of relations with Russia. Trump's visit to NATO headquarters in Brussels on May 25, 2017, achieved the first of these two goals and may portend the emergence of a Chinese–German alliance in the future. The second goal concerning Russia has been more elusive to achieve for a variety of reasons. Step three, according to Klare, was the withdrawal from the Paris Agreement. The rejection of the Paris pact is a move by Trump to achieve a number of "pro-carbon objectives" and obliterate incentives for renewable energy in an effort to advance "the creation of a new world order governed largely by energy preferences" (Klare, 2017b: 5). More evidence to support Klare's thesis emerged in December 2018 at the COP 24 in Katowice, Poland, when the Trump administration's "unapologetic defense of fossil fuels" and argument "that a rapid retreat from coal, oil and gas was unrealistic" found a "receptive audience" among other major fossil fuel producers such as Russia, Saudi Arabia, and Australia (Plumer and Friedman, 2018: A1).

Whatever larger structural forces drove the Paris decision, in the fall of 2017 the Trump administration took formal steps to eliminate the most significant component of the country's contribution to the agreement. On October 9, at an event in the coal country of eastern Kentucky, then EPA head Pruitt announced he would sign a proposed rule (to be filed in the Federal Register) to repeal the Clean Power Plan, Obama's signature climate policy (the courts have yet to rule on this proposed rule change). As the *New York Times* explained, "Eliminating the Clean Power Plan makes it less likely the United States can fulfill its promise as part of the Paris climate agreement to ratchet down emissions that are warming the planet and contributing to heat waves and sea-level rise" (Friedman and Plumer, 2017: 1). Repealing the Clean Power Plan would drive the final stake in the heart of the Paris Agreement. Some supporters of the fossil fuel industry urged the EPA to go even further and rescind the all-important Endangerment Finding itself, the foundational conclusion that forms the legal basis for greenhouse gas regulations at the agency. As its first year in office ended, the Trump administration appeared to have been highly successful in rolling back Obama's limited environmental legacy: rescinding fuel-efficiency standards, issuing new climate directives in the March 28 executive order, exiting the Paris Agreement, and then attempting to repeal the Clean Power Plan. Trump's success on the climate front was so overwhelming that he did not even attempt to interfere with the release of the first volume of the Fourth National Climate Assessment in early November 2017, even though the report completely contradicted everything the Trump administration has said about climate change from its first day in office. Produced by the U.S. Global Change Research Program, the Climate Science Special Report was unequivocal: climate change is real, it is caused by human activity, especially greenhouse gas emissions, and it is getting more dangerous every day. There is no convincing alternative explanation for global warming and climate disruption. The stark scientific warning in the National Climate Assessment stands in sharp contrast to Trump's climate crimes of denial, political omission, and the promotion of the greater extraction of fossil fuels.

During the second and third years of the Trump administration, the pace of its assault on climate policies slowed somewhat as other issues took center stage (North Korea, Iran, the Russian investigation by special counsel Robert Mueller). But Trump's "disdain for action on climate change" was in full display at the G7 summit in Canada in June 2018, where the president "skipped the G7's formal discussion on the global warming crisis" and the United States "refused to join in common statements by the other six nations reaffirming their commitment to the Paris climate agreement" (*Inside Climate News*, 2018: 1). Instead, the United States vigorously promoted fossil fuels as it had bizarrely done at the COP 23 in Bonn in December 2017 (and did again at COP 24 in Katowice, Poland, in December 2018 to much mockery). In addition, several important members of Trump's cabinet were replaced during the second year. Tillerson resigned as secretary of state in March 2018, and Pruitt was then forced to resign as head of the EPA in July due to mounting allegations and investigations of unethical behavior while in office. Mike Pompeo, a climate policy foe with close connections to the oil magnate Koch brothers, was picked to replace Tillerson as secretary of state. According to *Inside Climate News*, Pompeo's appointment "signals a hardening stance against international engagement on climate change" (Lavelle, 2018a: 1). Andrew Wheeler, a former coal industry lobbyist and congressional staffer, took over at the EPA following Pruitt's departure. Wheeler is committed to continuing Trump's extreme environmental regulatory rollback, but, with more experience in Washington, he is expected to be more careful and less inflammatory than Pruitt was. As one environmentalist put it, "He'll be Pruitt without the nasty smirk and the edge" (Lavelle, 2018b: 2). Yet despite these changes in personnel, no major changes in the direction of the administration's climate policies are to be expected during the remainder of Trump's time in office. What can be expected from the Trump administration is a continuing stream of anti-environmental policy decisions. The *New York Times* has counted eighty-five environmental rules that have been or are being rolled back under Trump (Popovich, Albeck-Ripka and Pierre-Louis, 2019). A running list of how the administration is changing environmental policy can also be found on the National Geographic website. And according to two research reports issued in early 2019, Trump's elimination of climate change rules, deregulation of energy corporations, and general environmental policy rollbacks will dramatically increase greenhouse gas emissions, cause thousands of premature deaths per year, and save the fossil fuel industry over $10 billion (Banerjee, 2019; Corbett, 2019; Cutler and Dominici, 2018).

In the face of mounting scientific evidence of climate disruption and massive ecological and social harms that result from the failure to mitigate carbon emissions, Trump and his administration, on behalf of the fossil fuel industry, have made a deliberate political choice to continue to inflict harm and put the world at catastrophic risk. As John Bellamy Foster (2017: 103) points out, these efforts "are necessary parts of an attempt by carbon capital to proceed undeterred with the burning of fossil fuels, as if this did not constitute a dire threat to the human species." These civilization-threatening actions, the promotion of greater fossil fuel extraction, the institutionalization of climate change denialism in the White House, the assault on Obama's climate policies, the abandonment of the Paris Agreement, and the increasing militarization of adaptation constitute blameworthy harms that are so terrifyingly reckless that they have led many commentators to label them as an extreme form of criminality. As David Bromwich (2019: xxxiii) argues concerning Trump's actions in relation to the global climate crisis, "More than cheating in an election or insulting traditional allies or degrading the norms of public speech in unheard-of ways, his denial of the existence of this more-than-national predicament should be counted the largest of his crimes."

CHAPTER 6

"Blood for Oil," Pentagon Emissions, and the "Politics of the Armed Lifeboat"

CLIMATE CRIMES OF EMPIRE

I WAS BORN in January 1951, the first month of the second half of the twentieth century, a half-century that witnessed massive social change. At the time of my birth, the "Great Acceleration" into the epoch of the Anthropocene was just beginning. This period of "unprecedented economic growth and environmental devastation" was transformative, "a turning point in Earth history" (Angus, 2016: 41). But a second and related major transformative process also began at the midpoint of the twentieth century: the full emergence of the "global American empire," coupled with an influential "culture of militarism" in U.S. society (Boggs, 2010, 2017; Freeman, 2012). The American empire and a dominant U.S. military institution would serve to magnify the economic and environmental effects of the Great Acceleration.

Like the "Great Transformation" of the nineteenth-century political economy that Polanyi analyzed in 1944, the two interrelated transformations in the second half of the twentieth century each had enormous impacts, both positive and negative, on the global economic and political order, the social structure and culture of the United States, and the personal lives of millions. Following historian William Appleman Williams in his classic *Empire as a Way of Life* (1980), I argue that the U.S. global empire, American militarism, limitless economic growth, an expanding consumer culture, and the increasing production of greenhouse gas emissions through the burning of fossil fuels all became "a way of life" during this half-century and served to not only sustain fossil capitalism at home and abroad but aggravate the problem of climate disruption as well.

Among the myriad social, political-economic, and ecological consequences of these two great transformative processes are three specific, interrelated, and often "unrecognized" blameworthy harms (crimes) that are connected to climate change: (1) a large-scale pattern of illegal imperial interventions and wars, many undertaken to secure access to and control over oil by the U.S. "warfare state" (Boggs, 2017; Dower, 2017; McCoy, 2017), which in turn, facilitated the greater extraction of fossil fuels and ensured rising carbon emissions during the Great Acceleration; (2) the deployment and operation of the American military during these imperial actions and wars, coupled with a supporting domestic and foreign military base structure (C. Johnson, 2004), that results in the release of huge quantities of greenhouse gases by the largest organizational polluter in the world, the U.S. Department of Defense (Sanders, 2009, 2017); and (3) the planning for and beginning execution of a "militarized" form of adaptation to climate disruption that involves responding to the problem with "the politics of the armed lifeboat" (Parenti, 2011: 11). These harmful state–corporate actions are climate crimes that can best be understood within the context of the development of the post–World War II global U.S. empire, the normalization of a culture of militarism within American society, and the resulting creation of a "treadmill of destruction" as a concomitant to the operation of the treadmill of production/high-carbon system (Clark and Jorgenson, 2012; Hooks and Smith, 2005). From this perspective, the interrelated climate crimes of U.S. military interventions to control oil, Pentagon greenhouse gas emissions that result from these wars and other operations, and the choice of a militarized form of adaptation to climate disruption can also be described as "crimes of empire" (Boggs, 2010).

A Short History of the American Empire and Its Crimes

Historian Alfred McCoy (2017: 40) reminds us that "empire is not an epithet but a form of global governance in which a dominant power exercises control over the destiny of others, either through direct territorial rule (colonies) or indirect influence (military, economic, and cultural)." According to various scholars, the United States has been an imperial project from its earliest years (F. Anderson and Cayton, 2005; Bacevich, 2002; Ferguson, 2004; Hahn, 2016; Immerman, 2010; Lens, 1971; Nugent, 2008), and "empire as a way of life" (Williams, 1980) is a strong determining structural and cultural factor in the U.S. propensity to commit state crimes (Boggs, 2010; Iadicola, 2010; Kramer, 2018). One important dimension of empire as a way of life is that efforts at imperial domination (crimes of empire) are almost always rationalized within a broad historical and cultural narrative often referred to as "American exceptionalism" (Appy, 2015; Dower, 2017; Fiala, 2008; Hodgson, 2009; Koh, 2003; Swanson, 2018). As American political activist Sidney Lens (1971: 1) noted, empires always seek to cover themselves in "the myth of morality," and the American empire in particular has formulated such myths "to assuage its conscience and sustain its image." Historian Carl Boggs (2010: 6) argues that in the United States, various myths of morality "permeate a political culture that gives policy-makers a relatively free hand to pursue geopolitical ambitions" and commit "crimes of empire."

Increasingly, these state crimes, and the deaths and destruction they cause, are also "normalized" by the development of a post–World War II American culture of militarism. As sociologist Michael Mann (2003: 16–17) defines the term, militarism is "a set of attitudes and social practices which regards war and the preparation of war as a normal and desirable social activity." It involves an extreme emphasis on the value of military institutions in society and the predominant use of military power to solve complex social and economic problems. Historian Andrew Bacevich (2005: 2) argues that "To a degree without precedent in U.S. history, Americans have come to define the nation's strength and well-being in terms of military prepared-ness, military action, and the fostering of (or nostalgia for) military ideals." Militarism operates in both the larger political system and national culture, as well as within specific government organizations such as the Department of Defense (Pentagon) and the Central Intelligence Agency (CIA). As Boggs (2010: 249) observes, state actions that may be "viewed from outside this paradigm [culture] as criminal or barbaric may seem rather normal, accept-able, even praiseworthy *within* it, part of a taken-for-granted universe of meanings." Sociologists refer to this phenomenon as the "normalization of deviance" (Vaughn, 1996).

Historian Walter Nugent (2008) argues that the United States has cre-ated three empires during its history. The first form of empire building involved continental expansion from 1782 to 1853, and the foundational crimes of the country were the violent imposition of chattel slavery on kid-napped Africans and the genocide of Native Americans. According to his-torian Steven Hahn (2016: 2), "the model of governance inherited from the British was empire," and "from the birth of the Republic the United States was a union with significant imperial ambitions on the continent and in the hemisphere, many pushed by slaveholders and their allies." Mythmaking was in evidence early on too, as when Thomas Jefferson referred to the United States in 1809 as an "empire for liberty" (Immerman, 2010: 5). The "settler-colonialism" (Dunbar-Ortiz, 2014) of this first period of expansion resulted in the mass slaughter of Indigenous peoples, the theft of their land, the extension of slavery, and an expansionist war against Mexico in 1846.

Next came the offshore empire building from the 1850s to 1917 that resulted in territorial acquisitions and formal colonies. During this second form of empire, an imperial war against Spain (1898) and other foreign interventions led to the territory and people of the Philippines, Cuba,

Puerto Rico, Samoa, Guam, and Hawaii being annexed and colonized. Theodore Roosevelt, about to become vice president in 1901, defended this "new American imperialism," arguing that "the expansion of America overseas was simply an extension of American expansion over the continent" (Judis, 2004: 61).

The third form of empire involved not the acquisition of territory per se but the extension of American political, economic, and military power around the world. Williams (1959) described it as "open door" imperialism. According to Nugent (2008: 306), this "new, virtual-global empire," which was "only embryonically evident during the interwar years of 1918–1939," emerged most dramatically after World War II, particularly during the Cold War, and it continues in strong form today through what has been called the "warfare state" (Boggs, 2017), the "Pax Americana" (Freeman, 2012), or the "corporate state" (Hedges, 2009). My focus in this chapter is this third form of the U.S. Empire and the climate crimes it fosters.

All three historic American empires described by Nugent (2008: xvi) have rested not only on an "ideology of expansion" but also on the cultural narrative of American exceptionalism that served to justify that expansion and the violent actions and other crimes that accompanied it at each stage. American exceptionalism generally portrays the United States as a nation of exceptional virtue, a moral leader in the world with a unique historical mission to spread "universal" values such as freedom, democracy, equality, popular sovereignty, and, increasingly, global capitalism. British historian David Ryan (2007: 119) argues that this mythic cultural conception "thoroughly informs U.S. constructs of its identity." American historian John Dower (2017: 120) refers to it as "the gospel" of American exceptionalism wherein "Americans surpass all others in virtue and practice." From John Winthrop's "City upon a Hill" sermon in 1630, to the Puritans about to embark from the *Arbella* to establish the Massachusetts Bay Colony, to publisher Henry Luce's influential notion of the "American Century" in 1941, the message "was and remains idealistic, generous, moralistic, paternalistic, patronizing, riddled with double standards and hypocrisy, and notably lacking in self-reflection or self-criticism" (Dower, 2017: 120). Every American president must pay extreme obeisance to this "gospel" or pay a political price, and it has served to rationalize and justify a long list of deadly imperial crimes. Criminologists sometimes refer to rationalizations and justifications for criminal behavior as "techniques of neutralization" (of guilt).

As the United States rose to ever greater power after World War I and then World War II, it also developed a self-image as a "reluctant superpower," a key theme within American exceptionalism that claims that the United States involves itself in world affairs only under duress and then always for selfless reasons (Bacevich, 2002). President Woodrow Wilson's famous claim that the United States must enter the Great War (World War I) because "the world must be made safe for democracy" exemplifies this narrative theme in action. The need to ensure that the United States could play a significant role in creating a new political and economic order out of the collapse of the Ottoman and Austro-Hungarian empires after the war was interpreted as selflessness rather than self-interest (C. Johnson, 2004). Such "idealism" would continue to inform American foreign policy for the rest of the twentieth century and into the next.

While isolationist sentiments at home would stymie Wilson's particular vision for a new political and legal world order, World War I did mark the emergence of a "new order of power" in the world that featured a "new centrality" for the United States. British historian Adam Tooze (2014: 6) argues that the United States emerged from the Great War "unscathed" and "vastly more powerful," a "novel kind of 'superstate' exercising a veto over the financial and security concerns of the other major states of the world." Although American political leaders were not yet committed to the full assertion of military power at this time, they did indirectly exercise forms of economic and political power that would eventually lead to the full-blown creation of the Pax Americana that still dominates the world today.

World War II would provide the United States with new and unique opportunities for various forms of empire building (Freeman, 2012). As the war progressed and it became clear that the United States would be able to exercise hegemonic power in the postwar era, American leaders began to plan for the construction of new global institutions that would greatly advance the country's political and economic dominance. As Howard Zinn (1980: 414) points out, "Before the war was over, the [Franklin Roosevelt] administration was planning the outlines of the new international economic order, based on partnership between government and business." The war, and this new international economic order, would finally lift the United States completely out of the Great Depression and establish it as both the world's dominant military power and the economic hegemon in charge of the key institutions of global capitalism, such as the International Monetary Fund, the World Bank, and the General Agreement on Trade and Tariffs. A larger and stronger global American Empire was emerging.

World War II also provided the United States with an opportunity to create a new political and legal order under international law. With the signing of the Atlantic Charter in 1941 by Roosevelt and Winston Churchill, the establishment of the United Nations and the adoption of its Charter in 1945, the prosecution of Nazi "war criminals" at Nuremberg that same year, and the passage of the Geneva Conventions in 1949, the United States and Great Britain "led efforts to replace a world of chaos and conflict with a new, rules based system" (Sands, 2005: xi). This new set of international political and legal norms was indeed a significant accomplishment, although always selectively applied. As Dower (2017: 21) observes, "As it transpired, none of the victor nations sitting in judgement in these early postwar years ever seriously regarded the laws they were inventing and applying to defeated enemies as being applicable to their own countries." In the United States, these new international institutions and laws were actively opposed at the time by many on the far right. Indeed, by the beginning of the twenty-first century, some American political leaders, particularly hard-core nationalists and neoconservatives, came to believe that the existence of such an international legal framework for judging state aggression, war crimes, and human rights violations was illegitimate if used to judge U.S. actions, such as the clearly illegal invasion of Iraq in 2003. As Dower (2017: 125) observes, "The mystique of exceptional virtue does not accommodate serious consideration of irresponsibility, provocation, intoxication with brute force, paranoia, hubris, reckless and criminal actions or even criminal negligence."

One of the most consequential results of World War II was the emergence of the phenomenon of "perpetual war" and a "warfare state" that has become the most significant feature of contemporary American politics. According to Boggs (2017: 3), the "warfare state" in the United States refers to "a broad ensemble of structures, policies, and ideologies: permanent war economy, national security state, global expansion of military bases, merger of state, corporate, and military power, an imperial presidency, the nuclear establishment, superpower ambitions." This ensemble led to massive "defense" spending and huge annual Pentagon budgets, a form of militarized state capitalism or military Keynesianism that would prop up the U.S. economy. This gargantuan warfare state has been "sustained and legitimated" not only by the myth of American exceptionalism but also by a deeply entrenched "culture of militarism" that is itself a product of World War II (Boggs, 2017). This warfare state has not only transformed American politics and culture; it has also resulted in a wide variety of state–corporate crimes, many of which can be described as crimes of empire.

There were three great, interrelated challenges to the American imperial project in the post–World War II era: the threat of independent nationalism in the so-called Third World (Global South), the continuing Cold War struggle with the Soviet Union, and the establishment and maintenance of access to and control of oil, particularly in the Greater Middle East. All three would fuel state criminality on the part of the United States and its client states. Nations on the periphery and semi-periphery of the world system, many of them former colonies of the world's wealthy capitalist nations, were to be limited to service roles in the postwar global capitalist economy, providing resources, cheap labor, and retail markets for consumer products and finance capital (Frank, 1969; Wallerstein, 1989). U.S. planners were concerned that "radical and nationalistic regimes," more responsive to popular pressures for immediate improvement in the living standards of their people than in advancing the economic interests of foreign capital, could become a "virus" infecting other countries and threatening the "overall framework of order" that the leaders of the corporate state in Washington had constructed (N. Chomsky, 2003). The Soviet Union, with its rival ideology and its own imperial ambitions, was accused of frequently provoking or providing assistance to these nationalistic movements. Thus, the U.S. military and the newly created CIA, often in direct violation of international law, engaged in dozens of foreign interventions around the world to overthrow such "threatening" governments or prop up friendly repressive client states that would serve American economic and geopolitical interests (Blum, 2004; Boggs, 2010, 2017; Dower, 2017; Kinzer, 2006; McCoy, 2017).

The "Vietnam Wars" from 1945 to 1990, as historian Marilyn Young (1991) dates them, were among the most consequential of the imperial interventions that took place during the postwar period. After World War II, the United States made a critical decision to support the attempt of France to reassert colonial control over all of Indochina. But after the French were defeated by Vietnamese (Viet Minh) forces at Dien Bien Phu in 1954, the U.S. government took over as the central imperial power in the region. From then on, U.S. leaders attempted to create and prop up unpopular (and often repressive) governments in South Vietnam and defend them against the military attacks of the internal National Liberation Front (Vietcong) and the North Vietnamese Army. After the questionable Gulf of Tonkin incident in August 1964 (and the congressional resolution authorizing military force that followed), American military involvement dramatically escalated. A devastating imperial war ensued, and the "invading" U.S. forces (which reached a peak of 475,000 by 1967) rained destruction on Vietnam and were often involved in widespread and horrific war crimes that led to civilian suffering on a "stunning scale" (Turse, 2013: 6). American historian Christian Appy (2015: xiii) argues that the brutal Vietnam War "shattered the central tenet of American national identity—the broad faith that the United States is a unique force for good in the world, superior not only in its military and economic power, but in the quality of its government and institutions, the character and morality of its people, and its way of life." American exceptionalism was challenged as never before by the U.S. state crimes in Vietnam, and American political and military leaders would struggle for years to overcome the "Vietnam Syndrome." Eventually the dreaded syndrome would be overcome, American exceptionalism would be revived, and the United States would continue to directly and indirectly engage in dozens of military operations and wars around the world, but now taking special care to limit the number of American troops killed or injured (Appy, 2015). A resurrected American exceptionalism would later play a significant role in what Bacevich (2016) calls "America's War for the Greater Middle East," an expansive war whose initial focus, at least, was on access to foreign petroleum.

EMPIRE, MILITARY POWER, AND CHEAP ENERGY: "BLOOD FOR OIL"

In addition to thwarting independent nationalism in the Third World and fighting the Cold War against the Soviet Union, the other key geopolitical concern for leaders of the U.S. Empire in the post–World War II period was access to and control over the vast supplies of petroleum and natural gas in the Greater Middle East. Not only was the global capitalist system "fueled and lubricated by oil" (Everest, 2004: 32); it was "a resource so vital to American prosperity that access to it must be protected at any cost, including the use of military force" (Klare, 2004: xiv). The availability of cheap petroleum became a critical national security issue and would eventually lead to direct military intervention in the Middle East, what Bacevich (2005) refers to as "blood for oil." I describe such wars as a climate crime of empire, which then shaped the further climate crimes of Pentagon carbon emissions and a militarized form of adaptation to climate disruption.

At one time, the Persian Gulf region contained 65 percent of the world's known oil reserves and 34 percent of its natural gas reserves (Everest, 2004). U.S. efforts to secure access to these fossil fuels started even before World War II ended. In February 1945, on his way home from the Yalta Conference with Churchill and Joseph Stalin, Roosevelt held a now-famous meeting with King Ibn Saud of Saudi Arabia. According to Bacevich (2005: 180), "Out of this meeting came an understanding: henceforth, Saudi Arabia could count on the United States to guarantee its security; and the United States could count on Saudi Arabia to provide it preferential treatment when it came to exploiting the kingdom's vast, untapped reserves of oil." From 1945 to 1979, this agreement allowed American leaders an opportunity to achieve their two major foreign policy goals in the Greater Middle East, political stability in the region and access to the vital resource of oil. Furthermore, with this arrangement they could accomplish these aims "in a way that minimized overt U.S. military involvement" (Bacevich, 2005: 180).

Similar geopolitical objectives were at stake in August 1953 when the CIA sponsored a military coup in Iran to overthrow the democratically elected government of Mohammad Mossadegh and return the shah, Mohammad Reza Pahlavi, to the Peacock Throne. Prime Minister Mossadegh had angered

London and Washington by nationalizing the Anglo-Persian Oil Company (later known as British Petroleum, BP). After the U.S.-engineered coup, the nationalization of Iran's oil industry was annulled, and the United States was able to muscle in on Iranian oil as "40 percent of the share of oil in the new consortium was given to five big American oil companies, and BP's share was reduced" (Dreyfuss, 2005: 109). The return of the shah and his extremely repressive rule provided the desired political "stability" for the region but also sowed the seeds of the 1979 Islamic Revolution in Iran, one of the key catalysts of the looming American war for the Greater Middle East that would be initiated in the 1980s.

As the social and political conflicts of the 1960s roiled American society, particularly the struggle over civil rights and the ordeal of Vietnam, it may have seemed that the issue of access to cheap foreign oil had declined in importance. Michael Klare (2004: 37) asserts, however, that "anxiety over the depletion of U.S. petroleum reserves and a new appreciation of the link between oil and war . . . were just as compelling to Washington in the postwar era" as they had been in 1945 when FDR met with King Saud. "If anything," Klare argues, "petroleum took on even more strategic significance as the United States and other major powers came to rely on cheap and abundant oil to fuel their booming economies." The spectacular economic growth during the Great Acceleration would not have been possible without Persian Gulf oil. Likewise, the Pentagon would not have been able to maintain its protective empire of bases and a strong military machine capable of "containing" the Soviet Union during the Cold War without this petroleum.

From 1946 to 1971, the annual consumption of petroleum products in the United States tripled, from 1.8 billion barrels to 5.4 billion barrels (Klare, 2004). This massive consumption of oil was accompanied, of course, by ongoing efforts to extract and market fossil fuels, rising carbon emissions, and global warming. But American consumers knew little or nothing about the serious problem of climate change at the time. Their major concerns were increased choices in the market, jobs and greater economic opportunity, and continued material abundance, which required "assured access to cheap oil and plenty of it" (Bacevich, 2005: 182). In 1970, however, domestic oil production hit a peak, and thereafter the United States had to depend more and more on foreign sources (Juhasz, 2006; Klare, 2004). As American dependence on Persian Gulf oil grew after 1970, U.S. political and military leaders would face new and complex challenges to maintain access and control of this critical resource. As Everest (2004: 32) noted, "Since the end of World War II, dominating the Middle East and controlling these vast oil supplies have been crucial to U.S. foreign policy under 11 [now 13] different presidents." In pursuit of these imperial goals, direct military intervention in the Persian Gulf region would eventually occur, particularly in Iraq in 1991 and 2003. Blood for oil, a climate crime of empire, would result.

Two different "oil crises" in the 1970s set the stage for a stronger and more direct American involvement in the Greater Middle East. The first oil crisis occurred after the start of the Arab–Israeli War in October 1973. After President Richard Nixon threw his support behind Israel, Iraq responded by nationalizing Exxon and Mobil's shares in the Basra Oil Company. More important, the Arab nations of the Organization of the Petroleum Exporting Countries (OPEC) imposed a full oil embargo against the United States. American consumers suffered as the giant oil companies, rather than accept lower profits, chose instead to "raise oil prices and force all of the sacrifices onto the American public" (Juhasz, 2006: 153). From October 1973 to February 1975, the price of gasoline increased 30 percent, and the price for home heating oil went up by more than 40 percent. The OPEC oil embargo shocked and concerned Americans as it revealed the growing U.S. dependence on foreign petroleum and the way that oil could be used as a political weapon. The public demanded answers from the giant energy companies and the government about the cost and availability of oil. Few were provided at the time, and the concerns would increase and the demands would grow louder in 1979 during the presidency of Jimmy Carter.

The second oil crisis of the 1970s began in January 1979 with the Islamic Revolution in Iran, as a broad-based popular movement organized by militant Shiite clerics overthrew the U.S.-installed government of shah Mohammad Reza Pahlavi. A virulently anti-American government led by the Ayatollah Ruhollah Khomeini replaced the shah and his support for U.S. oil policies in the Middle East. The new Islamic Republic of Iran canceled all of its contracts with the major U.S. oil companies. Once again, American consumers would suffer as the price per barrel of oil more than doubled from 1979 to 1981, and gas prices at the pump increased 150 percent (Juhasz, 2006). This economic shock was just beginning to set in when a major political earthquake also shook the country on November 5, 1979, when Islamic militants took over the U.S. embassy in Tehran and held fifty-two Americans hostage. The humiliating Iran hostage crisis would drag on for 444 days amid intense media coverage and inflict severe political damage on Carter.

Carter first responded to the oil crisis with a televised address on July 15, 1979, that is usually described as the "crisis of confidence" or "malaise" speech. In his address, Carter tried to argue that the reduced supply of oil itself was not the major issue; the nation's "true problems" were an addiction to oil, a crisis of confidence, and a collapse of moral values that had led Americans to worship self-indulgence and materialistic consumption. He claimed that "energy will be the immediate test of our ability to unite this nation," and he laid out a path to recovery from the oil crisis "that can help us to conquer the crisis of the spirit . . . rekindle our sense of unity, our confidence in the future, and give our nation and all of us individually a new sense of purpose." Bacevich (2016: 18) argues that "In content if not in delivery," this address "bears comparison with Abraham Lincoln at his most profound, Woodrow Wilson at his most prophetic, and Franklin Roosevelt at his most farsighted." Had the country taken Carter's wise counsel to heart and begun to make the lifestyle and social changes he suggested and endure the sacrifices he proposed, perhaps Americans would have been in a better position to respond to the coming climate change crisis. But the president's words and recommendations were attacked and mocked, and the speech was perceived to be a political flop, "emblematic of his perceived inability to lead" (Bacevich, 2016: 22).

The Iranian hostage crisis of November 1979 and the Soviet Union's invasion of Afghanistan in December of that year alarmed the country, further imperiling the Carter presidency heading into an election year. Carter himself, using uncharacteristic political hyperbole, declared that the Soviet move into Afghanistan was "an unprecedented act" and "the most serious threat to world peace since the Second World War." The president's critics "advanced an even more fevered interpretation" of the Soviet invasion and attacked Carter's "fecklessness" in the face of the enemy's aggression (Bacevich, 2016: 26). To have any hope of winning re-election, Carter had to rebut the impression that he was weak, ineffectual, and lacking a strong foreign policy. He had to show that he grasped the seriousness of the situation in the Middle East and could take effective action.

On January 23, 1980, Carter seized the opportunity presented by the State of the Union address to unveil a new forceful military policy that would eventually put the country on the path to blood for oil in the Persian Gulf. As Bacevich (2016: 280) points out, Carter, "who just months before had earnestly sought to persuade Americans to shake their addiction to oil now decried . . . the possibility of outsiders preventing Americans from getting their daily fix." In what came to be known as the "Carter Doctrine," the president proclaimed: "Let our position be absolutely clear. An attempt by any outside force to gain control of the Persian Gulf region will be regarded as an assault on the vital interests of the United States of America, and such an assault will be repelled by *any means necessary, including military force* [emphasis added]." This was a major change in policy with enormous implications. From 1945 on, the United States had relied on other countries like Saudi Arabia, Iran under the shah, and its client state in Israel to provide the political stability (and military might) in the region that would guarantee its "vital interests" (the free flow of oil foremost among them). In fact, this position had become enshrined more generally in the so-called Nixon Doctrine. First announced in July 1969, this policy was aimed primarily at Southeast Asia where the United States was attempting to extricate itself from Vietnam, but it had broader application. The Nixon Doctrine sought to shift the major burden of the fighting in various conflict situations to allied nations, supported with American economic and military assistance (Klare, 2004. But now Carter declared that the United States would be taking on the military burden itself, at least as far as access to oil in the Middle East was concerned. As Bacevich (2016: 29) notes, "implementing the Carter Doctrine implied the conversion of the Persian Gulf into an informal American protectorate. Defending the region meant policing it." Carter then took a concrete step toward implementation of this policy by ordering the creation of the Rapid Deployment Joint Task Force (RDJTF), a military command center that was capable of quickly injecting U.S. troops into the Greater Middle East during a crisis situation (Klare, 2004). The stage was set for more direct military intervention in the Persian Gulf region to protect American access to and control of oil.

Although Bacevich (2016: 245) convincingly argues that the Carter Doctrine "initiated the War For the Greater Middle East by tying the American way of life to control of the Persian Gulf" and its oil, the declaration itself had little immediate impact. Despite the stronger foreign policy stance and the creation of the RDJTF, Carter still lost the 1980 election to Ronald Reagan, and the full military implications of the Carter Doctrine would not be realized until the 1990s. During the 1980s, Reagan's foreign policy did not prioritize the Greater Middle East, and his occasional forays into the region seemed "almost as an afterthought" (Bacevich, 2005: 186). Instead, his administration was preoccupied with waging the Cold War more generally, presiding over a huge military buildup (including a resurgent nuclear arms race with the Soviet Union), and engaging in imperial interventions throughout Central America to roll back "communism." But Klare (2004: 36) argues that Reagan "fully endorsed the basic premise of the Carter Doctrine," and in 1983 he elevated the RDJTF from an ad hoc organization into a "full-scale regional headquarters, the Central Command." CENTCOM, as it is known, would be the military command structure with the enduring responsibility for waging America's war for the Greater Middle East. Reagan also "carried on Nixon's policy of arming our Gulf allies to the hilt," including the sale of Airborne Warning and Control System surveillance aircraft to Saudi Arabia, which, according to Klare (2004: 47), "exposed the flip side of American dependence on Saudi oil: American acquiescence to Saudi demands."

Reagan's sporadic and varied interventions in the Persian Gulf region in the 1980s, which also included sending Marines into Lebanon during the civil war there, ordering a retaliatory bombing of Muammar al-Qaddafi's Libya after a terrorist attack in Berlin, tilting toward Iraq and re-flagging Kuwaiti tankers to protect them in the Persian Gulf during the brutal Iran–Iraq war, and providing covert assistance to the Islamic mujahideen resisting the Soviet invasion of Afghanistan, did not amount to a coherent strategy to enforce the Carter Doctrine. But Reagan did attempt to overcome the Vietnam Syndrome, resurrect American exceptionalism, rebuild the Pentagon, and reorient it on the Islamic world, and his "inconclusive forays in and around the Persian Gulf . . . paved the way for still larger if equally inconclusive interventions to come" (Bacevich, 2005: 193). Still, through the first decade of America's war for the Greater Middle East, Bacevich (2016: 109) concludes that "the very nature of the problem that the U.S. military was expected to solve remained ill-defined"; therefore, "commitments tended to be both modest in scope and revocable." That would change in the 1990s and the American army would be assigned a larger, more central role in the Middle East after the surprising end of the turbulent Cold War.

The fall of the Berlin Wall in 1989 and the unexpected collapse of the Soviet Union in 1991 brought the Cold War to an end, presenting the United States with a new set of opportunities and challenges. With the Soviet Union out of the way and American military supremacy unrivaled, the "unipolar moment," as some called it (Krauthammer, 1991), had arrived. The goals of global market fundamentalism and U.S. imperial domination never seemed more realizable. American military power, already normalized as a primary tool to achieve global hegemony, could now be used with even more political impunity, whether it was punishing small neighbors such as Panama and Grenada for their failure to fall in line with U.S. interests or war in the oil-rich Persian Gulf region.

At first, the end of the Cold War produced a sharp struggle between rival factions of the ruling elite in the United States over how to capitalize on the opportunities afforded by the fall of the Soviet Union while deflecting political "threats" presented by the possibility of a "peace dividend" (a reduced Pentagon budget) or a new American isolationism. One group supported a globalist and internationalist approach, often referred to as "open door imperialism" (Bacevich, 2002; Williams, 1959), which was typical of the administrations of George H. W. Bush and Bill Clinton. The other faction, that included hard-core nationalists and neoconservatives, argued for a more nationalist, unilateralist, and militarist revision of American imperialism, in some ways a return to earlier models of neocolonialism. Both the elder Bush and Clinton viewed the United States as a global leader that should use its economic and military power to ensure openness and integration in the global capitalist system (Bacevich, 2002). Their foreign policy remained consistent with the form of informal imperialism practiced by the United States since the beginning of the twentieth century, stressing global economic integration through free trade and "democracy" (Dorrien, 2004; Williams, 1959). But neither of these two presidents shied away from the use of military violence (in violation of international law) when deemed necessary to accomplish American imperial designs, as the Gulf War of 1991 (under Bush) and the bombing campaigns in Iraq (and the former Yugo-slavia) in the mid-1990s (under Clinton) demonstrate (Bacevich, 2005; N. Chomsky, 2000).

The Persian Gulf War in 1991, known as Operation Desert Storm, rep-resented the first major implementation of the Carter Doctrine and dra-matically escalated America's war for the Greater Middle East. This military operation against Saddam Hussein's Iraq would ratchet up American impe-rial involvement in the region and set in motion social and political forces that would produce devastating long-term consequences. After the 1979 Islamic Revolution in Iran, the United States viewed Hussein as a regional strongman who might provide a geopolitical bulwark against the Iranian mullahs, and Reagan provided support for Iraq during its bloody and expen-sive war with Iran in the 1980s. That war left Baghdad deeply in debt to its small neighbor Kuwait, and when the Kuwaiti government refused to for-give the debt, Hussein began to eye the country's rich oil fields as a way to solve his financial problems. Reagan's successor, the elder Bush, and his administration struggled at first to develop a coherent policy toward Iraq, but when Hussein's army invaded Kuwait on August 2, 1990, the White House quickly concluded that "Iraq posed an indisputable threat to Ameri-ca's interests in the Gulf" (Klare, 2004: 49). Bush was particularly con-cerned about the threat Iraq posed to the oil fields and safety of Saudi Arabia, and he cited America's energy needs and dependence on Saudi oil as major justifications for his decision to use military force to drive Hussein from Kuwait. As Operation Desert Storm got under way in early 1991, the G.H.W. Bush administration offered additional rationales for military intervention such as the need to liberate Kuwait, deter international aggression, and elimi-nate Iraq's weapons of mass destruction. But as Klare (2004: 50) points out, "The record makes it clear, though, that the president and his senior associ-ates initially viewed the invasion of Kuwait through the lens of the Carter Doctrine; as a threat to Saudi Arabia and the free flow of oil from the Gulf."

Operation Desert Storm appeared to be a complete and decisive vic-tory. U.S. forces quickly ejected the Iraqi invaders from Kuwait and restored the emir to his throne. Bush, however, decided not to pursue the Iraqi army all the way to Baghdad and, despite losing the war, Hussein would survive in power. When, in the aftermath of the war, Hussein began to brutally suppress the Iraqi Shiites and Kurds who rose up in opposition to his dictatorial rule (at American instigation), the United States responded by sending troops to Northern Iraq, imposing harsh economic sanctions, and creating the existence of "no-fly zones" across much of the country (all of which did little to actually protect the resisting Iraqis). In a very significant development, American troops that had been "temporarily" placed in Saudi Arabia (despite the great reluctance of King Fahd) during the war to protect the oil fields there from invasion now were to remain in the Islamic holy land, a situation that was deeply offensive to many Muslims. The "blow-back" from this perceived defiling of holy lands by "infidels" would eventually manifest as terrorist attacks against U.S. "interests" abroad, the rise of Osama bin Laden and Al-Qaeda, and ultimately terrorist attacks on U.S. soil (Dreyfuss, 2005; C. Johnson, 2000).

The much-touted victorious outcome of the Persian Gulf War now devolved unsteadily into a policy of containment, the "containment" of the still "dangerous" Hussein. As Bacevich (2016: 133) concludes, "Viewed as 'a war for oil,' which indeed it was, Desert Storm produced a satisfactory yet imperfect outcome." Kuwait had been liberated, the Saudis no longer faced the prospect of an invasion, and the United States had enforced the Carter Doctrine (also overcoming the Vietnam Syndrome along the way). But the menacing Hussein remained in power, and the political stability of the Persian Gulf region remained precarious. The war for the Greater Middle East was just getting started.

Clinton inherited and then continued the Bush policy of containment in Iraq, and during his presidency, the United States was still effectively "at war" with Hussein. Some administration officials referred to the containment approach as "keeping Saddam in his box" (Rieff, 2003). Clinton enforced the "no-fly zones" in Iraq, maintained the devastating economic sanctions, supported the UN-imposed inspection program seeking to find and dismantle Iraq's weapons of mass destruction, and did not shy away from the use of violence to enforce the overall policy of containment. In December 1998, after another in a series of minor confrontations with Hussein, Clinton ordered a major four-day bombing campaign known as Operation Desert Fox. Despite the intensity of this particular attack, it was not an isolated occurrence. During the final two years of the Clinton presidency, the United States "bombed Iraq on almost a daily basis, a campaign largely ignored by the media and thus aptly dubbed by one observer Operation Desert Yawn" (Bacevich, 2005: 196). Despite the daily bombings and the severe sanctions inflicted on Iraq, a botched military operation in Somalia, cruise missile strikes against some terrorist training camps in Afghanistan, and the bombing of a Sudanese pharmaceutical factory allegedly involved in the manufacture of chemical weapons, Clinton's contributions to America's War for the Greater Middle East appeared moderate and restrained compared with the record of the neoconservatives and hard-core nationalists that followed him into office.

The two rival factions of the U.S. ruling elite continued to struggle over how to best capitalize on the opportunities offered by the fall of the Soviet Union. Clinton, like the elder Bush, supported the globalist and internationalist approach. The other faction, the hard-core nationalists and neoconservatives, argued for a more nationalistic, unilateralist, and militarist revision of American imperialism. And they were now poised to take power. The term "neoconservative" was first used in the early 1970s to describe a group of political figures and intellectuals associated with the Henry "Scoop" Jackson wing of the Democratic Party, but in reaction to the cultural liberalism and anti–Vietnam War stance associated with the George McGovern wing of the party, neoconservatives moved to the right, eventually joining the Republican Party (Dorrien, 2004). A number of neoconservatives held positions in the Reagan administration, often providing intellectual justification under the banner of American exceptionalism for that administration's policies of military growth and the rollback of, rather than coexistence with, the Soviet Union (Chernus, 2006). As the Cold War ended, the "neocons" and hard-core nationalists like secretary of defense Dick Cheney occupied positions of power within the G.H.W. Bush administration from which they forcefully promoted an aggressive post-Soviet neo-imperialism (Armstrong, 2002).

Clinton's election in 1992 had removed the neocons from positions within the U.S. government but not from policy debates. From the sidelines they generated a steady stream of books, articles, reports, and op-ed pieces in an effort to influence the direction of U.S. foreign policy. Throughout the Clinton years, the neocons continued to warn about new threats to American security, repeatedly calling for greater use of U.S. military power to address them (Bacevich, 2005; Chernus, 2006; J. Mann, 2004). One persistent theme in these writings was the need to eliminate Hussein's government from Iraq, consolidate American power in the Greater Middle East, and change the political culture of the region (Dorrien, 2004). Once again, American exceptionalism was key to the public justification of this approach. As British journalist Godfrey Hodgson (2009: 171) observes, "By the 1990s the background to the growing obsession with Iraq among neoconservatives was exceptionalist sentiment. Neither Saddam Hussein nor any other foreign leader must stand against the high historic mission of the United States to bring democracy to the Middle East." Such sentiments, of course, had often served as a cover for more critical economic and geopolitical interests such as access to cheap oil.

Neoconservatives subjected the Clinton administration to a barrage of foreign policy criticism, particularly with respect to Clinton's handling of the Middle East and Iraq. In early 1998, the Project for the New American Century (PNAC), a key neoconservative think tank, released an open letter to Clinton urging him to forcefully remove Hussein from power (Halper and Clarke, 2004; J. Mann, 2004). In September 2000, PNAC issued a report titled *Rebuilding America's Defenses: Strategy, Forces and Resources for a New Century.* This report called for massive increases in military spending, the expansion of U.S. military bases, and the establishment of client states supportive of American economic and political interests. The imperial goals of the neocons were clear. What they lacked was the political opportunity to implement these goals. Two unanticipated events would give them that opportunity.

In December 2000, the U.S. Supreme Court awarded the presidency to George W. Bush, despite losing the popular vote by over a half-million ballots. This odd political turnabout would restore the neocons to power, with more than twenty neoconservatives and hard-line nationalists being awarded high-ranking positions in the new administration (Dorrien, 2004). The Pentagon and the office of the new vice president, Dick Cheney, became unipolarist strongholds reflecting the long-standing working relationship between neoconservatives and hard-core nationalists like Cheney and the new secretary of defense, Donald Rumsfeld (J. Mann, 2004). Even though a stroke of good luck had placed them near the center of power, the neocons found that the new president remained more persuaded by "pragmatic realists" in his administration, such as Secretary of State Colin Powell, than by their aggressive foreign policy agenda (Dorrien, 2004). In their view, this was to be expected. The PNAC (2000) report had predicted that "the process of transformation is likely to be a long one, absent some catastrophic or catalyzing event—like a new Pearl Harbor." The unipolarists needed another stroke of good luck to implement their imperial agenda, which included war to control oil in the Middle East.

The September 11 terrorist attacks on the World Trade Center in New York and the Pentagon in Washington, DC (a form of "blowback"), presented the neocons and the hard-core nationalists with the "catalyzing event" they needed to transform their agenda into actual policy. The attacks were a "political godsend" that created a climate of fear and anxiety, which the unipolarists mobilized to promote their geopolitical strategy and mythic ideals to a president who lacked a coherent foreign policy, as well as to the nation as a whole (Chernus, 2006; Hartung, 2004). As former treasury secretary Paul O'Neill revealed, the first goal of the neocons in the second Bush administration had always been to attack Iraq and oust Hussein (Suskind, 2004) to resume America's war for the Greater Middle East. Again, they believed this would allow the United States to consolidate its imperial power in the strategically significant, oil-rich Middle East and, consistent with American exceptionalism, change the political culture of the region and spread "freedom" and "democracy" (Dorrien, 2004; Hodgson, 2009). And now they had their chance to accomplish these imperial goals.

On the evening of September 11, 2001, and in the days following the terrorist attacks, the neocons and hard-core nationalists in the Bush administration advocated invading Iraq immediately, even though there was no evidence linking Hussein to the events (Clarke, 2004; Woodward, 2004). After an internal struggle between the "pragmatic realists" led by Powell and the unipolarists led by Cheney and Rumsfeld, the decision was eventually made to first launch a general "war on terrorism" by attacking the terrorist group Al-Qaeda's home base in Afghanistan and removing that country's Islamic fundamentalist Taliban government (J. Mann, 2004; Risen, 2006; Suskind, 2006). America's War for the Greater Middle East was about to dramatically escalate. The unipolarists in the Bush administration were only temporarily delayed insofar as they had achieved agreement with the realists that as soon as the Afghanistan war was under way, the United States would begin planning an invasion of Iraq (Clarke, 2004; Fallows, 2004). By November, barely one month after the invasion of Afghanistan, Bush and Rumsfeld ordered the Department of Defense to formulate a war plan for Iraq (Woodward, 2004).

Throughout 2002, as plans for the war on Iraq were being formulated, the Bush administration made a number of formal pronouncements that demonstrated that the goals of the unipolarists were now the official goals of the U.S. government. In the January 29 State of the Union address, George W. Bush honed the focus of the "war on terrorism" by associating terrorism with specific rogue states such as Iran, Iraq, and North Korea (the "axis of evil") who were presented as legitimate targets for military action (Callinicos, 2003). In a speech to the graduating cadets at West Point on June 1, the president unveiled a doctrine of preventive war, a policy that many judged as "the most open statement yet made of imperial globalization" (Falk, 2004: 189). This was followed by a new National Security Strategy, which claimed not only the right to wage preventive war but also, again in the spirit of American exceptionalism, that the United States would use its military power to spread "democracy" and American-style laissez-faire capitalism around the world as the "single sustainable model for national success" (Callinicos, 2003: 29). As Indian writer and political activist Arundhati Roy (2004: 56) noted at the time, "Democracy has become Empire's euphemism for neo-liberal capitalism."

In the marketing campaign to build public support for the invasion of Iraq, the Bush administration skillfully exploited the political opportunities provided by the fear and anger over the September 11 attacks. By linking Hussein and Iraq to the wider war on terrorism, the government was able to

establish the idea that security required the ability to attack any nation
believed to be supporting terror, no matter how weak the evidence. This
strategy cynically obscured the real geopolitical and economic motives for
going to war, specifically the control of oil. As sociologist Michael Schwartz
(2008: 5) argues, the "control of oil was a prime goal of the invasion," with
the idea that this control would then lead to "the establishment of the
United States as the preeminent power in the Middle East." Scholar Ahsan
Butt (2019: 1) argues that the Bush administration invaded Iraq "for its
demonstration effect"; the war was motivated by a desire to (re-)establish
American standing as the world's leading power" after the humiliation of
September 11. These were the real, imperial goals of the unipolarists in the
Bush administration as they made their plans for war against Saddam Hus-
sein, which would become the centerpiece of the larger War for the Greater
Middle East.

On March 19, 2003, the United States and Great Britain finally launched
their long-planned, unprovoked, illegal invasion of Iraq and subsequently
inaugurated a formal military occupation of that once-sovereign nation. In
public, the Bush administration's legal and political justifications for this
attack continued to migrate from the eradication of weapons of mass
destruction to the advancement of democracy in the region to fighting back
terrorism. But economic and geopolitical motives, such as controlling Iraqi
oil, reconstructing Iraq's economy into a radical free market system, or reas-
serting hegemonic control of the Middle East, were publicly disavowed by
the administration as objectives of the invasion, despite overwhelming evidence that these imperial goals actually drove the war. As Federal Reserve
chairman Alan Greenspan observed in 2007, "I'm saddened that it is politically inconvenient to acknowledge what everyone knows, the Iraq war is
largely about oil" (T. Anderson, 2011: 231).

Despite Bush's claim of "mission accomplished" in Iraq on the aircraft
carrier USS Abraham Lincoln on May 1, 2003, the United States became
bogged down in a belligerent and deadly military occupation. Efforts to
remake the Iraqi government and policymaking structure were often frustrated by the Sunni–Shia civil war that resulted from the invasion. U.S. oil
companies would gain access to the world's second largest supply of petroleum, but the extremely onerous political and security conditions in the
country often made it very difficult for them to benefit from this control.
The initial war of aggression, followed by years of conflicted military occupation, subjected the Iraqi people to a "tidal wave" of death, destruction,
and misery (M. Schwartz, 2008). By the time Bush left office in January 2009, more than 4,000 American military personnel had been killed in
the war, and according to economists Joseph Stiglitz and Linda Bilmes
(2008), the total cost of the invasion and occupation to American taxpayers

was expected to eventually exceed $3 trillion. Furthermore, as Juan Cole (2019: 6) points out, by lying the country into a war for oil, George W. Bush also "helped put 5 billion metric tons of carbon dioxide a year into the atmosphere."

The situation in Iraq would improve only slightly as President Obama attempted to withdraw combat forces in August 2010, wind down the occupation, and transfer greater power and control back to the Iraqis. While Obama sought to end America's War for the Greater Middle East, in the final analysis, "ending the conflict eluded his grasp, in no small part due to actions on his part that actually expanded and thereby perpetuated it" (Bacevich, 2016: 295). Back in 2002, before the start of this war of aggression, Amr Moussa, head of the Arab League, had predicted that a U.S. invasion of Iraq would "open the gates of hell" (Engelhardt, 2018: 3). It was a prescient observation. This war opened the gates of hell not only in Iraq but throughout the Greater Middle East. The "permanent" American war in the region drags on to this day, with terrible repercussions for many, especially for Iraqis (Englehardt, 2018), but to the great benefit of the American military–industrial complex and the transnational oil corporations.

There is a vast literature on the American War for the Greater Middle East in general and the U.S. invasion of Iraq and its aftermath in particular, and fully examining that literature is beyond my scope here. What I have attempted to do is provide a short history of the global American Empire and demonstrate that, as the United States pursued its imperial goals in the post–World War II era, it engaged in a variety of interventions and wars to gain access to and exert control over Persian Gulf oil. Even though America's War for the Greater Middle East became more complicated over time and included other geopolitical goals, the centrality of oil to the initiation of this war cannot be doubted. And even though American dependence on foreign oil has recently declined, and the era of readily accessible oil and gas is ending, the historic role of the U.S. military in securing energy resources remains critically important. I argue that these resource wars, shedding "blood for oil" as Bacevich (2005) puts it, are not only crimes of empire; they are also climate crimes. Using military force to access and control oil helped fuel the Great Acceleration. It resulted in greater economic growth, consumer consumption, and environmental destruction. And it helped to produce climate catastrophe by locking in the relentless search for and extraction of fossil fuels, ensuring that cheap but dirty forms of energy would be marketed and burned, resulting in the release of greenhouse gases and the heating of the planet.

PENTAGON CARBON EMISSIONS AND
THE TREADMILL OF DESTRUCTION

The rise of the global American Empire after World War II, and the enormous expansion of its military institutions, capabilities, and actions, resulted in a wide variety of political, economic, and social changes around the world. The operation of this permanent U.S. warfare state, with its high-tech weaponry, distant military base system, and vast armada of transportation vehicles, also caused widespread global environmental harms. Fighting wars, of course, destroys human life and property. But in addition, warfare, preparation for war, and the maintenance of a large military institution all also have devastating impacts on the environment (Smith, 2017). Although this critical issue is often overlooked, even by environmental sociologists, research shows that the activities of the military establishment in most countries increases the ecological footprint of the nation-state and exacerbates environmental problems (Jorgenson and Clark, 2009). As environmental sociologist Kenneth Gould (2007: 331) asserts, "militarization is the single most ecologically destructive human endeavor." Worldwide, military operations consume huge quantities of nonrenewable energy and other resources and also generate massive amounts of toxic waste and other pollutants. German environmental analyst Michael Renner (1991: 132) argues that "the world's armed forces are the single largest polluter on earth." Even in the absence of armed conflict, the routine activities of most military organizations degrade the environment in a variety of ways, whether through ecological withdrawals or ecological additions.

The global U.S. empire (warfare state) maintains an exceedingly large military–industrial complex, spends billions of dollars more on "defense" than most of the rest of the world combined, engages in an active, aggressive, and interventionist foreign policy, rings the world with hundreds of military bases to facilitate these imperial interventions, sustains a powerful arsenal of nuclear weapons on land, sea, and in the air, has been bogged down in an endless war (primarily for oil) in the Greater Middle East since the early 1990s, and is beset with an extreme and influential culture of militarism. These military structures, activities, and operations not only generate a huge ecological footprint but also constitute what environmental sociologists John Bellamy Foster, Brett Clark, and Richard York (2010: 345–346) call "ecological imperialism," which they define as "ecological damage wrought by the robbing of third world countries of their resources and the destruction of their environment." Following Renner's observation, Sanders (2009: 21) contends that the U.S. military in particular is "the largest single source of pollution in this country and the world." The larger environmental impact of the U.S. Empire and its culture of militarism is an extremely important and understudied issue, yet an emerging body of social science research shows that "ecological degradation is a concomitant of militarism" (Jorgenson, Clark, and Kentor, 2010: 23), and the Pentagon is the largest and most impactful military institution in the world. Concerning this critical issue, the consumption of fossil fuels, and the resulting production of greenhouse gases by the American military, I argue that the carbon emissions of the U.S. Department of Defense (DOD) can be defined as a climate crime of empire and that this form of crime can best be analyzed through the theoretical lens of the concept of the "treadmill of destruction."

To the list of the core human drivers of global warming (carbon majors), therefore, we can add the large military establishments of the world, specifically the U.S. Department of Defense (DOD). It has been argued that the role of these military organizations has been "undertheorized," and the environmental threats they pose "underestimated," within environmental sociology and the treadmill of production perspective (Hooks and Smith, 2005). As a number of scholars have documented, the Pentagon is the largest single consumer of petroleum and other nonrenewable forms of energy in the United States and perhaps the world (Crawford, 2019; Klare, 2012; Renner, 1991; Sanders, 2009; Santana, 2002). This consumption directly contributes to "carbon dioxide emissions and the emission of other greenhouse gases known to impact global warming and climate change" (Jorgenson and Clark, 2009: 626). Indeed, as Neta Crawford (2019: 1–2) argues in a recent report, "The DOD is the world's largest institutional user of petroleum and correspondingly, the single largest producer of greenhouse gases in the world." Thus, the U.S. military can be added to the roster of major carbon criminals.

It is extremely difficult to measure with any precision the amount of fuel the Pentagon consumes or the quantity of greenhouse gases it emits. The estimates that have been derived are generally quite large, but the scholars who undertake these studies concede the difficulties of carrying out the measurements. Due to the culture of militarism in the United States, the DOD is often treated as a "sacred cow" and is not subjected to the same level of scrutiny and oversight that is common for other federal agencies. As one of the pioneering investigators of this topic, Senior Fulbright Scholar Barry Sanders (2009: 37) notes, "Try as hard as one might, a person cannot find a solid baseline for virtually any military subject, because in great part such certainty . . . does not exist. How much fuel does the military consume? No one knows for sure." And that includes the Pentagon itself. The lack of accountability and openness in this and other agencies of the federal government makes it difficult for scholars, let alone the American public, to obtain information about and develop an understanding of what the state is doing in our name. Despite the difficulties, Sanders, in his groundbreaking

2009 book *The Green Zone: The Environmental Costs of Militarism*, along with an intrepid group of environmental sociologists, have done an excellent job of providing us with a broad overview of the Pentagon's energy consumption and carbon emissions.

Research on these topics is also made more difficult because of the wide range of activities that the military performs and the enormous number of material objects that are necessary to support its missions. Preparations for warfare, as well as actual combat operations, require large numbers of personnel, lots of intense training, high-tech weaponry and other equipment, a large and varied fleet of transportation vehicles, research and development activities, and an ancillary infrastructure. Soldiers must be trained, outfitted, housed, and fed. Bases must be built, maintained, and target-hardened, often using vast amounts of concrete. Air strips and roads must be constructed, and supplies must be transported. Other transportation must be also provided, often to distant locations. Planes, helicopters, aircraft carriers and other battleships, tanks, and other armed vehicles must be developed, manufactured, and maintained. High-tech weapons, machinery, and other equipment must be produced, tested, and kept ready. The provision of all these things, of course, consumes massive amounts of oil, gas, and other nonrenewable resources. It also frequently generates hazardous waste and other forms of pollution. And, most important for this analysis, it produces great quantities of greenhouse gases that contribute to global warming.

Despite the difficulties of measurement, Sanders (2009: 50) concludes that the U.S. military is the largest single consumer of petroleum in the world, "using enough oil in one year to run all of the transit systems in the United States for the next fourteen to twenty-two years." He adds that not only does the Pentagon consume "inconceivable amounts of gasoline," it also consumes "one quarter of the world's jet fuel." Sanders (2009: 66–67) calculates that Americans consume roughly twenty million barrels of oil a day, with the U.S. military alone consuming more than one million barrels. Translated into gallons of fuel, he estimates that each deployed soldier consumes almost twelve times the amount that the average civilian uses (15 gallons of fuel a day to 1.3 gallons). Environmental sociologists Brett Clark, Andrew K. Jorgenson, and Jeffrey Kentor (2010), concur with other scholars that the Pentagon is the world's largest consumer of petroleum, and in a recent cross-national study they found that high-tech militarization, measured as military expenditures per soldier, increases the scale of energy consumption. They also found that "total energy consumption is positively associated with the relative size of military troops" (Clark et al., 2010: 38). The Union of Concerned Scientists (2014) has also provided an estimate that the U.S. military consumes 100 million gallons of fuel every year. As the UCS points out, that is enough fuel to drive the average SUV around the earth four million times. All of these findings support the assertion that militarization expands energy consumption.

The fact that high-tech militaries, like the Pentagon, consume large amounts of fossil fuels also means that they release large amounts of greenhouse gases and therefore contribute to the accumulation of carbon dioxide in the atmosphere (Crawford, 2019). Thus, high-tech militaries, in particular the DOD, can be identified as major human drivers of global warming that bear significant responsibility for climate disruption. As Sanders (2009: 22) puts it, "The military—that voracious vampire—produces enough greenhouse gases, by itself, to place the entire globe, with all its inhabitants large and small, in the most imminent danger of extinction." Sanders's data support the assertion that the U.S. military can be considered a major carbon criminal. Through its imperial actions around the world, it releases huge amounts of carbon dioxide that significantly contribute to global warming, a climate crime of empire.

How much carbon dioxide does the Pentagon release? As with the energy consumption of the military, it is difficult to measure. Using the figure of one million barrels of oil a day for U.S. military energy consumption, Sanders (2009: 68) calculates that "the combined armed forces sends into the atmosphere about 400 million pounds of greenhouse gases a day, or 200,000 tons. That totals 146 billion pounds a year—or 73 million tons of carbon a year." In 2009, this figure for Pentagon emissions (the emissions of a single organizational actor) represented a significant proportion, over 5 percent of the total 30 percent of the world's greenhouse gases traced to the United States. Jorgenson, Clark, and Kentor (2010: 22), in their panel study of carbon dioxide emissions and ecological footprints of nations, conclude that military expenditures per soldier and military participation rate "positively affect both total and per capita carbon dioxide emissions. Thus, it appears that, all else being equal, nations with more high-tech and labor intensive militaries emit relatively higher overall levels and greater intensities of anthropogenic carbon dioxide gas." In a follow-up study, Jorgenson and Clark (2016: 512) found that an increase in military expenditures as a percentage of gross domestic product also led to "a larger increase in consumption-based carbon dioxide emissions in OECD [Organization for Economic Cooperation and Development] nations than in non-OECD nations." As the largest high-tech military in an OECD nation in the world, with an empire of bases ringing the globe, a pattern of enormous consumption of fossil fuels, and a long history of imperial interventions and wars, the U.S. Pentagon is one of the largest (if not the largest) single-source contributors to greenhouse gas emissions since World War II.

In addition to directly contributing to the consumption of fossil fuels and the generation of carbon emissions, militarization also appears to influence the willingness and timing of nations to ratify international environmental treaties, such as the Kyoto Protocol (Givens, 2014). Even though the United States never ratified the protocol, during an early round of the negotiations American officials insisted on and achieved the ability to opt out of fully reporting on or taking action concerning the Pentagon's greenhouse gas emissions. This provision was then locked in through a National Defense Authorization bill in 1999. U.S. officials knew that the carbon emissions of the Pentagon were huge, and they did not want those emissions recognized or acted upon in any international agreement. The blanket exception for Pentagon emissions in international climate agreements would have ended with the Paris Agreement (Nelson, 2015), but Trump's attempt to withdraw the United States from this critical accord may continue to shield the Pentagon's substantial contribution to global warming from critical scrutiny.

As environmental sociologists began to analyze the general destructive impact of war and militarism on the environment, and more specifically the consumption of fossil fuels and the production of greenhouse gases by military institutions, many found that the treadmill of destruction theory was a useful perspective for developing an understanding of the relationship between the military and the environment. Treadmill of destruction (TOD) theory argues that there is a distinct expansionary dynamic associated with war and militarism that is rooted in the state and its geopolitical interests, which, though related, cannot be reduced solely to the capitalist treadmill of production. This theoretical approach was developed by Gregory Hooks and Chad L. Smith (2004: 558) in an attempt to "recast the environmental sociology literature by specifying the scope conditions under which a 'treadmill of production' and a 'treadmill of destruction' are applicable." While the treadmill of production perspective explains the accelerating impact of society on ecological conditions due to economic factors such as capitalist competition, the quest for profits, and the search for increased market share, TOD theory points to a different underlying logic: the military as a social structure and the imperatives of militarism. As Hooks and Smith (2005: 21) explain, "The 'treadmill of destruction' draws attention to a distinct expansionary dynamic that also generates additions to and withdrawals from the environment . . . expansionary dynamics associated with war and militarism; geopolitical competition and arms races."

Environmental sociologists who examine the ecological footprints of nations, the environmental impacts of militarization, the energy consumption of military institutions, and the carbon emissions they generate often draw on the TOD approach to provide an explanatory framework for their research—and they have found solid empirical support for the theory (Clark and Jorgenson, 2012; Clark et al., 2010; Givens, 2014; Jorgenson and Clark, 2009, 2016; Jorgenson, Clark, and Givens, 2012; Jorgenson et al., 2010). As

Clark and Jorgenson (2012: 566) note, "The treadmill of destruction perspective within sociology makes an important contribution highlighting how the expansionary practices of the military produce a system that is highly resource consumptive and waste generating—not to mention it creates distinctive forms of environmental inequality." Jorgenson, Clark, and Givens (2012: 331) also argue that "A full understanding of the human dimensions of global environmental change . . . requires an examination of the world's national militaries, and the treadmill of destruction perspective provides an appropriate analytical lens for such assessments." I concur with these scholars that TOD theory provides important insights into the climate crimes of military institutions, particularly the Pentagon, but would emphasize again that the state–corporate crime literature, the history of the global American Empire, and the political economy analysis presented throughout this book point to the importance of understanding the interrelationships between economic and political factors, the deep structural relation between corporations and the state. It is the dialectical relationship between private business interests (including fossil fuel corporations) and the American political system (including the military) that generates the historic pursuit of empire, produces specific imperial wars and military interventions, creates a deeply rooted and influential culture of militarism in U.S. society, and results in the various climate crimes of empire.

Militarized Adaptation: "The Politics of the Armed Lifeboat"

Another critical issue to examine concerning the relationship between the U.S. global empire (with its culture of militarism) and climate change is the political decision to adopt militarized forms of adaptation to the harmful ecological and social consequences of global warming. Recall that two of the central issues in the climate change literature are *mitigation* and *adaptation*. Mitigation refers to the necessity of a drastic reduction of carbon emissions, while adaptation is generally understood as "efforts made to reduce the vulnerability of human, natural and social systems to the impacts of climate change" (Buxton and Hayes, 2016: 11). In *Eaarth: Making a Life on a Tough New Planet*, McKibben (2010b) argues that, due to the climatic changes that have already been generated by global warming, we live on a planet today (which he calls Eaarth) that is significantly different from the one that existed at the end of World War II, prior to the Great Acceleration. Activists like McKibben, who accept that climate change is already under way and will greatly intensify in the future, contend that it is a moral imperative that we attempt to protect the vulnerable from harm and explore the least destructive ways to adapt to these catastrophic changes, while simultaneously seeking to mitigate the causes of global warming.

Specific forms of adaptation to climate disruptions can take one of two general paths. The first path seeks positive, progressive, cooperative, socially and economically just forms of adaptation that protect the vulnerable from harm, particularly those in the Global South, and assist them financially and technically in responding to life-altering changes. The other approach involves militarized, violent, and repressive forms of adaptation, Parenti's (2011) "politics of the armed lifeboat." In this case, adaptation is twisted and reinterpreted to mean protection for the wealthy and powerful in the Global North, rather than for the victimized and vulnerable in the Global South. I argue that the exclusion of just and cooperative forms of adaptation to climate change from political discourse and action, and the resort to militarized and violent forms of responding to the harmful effects of global warming, are both climate crimes insofar as they will bring predictable and avoidable harms to large portions of the human population, social systems, and ecosystems in order to benefit corporations and the elite in the richest and most powerful nations of the world.

The people and nations of the Global South will be the primary victims of these state–corporate climate crimes. According to the IPCC (2013), climate disruption will be far more devastating for populations in less developed countries as they depend more heavily on the environment for subsistence, already face problems of food insecurity, desertification, and limited access to potable water, often have low levels of arable land relative to population, and lack levels of technological development that could ameliorate the impacts of climate change. But the impacts of global climate disruption in less developed countries will extend well beyond the boundaries of devastated areas in the Global South. Insofar as human populations typically do not accept their demise passively, we can anticipate substantial climate-induced migration from these areas as the effects of global climate change deepen. These migrations pose additional risks and harms. In addition to the deep disruption to the lives of those who feel compelled to migrate due to climate-induced environmental changes, these movements carry a significant threat of violent conflicts as well (Crank and Jacoby, 2015; Parenti, 2011; Wallace-Wells, 2019). This potential for violent conflict, as well as the demand for economic resources and social services for climate refugees, is perceived by many in the Global North as a broad "security threat" that requires a resort to military methods to "control" borders.

Despite the increasingly negative social impacts, economic effects, and political threats related to global warming, genuinely progressive, fair, and cooperative adaptation to global climate change was not even on the agenda of the international political community until the G77—the group of developing nations of the world—demanded adaptation assistance and funding from the rich nations who are primarily responsible for the problem at the 2009 Copenhagen conference. These urgent requests, however, became a significant stumbling block to an agreement because governments of the Global North refused to acknowledge that they owed an ecological debt to the developing nations in the Global South, let alone that they should act to reduce this debt. The states (and corporations) that derive the most benefits from the global capitalist economy, and are at the same time most responsible for generating global warming, have generally refused to participate in any adaptation efforts that require "economic redistribution and development" or "a new diplomacy of peace building" (Parenti, 2011: 11).

According to environmental sociologists, reducing existing and future vulnerability to climate disruptions will require a wide variety of structural, institutional, and societal measures (Carmin, Tierney, et al., 2015). In particular, these sociologists highlight issues of providing adaptation finance, bridging climate and economic development, planning for disaster risk reduction, and the critical concern of managing human migration. While a number of efforts at technical adaptation to climate change are under way, primarily led by local governments (Flannery, 2015; Hertsgaard, 2011b) and "cool cities" (Barber, 2017) in the North, they are too few, too localized, widely scattered, and often underfunded. Moreover, many of these efforts do not address the significantly greater impact global warming will have on the human populations of the less developed countries of the Global South. In general, the overall political record of the nations of the Global North on adaptation issues is one of failure (Parenti, 2011, 2016a). Instead of planning for and adopting progressive, cooperative, and just adaptation policies, the world's most powerful states, the ones most responsible for global warming, have attempted to reframe climate change "from an environmental and social justice issue to a security issue" (Buxton and Hayes, 2016: 4) that usually demands a military response.

While the failure to adopt more cooperative, economically fair, and peaceful measures of climate adaptation can be defined as a crime of political omission, I argue that states in the Global North (particularly the U.S. global empire) are also guilty of the direct commission of violent crimes by responding to climate change with exclusionary immigration policies, expanded border enforcement, walls and surveillance, arrests and expulsions, political repression, and militarized policing methods (Buxton and Hayes, 2016; Miller, 2017, 2019; Parenti, 2011). This militaristic form of adaptation to the climate crisis, which often specifically targets climate refugees, is a blameworthy harm that can be defined as a climate crime of empire.

In *Tropic of Chaos: Climate Change and the New Geography of Violence*, Parenti (2011: 7) describes how "the current and impending dislocations of climate change intersect with the already-existing crises of poverty and violence," a collision of global poverty and violence with climate change that he calls "the catastrophic convergence." Among many harmful impacts, this "convergence" generates climate migrants or refugees, defined as "persons or groups of persons who, for reasons of sudden or progressive climate-related change in the environment that adversely affects their lives or living conditions, are obliged to leave their habitual homes either temporarily or permanently, and who move either within their country or abroad" (Environmental Justice Foundation, 2017: 6). The problem, as Parenti (2011) defines it, is that states in the Global North are responding to the catastrophic convergence (which they are significantly responsible for) and the climate migrations it causes primarily with militarism, violence, and repression.

It is difficult to measure the magnitude and severity of climate migration, what some have called *climigration* (G. White, 2011: 20), with any precision. More than two decades ago, the IPCC (1990: 2) warned that the "greatest single impact of climate change could be on human migration with millions of people displaced by shoreline erosion, coastal flooding and agricultural disruption." Oxford University professor Norman Myers predicted in 2005 that there would be fifty million climate refugees by 2010, an estimate that served as the basis for statements by the United Nations (Crank and Jacoby, 2015). A few years later those estimates dramatically escalated. Kramer and Michalowski (2012: 83) cite a 2009 report by the Asian Development Bank that concluded that in the Asia/Pacific region alone anywhere from 700 million to one billion people "will come under substantial pressure to migrate (temporarily or permanently, and internally or across borders)" due to climatological disruptions to shorelines and food systems. Parenti (2016a) cites a 2014 study from Columbia University's Center for International Earth Science Information Network that also projects that 700 million climate refugees will be on the move by 2050 (although most will not cross borders). The Environmental Justice Foundation (2017: 6) reports that in 2016 just one form of climate disruption, extreme weather-related disasters, "displaced around 23.5 million people" and that "since 2008 an average of 21.7 million were displaced each year by such hazards." Whatever the actual numbers of climate migrants, and the figures are highly contested (B. Hayes, Wright, and Humble, 2016; G. White, 2011), the governments and militaries of the Global North anticipate that climate-induced migration will increase dramatically, cause an "unimaginable" refugee crisis, and bring increased conflict and security threats (Carrington, 2016).

And there are some data to support that concern over the potential for climate change to cause violent conflict. According to estimates by political scientist Rafael Reuveny (2007), between 1960 and 1990 there were thirty-six violent conflicts resulting from or exacerbated by climate-induced migration due to increased competition for resources, intensified ethnic

tensions, inter- and intra-governmental distrust, and deepening sociopolitical fault lines. These data cover only the earliest possible impacts of global
climate change. Criminologists John Crank and Linda Jacoby (2015: xiii)
have further analyzed the connections between global warming, crime, and
violence, and they argue that "the impact of climate change on international security is likely to be substantial, with outcomes for nation-state
stability and security becoming increasingly unfavorable." As the catastrophic convergence analyzed by Parenti (2011) deepens, we can expect a
significant increase in climate migration–induced conflicts, violence, and
wars. And, in the absence of genuinely progressive, cooperative, and socially
just adaptations to global climate change, we can expect that militarized
adaptation will be embraced by the Global North. As Parenti (2011: 11)
warns, "One can imagine a green authoritarianism emerging in the rich
countries, while the climate crisis pushes the Third World into chaos.
Already, as climate change fuels violence in the form of crime, repression, civil
unrest, war, and even state collapse in the Global South, the North is responding with a new authoritarianism. The Pentagon and its European allies are
actively planning a militarized adaptation, which emphasizes the long-term,
open-ended containment of failed or failing states—counterinsurgency
forever." I would argue that the embrace of this authoritarianism, this militarized form of adaptation to climate-induced migration by the United
States and other countries of the Global North, is significantly related to the
history of the American Empire and its culture of militarism.

The development of the U.S. warfare and national security state, its
imperial interventions to secure geopolitical and economic interests, the
rise of the military–industrial complex, and the celebration of military values and institutions in the political culture did indeed create the conditions
for viewing climate refugees through the lens of militarism. Parenti (2011: 8)
references this history when he argues that "In the case of climate change,
the prior traumas that set the stage for bad adaptation, the destructive social
response, are Cold War–era militarism and the economic pathologies of
neoliberal capitalism." Throughout the post–World War II era, the pursuit
of empire, the confrontation with the Soviet Union, the attempts to secure
access to cheap and abundant oil, and the advancement of global capitalism
ended up distorting "the state's relationship to society—removing and
undermining the state's collectivist, regulatory, and redistributive functions, while overdeveloping its repressive and military capacities" (Parenti,
2011: 8). This has resulted in the political exclusion of progressive and just
adaptations to climate disruption and the adoption of border enforcement
and other forms of military security for dealing with climate refugees.

This political recasting of the problem of climate change adaptation can
be empirically observed through an examination of planning documents

and reports from the Pentagon and various national security think tanks (Bonds, 2015; Miller, 2017) and the actual border policies that have been implemented by the U.S. government in the past twenty years. One of the first reports to consider the national security implications of the impacts of climate change was a 2003 Pentagon-commissioned study titled *An Abrupt Climate Change Scenario and Its Implications for United States Security*. The authors, Peter Schwartz and Doug Randall (2003: 18), forecast a dark future of climate catastrophes, disruption, and conflict and then predicted that states like the United States and Australia would likely have to build "defensive fortresses" around their countries and strengthen borders "to hold back unwanted starving immigrants from the Caribbean Islands (an especially severe problem), Mexico, and South America." The idea that fortresses need to be built around countries in order to keep starving people out at gunpoint is a clear expression of the politics of the armed lifeboat. Many of the reports that followed argued that climate change would act as a general "threat multiplier." This concept first appeared in a 2004 report by a UN High-Level Panel on Threats, Challenges and Change but was given prominence in U.S. military circles through a 2007 document from the Pentagon-connected think tank the Center for Naval Analyses (CNA). In *National Security and the Threat of Climate Change*, the CNA (2007: 7) pointed out that, "unlike most conventional security threats that involve a single entity acting in specific ways and points in time, climate change has the potential to result in multiple chronic conditions, occurring globally within the same time frame." Another influential think tank report in 2007 also contributed to the emerging consensus among military officials in the United States that climate disruption was a threat to national security. *The Age of Consequences: The Foreign Policy National Security Implications of Global Climate Change* (Campbell, Gulledge, et al., 2007), produced by the Center for Strategic and International Studies and the Center for a New American Security, analyzed a number of scenarios related to climate disruption that may exacerbate a significant number of existing international problems and produce a variety of conflicts that could pose a national security threat to the United States.

Given the fact that so many conservative Republicans are climate change denialists, some environmental activists were happy to point out that the U.S. military, usually considered a conservative bastion, actually believes that global warming is real and needs to be addressed through government policy. As ammunition in political debates, the activists could argue that some conservatives do take the threat of climate disruption seriously and advocate political action to deal with the problem. These environmentalists hoped that the national security frame could be broadened and eventually push policymakers to seriously grapple with the causes of climate change (and consider mitigation strategies). But as Bonds (2015: 210) observes, "the record so far is that military planners will opt to sidestep this complicated political problem to instead focus on more technical questions about how to deal with the new difficulties and opportunities that will arise in a warming world." Bonds reviewed eighteen high-level U.S. military planning documents from 2007 to 2014 that reference climate change as a security threat and found that only two of these reports "even mention that Congress could act to reduce threats associated with climate change by significantly cutting greenhouse gas emissions." Although there is evidence that framing climate disruption as a national security issue has raised more general awareness of the problem of global warming and has had some influence on public opinion, "an overemphasis of climate change as a national security threat risks underappreciating the interdisciplinary and interdependent nature of successful adaptation solutions" (King, 2011: 3). Furthermore, Bonds (2015: 209) argues that when global warming is viewed through a national security frame, it provides "a new legitimation for U.S. global militarism, just at the time when the United States needs to shift public resources from funding soldiers and weaponry to instead building the green infrastructure that is required to meaningfully address the climate crisis."

Despite the hopes and concerns of climate activists and environmental academics, the U.S. government continues to define climate disruptions and conflicts more narrowly as national security threats that require a military or law enforcement response. In the 2008 National Defense Authorization Act, for example, Congress directed the Pentagon to consider the effects of climate change on national security in the Quadrennial Defense Review (QDR), a legislatively mandated review of DOD strategies and priorities. Both the 2010 and 2014 QDR robustly addressed the concern, again identifying climate disruption as a "threat multiplier" that must be taken seriously by the U.S. military (Miller, 2017; Parenti, 2011). Obama's first comprehensive National Security Strategy in 2010 also predicted that climate change will lead to new conflicts over resources and refugees and global suffering from the ecological impacts of global warming (King, 2011). And in January 2016, under the Obama administration, the DOD issued a very broad directive titled "Climate Change Adaptation and Resilience" that required all units and departments within the Pentagon to take climate change into account in their strategic planning (U.S. Department of Defense, 2016). But even the notion that climate change is a national security threat is too much for the extreme denialists in the Trump administration. Late in 2017 Trump dropped climate change from his administration's National Security Strategy (Borger, 2017), although immigrants would continue to be targeted and borders fortified for other political reasons.

The political efforts to define climate disruptions in general, and climate refugees in particular, as national security threats were reflected over the years in specific government policies that expanded border enforcement and criminalized immigrants. Border-fortification strategies, of course, are not new to the era of climate change. Throughout American history, restrictionist policies on immigration have been implemented, often motivated by xenophobia, racism, and smug nationalism or rationalized as promoting respect for the law, defending the "homeland" against terrorists, and protecting American workers. But in recent decades border-policing strategies have also been significantly influenced by the perceived "threat" of climate change. According to sociologist Timothy J. Dunn (1996), the militarization of the U.S.–Mexico border was already well under way in the 1980s and early 1990s. However, the hiring of new federal agents to work on the border, budget increases for agencies charged with border policing, and the deployment of new fortification methods and surveillance technologies were all ratcheted up during the second Bush and Obama administrations. As journalist Todd Miller (2017: 59) points out, despite all his campaign bluster about building "the wall" and deporting "illegal" immigrants, when Trump entered office in 2017, "at his disposal was the most massive border enforcement apparatus in United States history, built on turbo charge for more than 20 years." Miller goes on to note that Trump had at his command more U.S. Border Patrol agents than ever before in American history (around 21,000) and that the Customs and Border Protection, with more than 60,000 agents, was the largest law enforcement agency in the country. The 2017 border and immigration budget, including funding for the Immigration and Customs Enforcement agency, was $20 billion, a huge increase from 1990 (Miller, 2017: 59).

Before Trump even started his presidential campaign, this imposing border enforcement regime had already attempted to close off traditional crossing points into the United States by cutting through the desert landscape to erect hundreds of miles of walls and barriers. Razor wire, checkpoints, and armed patrols were in use. Surveillance technologies such as cameras, radar, and drones were deployed. Those caught crossing could be arrested without charges, placed in detention centers or relocation camps, and occasionally subjected to torture and other forms of violence. These efforts had deadly consequences. As historian and activist Aviva Chomsky (2018: 3) notes, "The significant fortifications already in place on the U.S.–Mexican border have already contributed to the deaths of thousands of migrants." Many of those who did make it across the increasingly militarized border were later subjected to arrest, detention, and deportation. Between 2009 and 2015, the Obama administration removed more than 2.5 million people through immigration orders, leading some immigration rights activists to refer to Obama as the "Deporter in Chief" (ABC News, 2016).

A significant feature of this border security system is the important role that private contractors play in its construction and operation. Taking off from the broader concept of the military–industrial complex (Hartung, 2011; Turse, 2008), this border apparatus is often described as the *security–industrial complex* (Buxton and Hayes, 2016; Engelhardt, 2014, 2018; Parenti, 2011). Tom Engelhardt (2014: 80) labels the private contractors who are deeply involved in the overall national security state as "warrior corporations." These security-oriented corporations provide a broad array of services and supports to the Pentagon in Afghanistan, Iraq, and other sites of imperial intervention, as well as aiding the federal government's harmful policing efforts along the U.S.–Mexico border. This nexus of state actors (military organizations and federal law enforcement agencies) and warrior corporations strongly influences the politics of the armed lifeboat that is directed at climate refugees. Parenti (2011: 14) observes that "As a politics of climate change begins to develop, this matrix of parasitic interests has begun to shape adaptation as the militarized management of civilization's violent disintegration." Given this intersection of business and government interests, militarized adaptation to climate disruption can be described as a form of state–corporate crime.

All of the climate crimes that involve the American military need to be placed in the context of the operation of the U.S. global empire. Whether we are analyzing imperial interventions and wars for access to and control over oil, the carbon emissions spewed out by the Pentagon, or militarized forms of adaptation to climate-induced migration and conflict, the role of empire remains paramount. Activist Harsha Walia (2013) calls the creation of a militarized security apparatus directed at climate refugees "border imperialism;" and in *Empire of Borders*, Todd Miller (2019) has recently analyzed the expansion of the U.S. border around the world. Miller (2018a: 3) also puts militarized adaptation in the context of empire when he argues that "The border has indeed become a place where the world's most powerful military faces off against people who represent blowback from various Washington policies and are in flight from persecution, political violence, economic hardship, and increasing ecological distress." Militarized adaptation to climate change is a harmful, unethical, and ultimately unworkable policy that favors the rich and powerful and turns the victims of empire into "offenders" and thus victimizes them yet again. It is a policy that disregards the larger structural causes of global warming and climate migration. It ignores the issue of mitigation. And even though the Trump administration has stripped the term *climate change* from its National Security Strategy, it has continued and intensified the overall hard-line border enforcement policy. During the 2018 midterm election campaign, in an effort to motivate his xenophobic base, Trump continually demonized a so-called caravan of

migrants traveling from various places in Central American through Mexico toward the United States to seek legal political asylum. In an expensive
political stunt, Trump even deployed active-duty troops to the border with
an authorization to use lethal force to stop desperate people fleeing climate
disruption and violence (much of this violence is, of course, related to previous U.S. interventions in the region). As Miller (2018b) pointed out, the
migrant caravan story during the 2018 midterm election was "a climate
change story." It is also a climate crime story.

CHAPTER 7

The "Climate Swerve"

HOPE, RESISTANCE, AND CLIMATE JUSTICE

We will not be able to prevent a good deal of very harmful climate change that is either already manifest or soon to strike us. But we have a genuine capacity for preventing the most extreme forms of catastrophe. That is why, from now on, our actions are always and never too late.
—Robert Jay Lifton, *The Climate Swerve: Reflections on Mind, Hope, and Survival* (2017: 148)

Yet, as deep as our [climate] crisis runs, something equally deep is also shifting, and with a speed that startles me. As I write these words, it is not only our planet that is on fire. So are social movements rising up to declare, from below, a people's emergency.
—Naomi Klein, *On Fire: The (Burning) Case for a Green New Deal* (2019: 21)

ON APRIL 28, 2017, just three months after Donald Trump was inaugurated as the forty-fifth president of the United States, my wife Jane and I drove 600 miles from our home in Kalamazoo, Michigan, to Gaithersburg, Maryland, a city that is 22 miles northwest of Washington, DC. Just as in 1971 when I was a college student, we were heading to the Capital to protest. After spending the night at a local hotel, Jane and I drove to the nearby Shady Grove Metrorail stop the next morning to catch the subway train into Washington to participate in an important political event. At each stop our car and most of the others began to fill up with excited and concerned citizens from all over the country, many carrying political signs, all headed into the city for what would turn out to be one of the largest climate change marches in U.S. history. "We Resist, We Build, We Rise" was the slogan the Peoples Climate Movement selected for the April 29 demonstration in Washington (and more than 300 other sites across the country and around the world), which was called to protest the environmental policies of the climate change–denying Trump administration. We exited the Metro train at Union Station in the heart of DC and headed toward Capitol Hill, the staging area for the march. A conservative estimate later placed the size of the demonstration at more than 200,000 people. As we marched in 90-degree heat (in April!) past the White House and around to the Washington Monument grounds, my wife carried a sign that read: "We're Marching for Our Grandsons' Future." One side of my sign pleaded to "Stop Climate Crimes" and the other called for "System Change, Not Climate Change." The marchers, the signs, the speeches, and the music that day all conveyed both resistance and hope—resistance to the continuing harmful actions of the fossil fuel corporations and the political omissions of the U.S. government and hope that the people of the world could still create a mass social movement powerful enough to compel leaders to reduce global warming and achieve social justice for those who experience the worst impacts of the climate crisis but have done the least to cause the problem.

On a personal level, our presence at the Peoples Climate March that hot April day in 2017, and the messages we carried, signified our grave concern with the critical issue of climate change. Jane and I were motivated to drive hundreds of miles to Washington, DC, and participate in this massive political protest because we truly do fear for the future of our three grandsons and want to protect them as much as possible from future climate disruptions that could affect their lives in very harmful ways. We would like the boys—Truman, Malcolm, and Calvin—to be able to grow up in a stable climate, as we were able to do, not having to worry about the ecological and social consequences of global warming. As Noam Chomsky (2013: 16) has observed, "We don't have to be lunatics who are willing to sacrifice our grandchildren so that we can have a little more profit."

While concern for our grandsons provided the primary motivation to join the resistance and march that day, the sign I carried provided, at least in a very truncated form, a hint at the criminological and sociological analysis that I have presented more fully in this book. I have argued that continuing to extract and market fossil fuels that release Earth-warming greenhouse gases into the atmosphere when burned, refusing to take the necessary political actions required to drastically mitigate carbon emissions and adapt to climate disruptions in socially just and progressive ways, promoting dishonest climate change denialism to block political measures on this critical issue, and engaging in various climate harms related to the pursuit of empire all constitute ongoing climate crimes that must be resisted, reduced, or prevented. I also take the position that the carbon criminals responsible for these state–corporate crimes should be held legally and morally accountable for the harms they inflict. And since climate crimes are rooted in the social formation of neoliberal global capitalism and its destructive treadmill of production, I argue that for the mitigation of carbon emissions to occur and climate justice to prevail, the political–economic system must be challenged, effectively regulated, and ultimately restructured. This book is an attempt to elaborate on those short, clipped messages my wife and I displayed on our signs as we marched for climate justice and our grandchildren that day in Washington.

"THE POSSIBILITY OF HOPE": SPECIES AWARENESS AND THE CLIMATE SWERVE

But can climate crimes be stopped? Can society really do anything to halt global warming and reduce the risk of further climate disruption? Do we have any hope of drastically cutting greenhouse gas emissions, transitioning to a system of clean, renewable energy, and adapting to the harmful climatic changes that are already occurring in a way that is just and cooperative? Can a criminological and sociological analysis provide any assistance to the political effort to resist climate crimes, hold carbon criminals accountable, promote system change, and achieve climate justice? Given the mounting and extremely alarming scientific evidence of catastrophic climatic harm, it is easy to become pessimistic and resigned to our fate when considering these questions. Many thoughtful people who have analyzed the problem have abandoned hope for a solution. In the first volume of his massive two-volume tome *Carbon Ideologies*, American journalist William T. Vollmann (2018: 3) begins by stating that he will not offer any solutions to save the world as we know it from climate disruption because "nothing can be done to save it; therefore, nothing need be done." In the opinion of Nathaniel Rich (2018b: 1), that makes *Carbon Ideologies* "one of the most honest books yet written on climate change." In his thought-provoking book *Reason in a Dark Time*, subtitled *Why the Struggle against Climate Change Failed—and What It Means for Our Future*, philosopher Dale Jamieson (2014: ix) argues for "realism" and declares, "We are stuck with climate change. This book is about what it is, why we are stuck with it, and what we can learn from our failures to get out of the ditch." Adding to the pessimism of these thinkers, the landmark report from the United Nations Intergovernmental Panel on Climate Change warned in early October 2018 that due to the accumulating scientific evidence of accelerating climate disruptions, urgent and unprecedented actions to transform the world economy were necessary to limit global warming to 1.5 degrees Celsius (Davenport, 2018). This "very grim forecast" as Bill McKibben (2018b) called it, suggested that the international community had only around twelve years to avert further catastrophe, deepening the despair among some environmental activists and other concerned citizens (Watts, 2018).

Yet, even in the face of dire scientific reports and grim forecasts, I argue that we should not succumb to resignation. In this final chapter, I want to make the case for fostering hope, engaging in critical analysis of our current ecological crisis, and participating in political and social actions to resist the climate crimes of the powerful and strive for social justice for those most afflicted by the catastrophic effects of global warming. Like Noam Chomsky (2017), I believe that we should choose "optimism over despair" and maintain what Howard Zinn (1994) called "the possibility of hope." I am not a "Pollyanna." We do indeed face a very dangerous and uncertain situation, and the October 2018 IPCC report indicates that the window of opportunity to take meaningful action is rapidly closing. Australian scientist Tim Flannery (2015: ix) has argued that "if we are to have real hope, we must first accept reality"; we must understand the hard facts about "our climate predicament." I, too, believe that we must unflinchingly confront what American journalist Chris Hedges (2018) calls "the greatest existential crisis of our time," the "tragic reality before us." As we face that reality, we must understand that there are no easy answers or simple solutions to the climate crisis. There is no quick fix for our predicament. It is already too late to stop certain forms of dangerous climate disruption from occurring—too late to prevent a large amount of human suffering due to global warming. But as famed American psychiatrist, historian, and public intellectual Robert Jay Lifton (2017: 148) points out in this chapter's epigraph, "it is always and never too late" to take meaningful action on harmful climate change. Yes, a great deal of pain and suffering is now inevitable. It cannot be prevented at this late hour. In that respect, what we do now is *always too late*. But Lifton argues that we still have "a genuine capacity for preventing the most extreme forms of catastrophe." In other words, it is *never too late* to do what we still can to maintain a viable climate and reduce human suffering. That perspective, that hope, guides the analysis and advocacy in this chapter.

Not only does Lifton (2017: ix) insist that it is never too late to act to reduce the most extreme forms of climate catastrophe; he also argues that around the world today there is an "evolving awareness of our predicament," an awareness that he calls the "climate swerve." The "climate swerve," he declares, "creates a mind-set capable of constructive action, and is a significant source of hope." The Paris Climate Conference in 2015, he asserts, "was a stunning demonstration of universal awareness of the danger of global warming" and a "display of species awareness" that is "unprecedented, and holds out the possibility that we humans may extricate ourselves from extreme climate catastrophe" (Lifton, 2017: xi). While there is "profound resistance" and at times a "vicious backlash" to the emerging climate swerve, especially as manifested in the extremely reckless and reprehensible political actions of the Trump administration, Lifton (2017: 154) still believes that our evolving species awareness can take us on "a survivor mission of preserving our habitat and embracing genuine forms of adaptation," a mission in which we can "reassert our larger human connectedness, our bond with our species." This more "formed awareness" and the hope it provides is a necessary, but not sufficient, condition for resisting climate crimes and achieving climate justice. Evolving awareness must be translated into political action by, among many others, what Lifton (2017: 100) calls "witnessing professionals" engaged in "advocacy research" and taking "necessary measures" to protect people and stave off climate disaster. This notion is similar to my call for public criminologists to speak in the prophetic voice about global warming (Kramer, 2012). Rob White (2018: 149) also argues that criminologists "need to engage ourselves as public intellectuals and in political action, to assume the mantle of stewards and guardians of the future, and to prioritise research, policy and practice around climate change themes." My hope is that the criminological perspective, theoretical explanations, and "advocacy research" presented in this book can help promote the climate swerve and assist other witnessing professionals and climate activists in their attempts to mitigate the worst suffering, reduce climate crimes, hold carbon criminals accountable, and restructure the global political economy to achieve climate justice. As the Czech dissident, writer, and statesman Vaclav Havel has argued (quoted in Jamail, 2019: 220), "Hope is not the conviction that something will turn out well, but the certainty that something is worth doing no matter how it turns out."

Targeting Those Responsible for Climate Harms

Analyzing global warming through the lens of the concept of state–corporate crime not only helps us understand the broader social forces shaping climate crimes; it also provides a framework within which we can advance the larger political project to reduce or prevent these blameworthy harms. To resist climate crimes, we must correctly identify those who are most responsible for the ecological and social destruction that is occurring and target them for specific political and legal actions. We need to have a clear understanding of who the carbon criminals are and what needs to be done to reduce or prevent their ongoing harmful behavior. As the evidence I have presented shows, corporations in the fossil fuel industry, energy trade associations, many conservative think tanks and foundations, specific U.S. governmental agencies, and the U.S. government in general, along with the international political community as a whole, have all engaged in various forms of climate crimes and can be defined as carbon criminals. These organizational actors (and some specific individuals who occupy powerful positions within organizational structures), enmeshed in a larger global capitalist political economy, must be the intended targets for social, political and legal actions if the human species is to have any hope of avoiding extreme climate catastrophe and achieving some form of climate justice.

While the consumption choices and practices of individuals and households are also a major contributor to anthropogenic climate change (Ehrhardt-Martinez, Schor, et al., 2015), and individual consumers should not be left off the hook for their role in causing global warming, reducing the ecological footprint of people at this level, important as that is, will not be enough to stave off climate disruption at this late date. It is simply inadequate, for example, to first and foremost encourage people to alter individual behavior like changing light bulbs, as Al Gore did at the end of his award-winning 2006 documentary *An Inconvenient Truth*. Not only is that prescription "completely incommensurate with the scale of the problem" (Mark, 2018: 14), it fails to recognize that "households are relatively passive actors in a system in which greenhouse gas emissions are structurally driven by the decisions of business and government" (Ehrhardt-Martinez et al., 2015: 95). It is unacceptable to suggest that we are all somehow equally responsible for climate change or to blame "human nature" in general for our inability to address global warming, as Nathaniel Rich (2018a) appeared to do in his extensive *New York Times Magazine* article, "Losing Earth." In a critique of Rich's argument, Naomi Klein (2018a: 7) repeated her assertion from *This Changes Everything: Capitalism vs the Climate* (2014) that we failed to take lifesaving climate action in the 1980s not because of human nature but instead due to "an epic case of bad timing." This was the time period, she points out, when "the global neoliberal revolution went supernova, and that project of economic and social reengineering clashed with the imperatives of both climate science and corporate regulation at every turn." In another critique of Rich's article, the editor-in-chief of *Sierra Magazine*, Jason Mark (2018: 14) called out the "greed and duplicity" of corporate polluters and echoed the criminological perspective of this book: "Climate change isn't so much a failure of human nature as it is the predictable result of a number of corporations putting their profits ahead of humanity's future and the planet's well-being. A growing number of activists have grasped this fact and are saying with new clarity and force, that global warming is a *crime* [emphasis added] perpetrated by a group of (what else to call them?) corporate villains." To his credit, when Rich (2019) expanded his article into a book and extended his analysis to the present in a new afterword, he did identify fossil fuel corporations, climate change deniers, and government officials who are all making the crisis worse as "villains" whose actions are "sociopathic" and constitute "crimes against humanity."

Targeting those who are responsible for climate crimes, the "villains" as Rich puts it, is important. Mark (2018: 14) asserts that it is "a proven axiom of social change" that to solve a problem, "you first have to name a perpe-trator." That is what I have tried to do in this book. Naming carbon crimi-nals (the perpetrators), empirically documenting their climate crimes, and developing a theoretical understanding of their harmful organizational actions under global capitalism are important steps that public criminolo-gists can take to speak in the prophetic voice about global warming. But, as I argued in chapter 1, speaking in the prophetic voice also requires crimi-nologists to go beyond academic tasks and become engaged in a broader array of public conversations and political actions. The field of criminology in general has always been a "policy science," as proclaimed in one of the major journals of the American Society of Criminology, *Criminology and Public Policy*. Just as criminologists are expected to propose, analyze, and often become advocates for policies, program, laws, governmental actions, and societal efforts to control, reduce, and prevent conventional forms of criminal behavior (street crimes), criminologists should also be expected to do the same with regard to state–corporate crimes in general and climate crimes in particular. Thus, in this final chapter, I will describe and advocate for specific governmental and societal actions that can be directed at the carbon criminals I have identified, in an effort to reduce, control, and prevent the ongoing climate crimes they perpetrate.

PUBLIC EFFORTS AND STATE ACTIONS TO RESIST CLIMATE CRIMES

Following the description and analysis of climate crimes presented in this book, and after an examination of the literature on the major policy goals of climate change mitigation and adaptation, a broad climate action plan can be identified. The primary objective of a general mitigation plan must be to rapidly decarbonize the high-carbon social system currently in place. The road to global decarbonization must involve not only changing the behavior of the "interlocking carbon interests" (Urry, 2011) that domi-nate the world's economy but also expanding renewable energy (Rock-ström, Gaffney, et al., 2017). The first step in this climate action plan is to drastically cut greenhouse gas emissions. After decades of political failure, governments must finally act and impose a "fossil freeze" (Brecher, 2017). Exploration for new sources of fossil fuels cannot be permitted. The climate crime of continued extraction must be stopped. The remaining deposits of coal, oil, and gas have to be kept in the ground. To kick-start this process, many climate activists argue that some form of carbon pricing must be adopted (Rabe, 2018). Furthermore, there must be a rapid phase-out plan for existing fossil fuel reserves and processing facilities, what some call "managed decline" (Lofoten Declaration, 2017). In addition to legally man-dated reductions in fossil fuel production and carbon emissions, a number of analysts also advocate for massive state investment in controversial technologies of carbon capture and sequestration. While some climate experts have serious concerns about such proposals for "negative emissions," both the UN Environment Programme and the 2018 IPCC report argue that in order to meet the goals of the Paris Agreement, some removal of carbon dioxide from the atmosphere will also be necessary (Buck, 2018; Kolbert, 2017; Shankman, 2018).

Second, at the same time that the existing fossil fuel infrastructure is being shut down, there must also be a rapid shift to clean, renewable sources of power, what Lester Brown (2015) of the Earth Policy Institute calls the "Great Transition." As a group of European researchers argue, "global decarbonization must involve renewable energy" (Rockström et al., 2017: 1,269). A considerable amount of evidence suggests that a major clean energy transformation is currently under way and that 100 percent renewable energy for all is achievable by 2050 (McKenna, 2018; McKibben, 2017a, 2019a; Queally, 2015; Teske, 2015). As Brown (2015: 135) points out, government policy instruments are "an important component of the energy transition." States will have to do much, much more than they are currently doing to promote renewables. The Great Transition will require public investment, research and development, and planning and coordination. It will also require stronger global governance. The international political community through the UNFCCC process will have to play a major role in pushing the clean energy transformation forward.

The third element of a general climate action plan concerns just adaptation. To achieve climate justice, states will have to plan for and implement forms of fair, cooperative, and progressive adaptation to climate disruptions. The barbarism of militarized adaptation must be disavowed, and in its place the Global North will have to provide critical technological assistance and generous financial aid (perhaps expanding the current Green Climate Fund) to help foster adaptation and build resilience in the Global South (M. Robinson, 2018). Massive social, material, and expressive support will also have to be provided for climate refugees forced to flee (B. Hayes, Wright, and Humble, 2016). Responding to the refugee crisis in a just way will require political will, public planning, and massive state funding.

These three broad elements are widely recognized, feasible, and the subject of much critical analysis. The required actions are, for the most part, the opposite of the climate crimes examined in this book. But how can such an action plan be implemented? Or, to consider it from a more criminological perspective, how can social and political efforts to reduce, control, and prevent climate crimes be put into effect? To provide a framework for thinking about the societal efforts needed to deal with climate crimes and enact a climate action plan, I return to Agnew's integrated definition of crime.

Recall that Agnew (2011b) identified three core characteristics of "crime": morally blameworthy harms, public condemnation, and state sanction. In the preceding chapters, I have described and analyzed a number of blameworthy ecological and social harms that I have called climate crimes, perpetrated by corporate and state actors I have labeled as carbon criminals. In presenting these analyses, I have occasionally touched on issues concerning public attitudes and reactions to climate harms and examined the general failure of states to take appropriate mitigation and adaptation actions in response to these harms (political omission). Putting aside Agnew's important concept of unrecognized blameworthy harms (blameworthy harms that are not strongly condemned by the public and sanctioned by the state), I believe that, from a policy perspective, it is useful to further explore the issues of public condemnation and state sanction as they relate to carbon criminals and climate crimes: What political efforts have been or could be undertaken by a broad-based "public" concerning these crimes and criminals, and what actions have been or could be taken by the "state" in response to them?

According to Agnew (2011b: 36), public condemnation can be conceived as having a number of dimensions: To what extent do people believe certain acts to be harmful and blameworthy; evaluate these acts as wrong or immoral; and experience anger, disgust, or other negative emotional reactions to these actions; and finally, how strongly are they inclined to sanction those who commit the acts? While these are all important issues, I will expand this focus by exploring the additional dimension of public actions taken as an additional measure of public condemnation. To what extent do individuals or groups act on their beliefs, evaluations, emotional reactions, or inclinations to sanction? To what extent do people in the public sphere engage in political efforts to control, reduce, or prevent the blameworthy harms of fossil fuel corporations and government agencies? To what extent do they develop and advocate for elements of a broad-based climate action plan? To what extent do people contribute money to or volunteer with political campaigns; vote for candidates and ballot proposals; send written communications and make phone calls to elected officials; meet directly with political leaders to lobby them; organize teach-ins, panel discussions, or other forms of public education and discussion; participate in protests; go on strike; engage in acts of civil disobedience; or join and work with social movement organizations? What specific political efforts have individuals or groups become publicly involved with related to the issue of climate change?

Agnew's third core characteristic—state sanction—concerns specific laws (criminal or civil), the severity of legal penalties, and the certainty of punishment (the "law in action" as opposed to the "law on the books"). From this narrower legal starting point, I also argue for a very broad expansion of the concept of state sanction to include any and all forms of state and governmental action, targeted at corporations in the fossil fuel industry or related more generally to the mitigation of and adaptation to climate change. This would include all state legal definitions, criminal prosecutions, forms of civil litigation, regulatory enforcement actions, judicial rulings, other types of domestic legislation, public investments, social welfare programs, and democratic mechanisms for social and economic planning and coordination. It would also comprise treaty negotiations and other forms of participation in interstate accords under international law, including trade and other economic agreements as they relate to global warming. What can government agencies or nation-states do to help mitigate carbon emissions, plan for and facilitate the transition to clean and renewable energy and more sustainable consumer choices and practices, and adapt in just and cooperative ways to climate disruptions? For those climate activists who argue against piecemeal, market-based climate policies and for a revolutionary restructuring of the global economic system—a move from capitalism toward democratic socialism, participatory economics, or ecosocialism—taking state power in some form will be essential to that structural transformation (Aronoff, 2018).

Agnew's (2011b) conceptualization of the core characteristics of crime raises an important question concerning the relationship or connection between the public condemnation of blameworthy harms and the possibility of state sanction of these harms. He argues that public condemnation may result in severe state sanctions, especially with "Core Crimes" (more traditional forms of criminal behavior). But he also conceives of disconnections that result in "Public-Only Recognized Harms" or "State-Only Recognized Harms." In my more expansive use of Agnew's concepts, it is important to highlight the relationship between actions in the public sphere and actions by the state. Many public efforts concerning climate change are directed at institutions of political governance in an attempt to force them to act. And the sanctioning of carbon criminals or the implementation of a climate action plan generally depends on state actions. While groups or organizations in the public sphere can take actions that independently target the fossil fuel industry or assist the transition to renewable energy, in most cases the public is demanding that the government "do something" about climate change. The core characteristics of public condemnation and state sanction can be interrelated in this area. Although research shows that the public generally has little ability to affect state decision-making under current political arrangements, I will examine those efforts undertaken by public sector actors to try and influence the state on the climate issue.

Calling attention to this potential interrelationship emphasizes once again the centrality of the state. As I noted earlier, Giddens (2011: 94) has asserted that nation-states "must take the lead in addressing climate change. . . . The state has to be the prime actor." Parenti (2011: 226) agrees, arguing that "the climate crisis is not a *technical* problem, nor even an *economic* problem: it is fundamentally a *political* problem." In a 2012 article titled "Why Climate Change Will Make You Love Big Government," Parenti points out that "Big Storms require Big Government." He contends that the state "cannot be avoided" in the broader effort to deal with global warming. This is true not only for conservative storm victims and liberal environmentalists but also for those progressive activists who advocate for system change. Parenti (2016b: 182) argues: "For Left politics to become effective, especially in the face of the climate crisis, they must come up with strategies that engage and attempt to transform the state. The idea of escaping the state is to misrecognize the centrality and immutably fundamental nature of the state to the value form and thus to capitalist society." Many Marxist scholars have emphasized not only the contradictory nature of the state's institutional power base but also its place as a site of social struggle that can occasionally be mobilized by powerless groups in political conflicts (Chambliss, 1989; Coleman, Sim, et al., 2009; R. White, 2018). Although governments are also involved in climate crimes like subsidizing the energy, automobile, and agricultural industries (resulting in the continuing extraction of fossil fuels and rising carbon emissions) and failing to act to mitigate and adapt to the climate crisis (political omission), state action is still necessary to avert disaster. As Rob White (2018: 148) concludes, "In the end, effective intervention around climate change frequently demands action in collaboration *with* and *against* the state." To control and prevent climate crimes, and implement a climate action plan, the public sphere will have to direct efforts against the fossil fuel industry and the state, contend for and capture political power in some form, and then use the power of the state to reduce carbon emissions, transition to clean energy, and provide climate justice.

A realistic analysis of this project, and the general relationship between the public sphere and the ensemble of governmental institutions we call the state, raises a number of serious issues including the extensive history of political omission concerning climate change. Despite various public-sphere efforts in the past, the U.S. government in particular has exercised the second dimension of power—the mobilization of bias—to avoid placing climate change on the political agenda and to avoid making mitigation and adaptation decisions. Non–decision making on the climate crisis reflects a more general and critical problem with the American political system. As Indian writer Amitav Ghosh (2016: 130) argues, "the public sphere's ability to influence the security and policy establishment has eroded drastically." This observation is supported empirically by an important study testing four major theories of American politics. Using a unique data set that included measures of key theoretical variables for 1,779 policy issues, political scientists Martin Gilens and Benjamin Page (2014: 564) found that "economic elites and organized groups representing business interests have substantial independent impacts on U.S. government policy, while average citizens and mass-based interest groups have little or no independent influence." These sobering findings confirm what many political activists have long known due to their own experiences: the majority does not rule. And, since at least the Reagan era, neoliberalism and corporate power have held the state hostage. As Gilens and Page's research demonstrates, citizens who advocate for policy changes are generally not successful if those policies are opposed by corporate interests. This has certainly been the case concerning the climate crisis over the years. And more recently, with the neoliberal Republican Party in control of both houses of Congress and the climate change–denying Donald Trump as president, this has been a serious obstacle to protecting even the more limited Climate Action Plan proposed by Barack Obama. With the more supportive Democratic Party gaining control of the U.S. House of Representatives in the 2018 midterm elections, Democrats may be able to "turn a spotlight on climate change and Trump's retreat from responsibility" (Lavelle, 2018c: 1). But even though climate change may be back on the political agenda, and the House may be able to provide some measure of protection for Obama administration policies, this is not likely to lead to any major climate initiatives since the fossil fuel industry remains powerful and the Republicans still control the Senate. The climate justice movement has its work cut out for it.

Despite the enormous political barriers that exist in the U.S. political system, a large and growing number of public-sphere efforts (and even a few state actions) to deal with the climate crisis have occurred. A full description and analysis of these efforts would require a separate book. In the remainder of this chapter, however, in an effort to foster hope, encourage resistance to ongoing climate crimes, and advocate for climate justice, I will highlight and briefly examine just some of the most important public-sphere efforts and state actions that have been taken or could be taken to resist climate crimes, implement a climate action plan, and advance the climate swerve. Some of the climate actions to be discussed are targeted at fossil fuel corporations, while others are directed at institutions of political governance. In examining these efforts, we should remain alert to the issue of the potential relationship between public condemnation and state sanction.

PUBLIC OPINION, BLOCKADIA, AND
FOSSIL FUEL DIVESTMENT

Since public support can have a large impact on shaping social and political responses to climate change, it is important to consider public opinion on a range of policy issues related to the crisis. Sociologists Rachael L. Shwom, Aaron M. McCright, and Steven R. Brechin (2015:

269–270) use the term *public opinion* to refer to "beliefs, attitudes, policy support, and behavioral intentions of people and groups within a particular geographic location." Providing an overview of mass opinion on climate change in the United States, as well as of variations in individuals' views on the topic, they caution that "public opinion on climate change is multidimensional, dynamic, and differentiated." After reviewing the available survey research data, Shwom, McCright, and Brechin (2015) conclude that "public awareness" of climate change in the United States has been increasing since the 1980s, as has "public knowledge" of the issue, although not as quickly. Despite their general awareness of climate change, levels of belief in the reality and human causation of global warming among Americans has fluctuated across recent decades. Shwom, McCright, and Brechin (2015: 277) point out that "Unlike most other industrial nations, the United States consistently has had a segment of the population either deny the reality or anthropogenic cause of climate change." They also note that citizens in the United States worry less about climate change than a host of other issues but that "Despite the low levels of relative concern, consistently more than 50 percent of Americans support a variety of energy and climate change policies." Although there is often a gap between what people say and what people do, some public support for climate action exists. The question is: How strong is that support and can it pressure government to act?

Activist Jeremy Brecher (2017) has also reviewed public opinion on climate change. Drawing on the work of the Yale Program on Climate Change Communication, he points out that 65 percent of registered voters in the United States support setting strict carbon emissions limits on coal-fired power plants, and 62 percent of Americans said it was important that the world reach an agreement to limit global warming at the 2015 Paris Climate Conference. Brecher also cites evidence that Americans favor more research into renewable energy, support regulating carbon dioxide as a pollutant, and would like to see a carbon tax imposed on fossil fuel companies. In an August 2018 report, the Yale Program on Climate Change Communication found continued and growing majority support for these and other climate change–related policies. The program's "Yale Climate Opinion Maps 2018" (Marlon, Howe, et al., 2018) also reported that 68 percent of Americans think corporations should do more to address global warming, and 62 percent think that Congress should do more to address the issue. In addition, a major CBS News poll in September 2019 found that two-thirds of Americans believe climate change is either a "crisis" or a "serious problem," with a majority "wanting immediate action to address global heating and its damaging consequences" (Milman, 2019: 1).

These results suggest that public support for climate action not only exists, but it appears to be growing. Brecher (2017: 54) cautions, however, that even though a majority of Americans are concerned about global warming and are in favor of clean energy, most still support an "all of the above" energy policy that would result in the continued extraction and burning of fossil fuels. Thus, he argues that "a central goal for the climate insurgency must be to move the public to the conviction that the greenhouse gases released by the burning of fossil fuels are the principal cause of climate change" and "persuade the majority of Americans who are worried about climate change that the elimination of fossil fuels is not only desirable but necessary" (Brecher, 2017: 54).

Many climate activists would argue that it is the job of environmental groups to "move the public" to the conclusion that carbon emissions are the cause of global warming and to "persuade" a majority of Americans to eliminate fossil fuels. As I noted in chapter 5, however, the activities of mainstream "green" groups concerning the climate change issue have generated considerable controversy. Recall that Skocpol (2013) argued that the failure to enact climate change legislation during Obama's first term was the fault of the mainstream U.S. environmental movement. Focusing specifically on the defeat of the cap-and-trade bill in 2010, she asserted that environmental groups had tried to play a bipartisan, "inside the Beltway" grand strategy that failed to appreciate the growing political polarization of the climate change issue in Congress, a result of intense efforts by the Tea Party and the climate change denial countermovement. Skocpol faulted environmentalists for failing to build a broad popular movement that could create the necessary political pressure on Congress for a change in climate policy.

Klein (2014: 195–196) has also criticized mainstream environmental groups. She argues that these groups have become "entangled" with the very same fossil fuel corporations that are responsible for the climate crisis, asserting that "the painful reality behind the environmental movement's catastrophic failure to effectively battle the economic interests behind our soaring emissions" is that "large parts of the movement aren't actually fighting those interests—they have merged with them." Klein contends that funding from the fossil fuel corporations and the adoption of the values of centrist policy foundations caused mainstream green groups to advocate for "market-based" climate solutions and support free trade agreements that contained only token environmental protections. In her view, this "reform environmentalism" undercuts the potential "to build a mass movement capable of taking on powerful polluters" (Klein, 2014: 199). Brulle (2013: 174) concurs, observing that "Rather than engaging the public, reform environmentalism focuses debate among experts in the scientific, legal, and economic communities. It may provide technical solutions to specific problems but it neglects the larger social dynamics that underlie environmental degradation."

Due to these critiques, and the repeated political failure of reform environmentalism, some climate activists began to develop a different vision for climate politics that included new targets, strategies, and tactics (Hadden, 2017). Grassroots environmental organizations—such as McKibben's 350.org, attempting to mobilize greater public-sphere support and build a broader movement to pressure the state to take climate action—came to the forefront. After struggling for years to figure out where to "place the bulls-eye," as Mark puts it, these more radical climate activists began to target fossil fuel infrastructure projects such as the Keystone XL and Dakota Access pipelines. Many of these "infrastructure fights" over pipelines, oil trains, or fracking projects are directed at fossil fuel companies, while some target both energy corporations and government agencies. All of them provide something that climate campaigners have long been in search of: "tangible local targets that ordinary citizens can focus on" (Mark, 2018: 16). Klein (2014: 294–295) refers to the "climate warriors" engaged in these infrastructure fights as "Blockadia," noting that "resistance to high-risk extreme extraction is building a global, grass-roots, and broad-based network the likes of which the environmental movement has rarely seen." She argues that "Blockadia is not a specific location on a map but rather a roving transnational conflict zone that is cropping up with increasing frequency and intensity wherever extractive projects are attempting to dig and drill, whether for open-pit mines, or gas fracking, or tar sands oil pipelines." Blockadia is a form of active resistance to ongoing and immediately visible climate crimes.

The climate warriors flocking to the barricades of Blockadia are often young people who distrust mainstream environmental groups, as well as members of a "reinvigorated Native American sovereignty movement," which has been involved in battles over projects such as the Dakota Access pipeline on the Standing Rock Reservation (Mark, 2018). First Nations groups in Canada have led the fight against tar sands mining in Alberta, and other Indigenous people's groups have also been at the forefront of fossil fuel infrastructure fights around the world (Klein, 2014). The Blockadia resistance movement has frequently utilized civil disobedience as a tool, as 350 .org did in the battle against the Keystone XL pipeline. Many engaged in these protests have subsequently faced legal sanctions. A group of climate activists protesting fossil fuel mining in Minnesota, accused of shutting off a tar sands pipeline valve, used the "necessity defense" to justify their "criminal" actions, which the judge allowed (McKenna, 2017). We can expect that this defense will be used more and more in the future as the Blockadia movement grows and becomes more militant (Germanos, 2018b).

Another important public sphere effort to combat climate change is the fossil fuel divestment campaign. This movement to persuade institutional investors to remove fossil fuel investments from their portfolios and endowments was launched shortly after McKibben's "Global Warming's Terrifying New Math" article went viral in 2012. Echoing the divestment campaign against South African apartheid a generation earlier, students at hundreds of campuses around the world began demanding that their universities commit to divesting from fossil fuels. The fossil fuel divestment movement has spread to thirty-seven nations around the world and has become "the largest anticorporate campaign of its kind in history" (McKibben, 2016b: 14). A 2018 report from Arabella Advisors calculated that investors with more than $6 trillion in assets have committed to divest from coal, oil, and gas, an increase of a trillion dollars from the previous report in 2016 (Carrington, 2018c); and in 2019 McKibben (2019b: 196) reported that the number had grown to nearly $8 trillion.

The sell-off of fossil fuel investments has, perhaps surprisingly, been led by the insurance industry, with more than $3 trillion divested. Other major institutional investors who have joined the movement include the University of California system, the World Council of Churches, the Rockefeller Family Fund, and dozens of other universities and Christian denominations (McKibben, 2018a). In July 2018, Ireland became the first nation to divest, and in September of that year, New York City mayor Bill de Blasio and his London counterpart, Sadiq Khan, announced that they were divesting their public pensions and other assets from fossil fuels, challenging other major cities to join them in this effort (de Blasio and Khan, 2018; Kusnetz, 2018a). Academic research cited by McKibben (2018a, 2019b) has demonstrated that these fossil fuel divestment decisions have had several powerful effects. First, the campaign has thrust the issue of climate change, and the oil industry's culpability for the crisis, squarely into the political mainstream in such a way as to alter the political debate. The second finding is that the divestment movement has actually begun to cost the fossil fuel industry real money. The oil companies themselves have started to admit that divestment poses a "material risk" to their business (Carrington, 2018c). While the proponents of divestment may have hoped to someday affect the business decisions of the fossil fuel corporations, their primary goal in launching the movement was to place a morally stigmatizing label on the industry: to publicly proclaim, as McKibben (2012a: 7) did, that the fossil fuel industry "has become a rogue industry, reckless like no other force on Earth. It is now Public Enemy Number One to the survival of our planetary civilization."

More recently, McKibben (2019c: 4) has proposed "extending the logic of the investment fight one ring out, from the fossil fuel companies to the financial system that supports them." McKibben suggests that the banking, asset management, and insurance companies who fund the fossil fuel industry that is wrecking the planet could be targeted and pressured to change their destructive financial practices. He points out that corporations like JP Morgan Chase, Blackrock, and Chubb control huge amounts of money and routinely invest enormous sums in dirty energy companies responsible for carbon emissions. Slowing the flow of money to the fossil fuel industry could also be a major means of slowing the release of greenhouse gases into the atmosphere. McKibben (2019c: 18) predicts that an effort to shame these corporations and convince them to end the financing of global warming could "become one of the final great campaigns of the climate movement—a way to focus the concerted power of any person, city, and institution with a bank account, a retirement fund, or an insurance policy on the handful of institutions that could actually change the game." Blocking money from flowing to the fossil fuel industry may be even more effective than blocking extraction projects with bodies.

CLIMATE LITIGATION AND CARBON PRICING

Both the Blockadia activists and the divestment campaign target fossil fuel companies directly in their efforts to keep oil, coal, and gas in the ground and reduce greenhouse gas emissions. Other climate advocacy and environmental groups attempt to put political pressure on governments to take actions against the industry to force a decrease in the production of fossil fuels. These public-sphere organizations seek state interventions to stop the extraction of fossil fuels and lower carbon emissions. They promote governmental actions that can not only reduce the demand for dirty energy but also stop the expansion of fossil fuel projects and "manage the decline" (Lofoten Declaration, 2017) of existing production in order to achieve the climate goals set in Paris in 2015. These climate groups encourage the state to use a variety of tools to accomplish these goals, including corporate regulation, civil litigation, the withdrawal of government subsidies to the fossil fuel industry, and various forms of carbon pricing.

Many environmental organizations, climate activists, and even governmental entities have turned to specific legal actions and the courts to resist climate crimes and "bring climate wreckers to account" (Mark, 2018: 17). For many in the climate justice movement, the law has become the weapon of choice in the fight to limit anthropogenic global warming and compensate its victims. Regulatory efforts and civil litigation represent "a new front of climate action, with citizens aiming to force stronger moves to cut emissions, and win damages to pay the costs of dealing with the impacts of warming" (Carrington, 2018b: 1). James Hansen argues that a "litigate-to-mitigate campaign is needed alongside political mobilization because judges are less likely than politicians to be in the pocket of oil, coal and gas companies" (quoted in Watts, 2017c: 1). And there has been a "growing wave" of legal actions concerning climate change in recent years (Irfan, 2019). The

Sabin Center for Climate Change Law at Columbia University Law School has logged more than 1,000 lawsuits that have been brought across the globe (Carrington, 2018b: 2). Some of these suits were initiated by public-sphere environmental groups, some by various levels of government within the ensemble of institutions that make up the state. Some of these cases target fossil fuel corporations and some target national governments. All demand that the state's legal institutions act to protect the environment and hold carbon criminals accountable. These lawsuits "could lead to multibillion-dollar payouts, and force an unwilling government to make cutting greenhouse gases a central priority" (Irfan, 2019: 17). Whatever the specific targets and requested sanctions and remedies, this dramatic "surge" of litigation has emerged as one of the most important features of the climate swerve (Kusnetz, 2019).

Legal actions directed at companies in the fossil fuel industry can take three different forms: criminal prosecution, regulatory law enforcement, or civil litigation. Criminal prosecutions are generally not used in the attempt to mitigate greenhouse gas emissions, although Rob White (2018: 117) argues that reducing emissions is such a central element of any climate action plan that such efforts "could and arguably should involve the criminalization of carbon emissions and the forced shutdown of dirty industries." The International Criminal Court (ICC) took a tentative step in that direction in 2016 when, in a change of focus, it announced that it would widen its remit and begin to prioritize crimes that result in the "destruction of the environment," "exploitation of natural resources," and the "illegal dispossession" of land. One international criminal law attorney suggested that "the new ICC focus could open the door to prosecutions over climate change" (Vidal and Bowcott, 2016: 1).

While criminal prosecutions of climate crimes are likely to remain rare, regulatory actions concerning environmental issues are more common, albeit with questionable effectiveness. Wood (2014: xvi) provides a thorough review and critique of environmental regulatory law, especially as it relates to climate change, and concludes that "across nearly all fronts of ecological assault, environmental law has failed in its basic purpose to safeguard natural resources," including the atmosphere. And the federal regulatory system has not lacked an opportunity to act. As I noted previously, in a significant development related to climate change policy, the U.S. Supreme Court ruled in 2007 that the Environmental Protection Agency does have the legal authority under the Clean Air Act (CAA) to regulate carbon emissions (*Massachusetts v. EPA*)—but the agency generally did not use that authority. Obama, however, blocked by congressional Republicans from pursuing legislative policies to deal with global warming, took executive action and ordered the EPA to move ahead with a carbon emissions plan.

The agency first issued an Endangerment Finding (required under the CAA) and then took steps in 2015 to begin implementing Obama's Clean Power Plan aimed at reducing emissions from coal-fired power plants. Although the Clean Power Plan was criticized by some climate activists as not going far enough, Republicans denounced it as a "war on coal," and the anti-environmental Trump administration has taken steps to eliminate the plan (and roll back other regulatory efforts as well). As long as Trump remains in office, the regulatory path to climate action will be completely blocked.

Since criminal prosecutions are seldom pursued and regulatory efforts have generally been stymied or proven ineffective, climate activists have increasingly turned to civil litigation as the primary legal tool in their efforts to force fossil fuel corporations to cut emissions and pay the costs of the ecological and social damage they cause. Environmental groups, plaintiffs seeking damages for injuries, state attorneys general, cities, one U.S. state, and several national governments have all filed lawsuits to block extraction projects, reduce carbon emissions, or obtain financial compensation for victims of climate harms. In a global review of "the status of climate change litigation," the UN Environment Programme (UNEP, 2017: 8) asserted that "litigation has arguably never been a more important tool to push policymakers and market participants to develop and implement effective means of climate change mitigation and adaptation than it is today." UNEP (2017: 10) pointed out that, as of March 2017, climate change cases had been filed in twenty-five countries, with more than 600 cases being initiated in the United States alone. While the majority of these lawsuits, like *Juliana, et al. v. United States* target national governments, fossil fuel corporations have been targeted as well.

In a number of those suits, plaintiffs have sought (unsuccessfully so far) to establish that the carbon emissions of particular fossil fuel companies have caused them specific harms that require compensation. In these legal actions, "plaintiffs grounded their claims against these private companies in a theory of public nuisance under federal common law" (UNEP, 2017: 19), which the courts have rejected. As climate science advances, and scientists improve their ability to attribute specific harms to climate-related disruptions and events (source attribution), these suits will become more numerous and likely more successful (Leber, 2019).

The state attorneys general who have filed lawsuits (New York) or are considering filing (Massachusetts) against a number of fossil fuel corporations, including ExxonMobil, are on more solid legal ground. These state law enforcement officials, joined by several groups of investors and the federal Securities and Exchange Commission, argue that the energy companies have engaged in fraud by lying to the public and their investors about what they knew about the risks of climate change and how such risks could negatively affect the fossil fuel industry in the future. ExxonMobil and the other energy giants have fought back ferociously against these investigations, and it is not clear how they will be resolved. However, one former oil company executive believes that these "legal actions add a further dimension to the pressure for change in an industry that has begun to accept the need to reinvent itself" (Carrington, 2018b: 3).

Beginning in July 2017, several local governments in the United States have also tried to increase the pressure on the fossil fuel industry to change its business practices through climate liability litigation. New York City, San Francisco, and Oakland are among the dozen cities, counties, and municipalities that have filed civil lawsuits against various oil and gas companies attempting to recover costs for dealing with climate impacts like sea-level rise and extreme heat (Hasemyer, 2018a). One attorney, representing several smaller cities and counties in California, asserted, "Our lawsuits are tort cases seeking money from companies based on wrongful behavior. All of these cases are about remedying the incredibly large cost that public entities are incurring from past behavior" (Drugmand, 2018: 2). These legal actions are targeted at the same kind of corporate wrongdoing that was the basis for successful litigation against the tobacco industry, including a major settlement in 1998 (Oreskes and Conway, 2010). Like the tobacco companies, fossil fuel corporations have knowingly sold a harmful product while denying or downplaying the risk that product produces for the general public. In June 2018, a federal judge dismissed the San Francisco and Oakland cases saying that, although the dangers of climate change are "very real," the issue should be solved by Congress or handled by the regulatory process at the EPA (Hasemyer, 2018a). In July of that year, the New York City case was also tossed out on similar grounds. Despite these dismissals, climate liability lawsuits against fossil fuel companies continue to be filed. In July 2018, Baltimore became the latest city to sue for climate damages, this time in state court, not federal court (Dorroh, 2018a). And that same month, Rhode Island became the first U.S. state to file a climate suit versus the fossil fuel industry, targeting twenty-one major oil companies "for knowingly contributing to climate change, and causing catastrophic consequences to Rhode Island, our economy, our communities, our residents, our ecosystems" (Corbett, 2018b: 1). More climate liability litigation by U.S. cities and states is sure to follow.

Throughout the rest of the world, one of the emerging trends in climate change litigation concerns the question of whether the extraction and production activities of the transnational fossil fuel corporations, the carbon emissions they generate, and the climate harms they cause violate the human rights of people under international law. While the core international human rights treaties do not explicitly recognize the right to a stable climate, it has long been understood that forms of environmental destruction can undermine the effective enjoyment of many important human rights that are enumerated in these laws, including the right to a clean environment (UNEP, 2017). Philosopher Simon Caney (2010: 166) has argued that climate change jeopardizes three key human rights: "the human right to life, the human right to health, and the human right to subsistence." He also argues that if climate change does violate human rights, then compensation is due to those who have been victimized. Based on assertions of this kind, many lawsuits have been filed around the world that allude to international human rights and obligations (UNEP, 2017).

The UN Human Rights Council has issued numerous resolutions recognizing the harmful effects of climate change on human rights. And, in March 2018, the United Nation's Special Rapporteur on Human Rights and the Environment issued a report that outlines a legal framework for the international community to recognize a healthy environment as a human right. According to Jennifer Dorroh of *Climate Liability News* (2018b: 1), "The report could bolster the arguments of citizens seeking accountability for ecological harm in human rights courts around the world." Cole (2015) takes the argument a step further by arguing that the United Nations needs to establish by treaty an international climate court where torts and human rights violations can be decided. According to Rob White (2018: 122), this climate litigation and other calls for legal action and human rights inquiries concerning global warming reaffirm that the legal arena serves as "a major source of contestation and political struggle."

One of the most intriguing ideas for creating a new legal mechanism to deal with global warming is to make ecocide an international crime. The concept of ecocide emerged in the 1970s, but it was given new prominence in April 2010 when UK lawyer Polly Higgins submitted a formal request for an international law of ecocide to the United Nation's Law Commission (Higgins, Short, and South, 2013). Higgins proposed an amendment to the Rome Statute (the statute that established the International Criminal Court), stating that "Ecocide is the extensive damage to, destruction of or loss of ecosystem(s) of a given territory, whether by human agency or by other causes, to such an extent that peaceful enjoyment by the inhabitants of that territory has been severely diminished." According to Higgins (2010, 2012), this proposal, if enacted, would make ecocide the fifth international "crime against peace," joining genocide, crimes against humanity, war crimes, and crimes of aggression. Ecocide should also be considered another form of climate crime, and the urgency and impetus for this concept to be officially recognized as a bona fide international crime have been heightened by the woefully inadequate responses by governments, individually and collectively, to global warming (R. White and Kramer, 2015b). The prospects for the inclusion of ecocide as a new international crime are not promising at this time, but as Rob White (2018: 122) points out, "New concepts of harm, as informed by ecological sciences and environmental values, will inevitably be developed as part of any legal reform process, and these can contribute towards climate justice ends."

While many environmental activists place their hopes for climate justice on legal actions and court rulings that could hold fossil fuel corporations accountable and provide financial compensation for the victims of climate disruptions, others continue to make the case for state policies that rely primarily on market mechanisms. Some climate groups argue, for example, that to slow greenhouse gas emissions the U.S. Congress needs to pass legislation that creates some form of carbon pricing. The "cap-and-trade" bill proposed during Obama's first term was one such effort. Another carbon pricing proposal, championed by the influential Citizens' Climate Lobby, suggests imposing a tax on emissions and then rebating the revenue back to citizens in equal proportion. A carbon tax places a levy on fossil fuels at the point where they are extracted or imported and would be paid by the oil, coal, and gas companies. Such a tax attempts to attach the ecological and social consequences of global warming to the fossil fuel production sources that are driving those consequences. As Irfan (2018: 2) puts it, "Ideally, it would push economies toward sustainability by making dirtier energy sources and industries more costly relative to their alternatives." And since the fossil fuel corporations would pass the increased cost on to consumers, most carbon tax proposals include a provision for providing a carbon dividend back to the people from the substantial revenues that might be raised by such a tax. Socialists are generally skeptical of such market-based policies, but some argue that a carbon tax could be a complement to other more progressive climate policies—one tool among many that could be used in the fight to limit global warming (Fremstad and Paul, 2018).

But there are three major concerns regarding the carbon tax proposal. First, although such a tax has wide support among environmentalists and the general public, it has long encountered what American political scientist Barry Rabe (2018: xvi) calls "political stumbling blocks" in many places around the world. In a comprehensive analysis, he examines a few examples of when carbon taxes work but more generally why they fail. In the United States, congressional Republicans still oppose the idea, and in July 2018, the House passed a resolution that denounced the idea of a carbon tax by a vote of 229–180, with Republicans arguing that such a tax would be detrimental to the U.S. economy (Lavelle, 2018d). As Rabe (2018) observes, the implementation of carbon pricing in any form requires building an "enduring political base," a base that still does not exist at the federal level in the United States.

The second concern is that most of the carbon tax proposals do not take into account the true "social cost" of carbon and thus set the tax too low. The social cost of carbon is a policy tool that attaches a price to the long-term economic and social damage caused by each ton of carbon dioxide. Irfan (2018: 2) points out that how high you set the carbon tax is a function of the empirical value you place on the social cost of carbon. Obama set the social cost at $50 per ton of carbon dioxide, while under Trump it has been estimated to be between $7 and $1. The Trump estimates are so low because the current EPA has focused only on damages from climate change that would occur within the borders of the United States and also put less weight on the climate harms that would afflict future generations by setting a high "discount rate" (Plumer, 2018b: A16). The discount rate is how economists value costs and benefits across time, with a high rate reflecting a concern with present economic costs and benefits, placing less value on future costs. William Nordhaus, a 2018 Nobel laureate of economics, advocates for a high discount rate and therefore sets the social cost of carbon at $30 per ton (Battistoni, 2018). However, if the median social cost of carbon is $417 per ton, as one recent study determined (Ricke, Drouet, et al., 2018), and the carbon tax is set at a very low $30 per ton, then the tax will not accomplish its stated objectives. In glaring contrast to Nordhaus and other mainstream economists, the 2018 IPCC report suggests raising the social cost of a ton of carbon to as high as $5,500 by 2030 to keep overall global warming below 1.5 degrees Celsius (Plumer, 2018a).

The third and greatest concern with a carbon tax is that fossil fuel corporations will oppose any such proposal unless the industry is able to shape it to their benefit, which they are trying hard to do. Americans for Carbon Dividends is a fossil fuel industry–backed organization whose mission is to build political support for the Baker-Shultz Carbon Dividends Plan. Named for its lead authors, former secretaries of state James A. Baker III and George P. Shultz, the plan proposes a gradually rising carbon tax that starts at $40 per ton and includes a provision for returning 100 percent of the revenues earned back to American citizens in the form of a monthly dividend payment. Aside from the fact that the tax is too low, there are other, more dangerous provisions in the proposal as well. The Baker-Shultz Plan, unveiled in 2017 by the Climate Leadership Council, a lobby group whose founding member organizations include ExxonMobil and other oil companies, also includes what it calls "regulatory simplification," a rollback of carbon emissions standards. And even more alarming, the plan also proposes an "end to federal and state tort liability for emitters." In this case, gaining a (minimal) carbon tax would result in losing the ability to regulate and the right to sue. As Karen Savage (2018b: 1) of *Climate Liability News* points out, "That means the federal government would no longer have the ability to regulate climate pollutants and courts would have no role in determining whether industry should pay for impacts already caused by fossil fuel–driven global warming." This preemptive "de-criminalization" of climate crimes could be described as a giant "get out of jail free card" for the fossil fuel industry, and ExxonMobil has already announced that it will pledge $1 million to the Americans for Carbon Dividends campaign to push the Baker-Shultz Plan through Congress (Savage, 2018a). This dangerous proposal must be opposed and the door to climate change litigation against fossil fuel companies left open.

STRONGER GLOBAL GOVERNANCE, DECARBONIZATION, AND JUST ADAPTATION

The failure of nation states to take climate action—the crime of political omission—has led some public-sphere groups to file lawsuits against national governments in an attempt to force them to act. While most climate litigation has been directed against fossil fuel corporations, some of the most significant legal cases related to climate change are those that target governmental institutions, such as the *Juliana, et al. v. United States* case. While a trial date in the *Juliana* case had still not yet been set by the end of September 2019, two landmark cases from Europe, *Urgenda Foundation v. the State of the Netherlands* and *Friends of the Irish Environment CLG v. Fingal County Council . . . and Ireland and the Attorney General*, have already rendered decisions that citizens do have a legal right to a safe climate and that governments do have a duty to protect the environment for their people.

In June 2015, in response to a lawsuit filed by an environmental group, the civil section of the District Court of the Hague ruled that the government of the Netherlands had "breached the standard of due care" by implementing a carbon emissions policy that was "insufficient to avoid dangerous climate change and was therefore unlawful" (de Graaf and Jans, 2015: 517). The court declared that the Dutch government had a "duty of care" to protect the environment and ordered it to cut emissions of carbon dioxide by 25 percent, compared with 1990 levels, by 2020. The plaintiff in the case, the Urgenda Foundation, cited the European Convention on Human Rights as a basis for its complaint and presented evidence that the government was not moving quickly enough to reduce greenhouse gas emissions and head off dangerous climate disruptions. In October 2018, the Hague Court of Appeal denied the Dutch government's request to overturn the lower court ruling. In a statement, the court said: "Considering the great dangers that are likely to occur, more ambitious measures have to be taken in the short term to reduce greenhouse gas emissions in order to protect the life and family life of citizens in the Netherlands" (Associated Press, 2018: 1). Similar cases are pending across the globe, and the Urgenda Foundation has proclaimed that the Dutch court's ruling puts "all world governments on notice" (Nelson, 2018: 3).

Another important legal judgment related to climate change has come from Ireland. On November 21, 2017, the Irish High Court recognized a constitutional right to a safe climate and environment (Ye, 2017). In a case brought by Friends of the Irish Environment (FIE), a nonprofit organization that advocates for sustainable planning and environmental justice, the High Court ruled that a constitutional right to environmental protection "that is consistent with the dignity and wellbeing of citizens at large is an essential condition for the fulfillment of all human rights" (Sargent, 2017: 1). According to *Climate Liability News*, the decision could force the Irish government to take stronger and more effective actions to meet its climate change goals (Savage, 2017). And FIE expects the judgment to have "profound implications beyond the scope of this case," arguing that "The state now has a duty to protect the environment in a way that is consistent with this newly established right." The Irish Green Party concurred, asserting that this "historic" ruling "will aid in future cases to hold the Government and State accountable for their responsibilities in environmental protection and tackling climate change" (Sargent, 2017: 2).

These court rulings constitute an important component of the climate swerve that Lifton (2017) has detected: the more "formed awareness" of the dangerous predicament we face. The decisions establish that citizens have legal rights to a clean and safe environment, that governments have a legal duty to protect those constitutional and public trust rights, and that states must remedy environmental harms and take stronger actions to implement a climate action plan that will limit global warming and protect the public (including future generations) from climate disruption. Courts are requiring governments to make deeper cuts in greenhouse gas emissions, assist the transition to clean energy, and adapt to climate harms in ways that protect human rights and do not cause further victimization. In other words, these judicial rulings are pushing states to implement important elements of a broad-based climate action plan. The combination of continuing political pressure from climate movement organizations, and the various forms of climate litigation that have been pursued, provide, as Lifton (2017) puts it, some "hope" that "constructive action" to address the climate crisis will be taken as the climate swerve evolves.

But no climate action plan can be implemented unless a new and much more potent global climate accord is reached. Stronger global governance is critical to climate change mitigation and adaptation efforts. While the nation-state remains essential to effective climate action, states must act together as part of the international political community if the world is to have any hope to limit global warming and achieve climate justice. As Ciplet, Roberts, and Khan (2015: 251) point out, "climate change is inherently a transnational problem in its causes and consequences," and "purely national forms of advocacy are no longer sufficient because the problems we face have taken on particularly global dimensions." Rob White (2018: 151) agrees, asserting that "The global nature of the problem—climate change—means, however, that inevitably our collective survival will require planetary cooperation and worldwide action." Yet planetary cooperation and worldwide action have often been in short supply during the UNFCCC negotiation process over the years, with the United States frequently impeding the process. Then in 2015 came the breakthrough at Paris. The Paris Agreement was the first comprehensive climate pact that the international community was able to achieve since the more limited Kyoto Protocol. It was an enormously important accomplishment but, at the same time, badly flawed and ultimately inadequate to the task of limiting global warming to 2 degrees Celsius (the original objective), let alone 1.5 degrees Celsius (the new, more stringent target that has emerged out of the negotiations).

Since the Paris Agreement is based on nationally determined and voluntary reductions in greenhouse gas emissions, there are no legally binding targets. In an important advance, however, all nations—not just the developed countries of the Global North—did agree to mitigate emissions. The concept of "common but differentiated" responsibility took on a new meaning: each nation agreed to calculate its own Intended Nationally Determined Contribution to the reduction of global carbon emissions and then develop its own voluntary methods for meeting this goal. While these mitigation plans are not legally binding, each state is required to publicly report on its progress in making the pledged cuts, and a review process has been set up to monitor emissions reductions and request adjustments as necessary. But there are two major problems with this agreement: first, the promised contributions to carbon emission reductions are woefully inadequate; second, the pledges are voluntary, not legally binding. In other words, Paris will not get us where we need to go. Critics of the agreement have pointed out that the emissions cuts promised in the accord would still result in an average rise in global temperatures of around 3.5 degrees Celsius, which would in turn result in "beyond catastrophic" climate disruptions. And while it was clear at the time that the emissions reduction commitments had to be voluntary ("should" be reduced, not "shall" be reduced) because of the negative political situation that Obama faced in the United States (with climate change–denying Republicans in control of the Senate), I argue that the pledges will have to be made legally binding ("shall" be reduced) in the future to be effective in the long run.

The inadequacies of the Paris Agreement were made even more painfully obvious with the release of the IPCC special report on global warming in October of 2018, which was commissioned by world leaders to explore the impacts of a rise in global temperatures of 1.5 degree Celsius above the pre-industrial average and related global greenhouse gas emission pathways. Recall that a reference to this more stringent threshold was inserted into the final document at Paris at the insistence of a number of countries, most from the Global South (D. Green, 2016). The landmark IPCC 1.5 report was written by ninety-one scientists from all over the world who analyzed more than 6,000 recent scientific studies related to global warming. Their conclusion was that "Without a radical transformation of energy, transportation and agriculture systems, the world will hurtle past the 1.5 degrees Celsius target by the middle of the century" (Berwyn, 2018b: 1). The IPCC asserted that the world was on the brink of failure to control global warming and that "urgent" and "unprecedented" changes are needed to reach the target, changes for which "there is no documented historic precedent" (Mooney and Dennis, 2018: 1). To keep warming under this threshold, the report found that global carbon emissions would have to be cut 45 percent by 2030 and reach net-zero emissions by 2050. This finding led some media commentators to conclude that the world has only twelve years to manage climate change before catastrophe is assured (Mooney and Dennis, 2018; Watts, 2018). Without a drastic transformation of the world economy, the IPCC predicts that by 2040 coastlines will flood, droughts will intensify, coral reefs will die, wildfires will increase, and poverty and food shortages will worsen, inflicting damages estimated at $54 trillion (Davenport, 2018: 1). These findings stress the urgency of limiting global warming to 1.5 degrees Celsius and reveal that the previously accepted threshold of 2 degrees Celsius "is now officially recognized as being catastrophically dangerous" (Athanasiou, 2018: 2).

As alarming as it was, the 2018 IPCC report probably underestimates climate dangers, as have previous IPCC assessment reports. Since these documents must be approved by all 195 member countries in a politicized process, they are generally more conservative in their findings and predictions. On top of that built-in conservatism, the writers of the 2018 report made no mention of three important issues. First, the IPCC does not discuss "the grim consequences of potential tipping points that could trigger conflicts over resources and mass migration" (Berwyn, 2018b: 2). The potential for "tipping elements," such as the loss of the Arctic-ice albedo effect, methane release from melting permafrost, and accelerating deforestation, to create runaway global warming that is out of human control has long been recognized. But a 2018 study published in the *Proceedings of the National Academy of Sciences of the United States of America* warned that the world may be just

1 degree Celsius of warming away from the risk that these dynamic, self-reinforcing feedback loops could push the earth system past a threshold that would result in continued warming on a "Hothouse Earth" pathway, even as human emissions are reduced (W. Steffen, Rockstrom, et al., 2018). Apart from the dangers of these tipping points, the IPCC report also fails to discuss the potential for the mass displacement of people and the problems that the developed nations will be confronted with as they are forced to deal with large numbers of climate change–driven refugees. A militarized response to climate refugees constitutes a form of climate crime, and justice demands that this type of violent and unfair adaptation be avoided. The final critical omission of the report was the failure to discuss the need for financial transfers to the Global South to assist those nations as they attempt to deal with climate disruptions, "transfers to secure the equity of the transition" that will be necessary to achieve climate justice (Athanasiou, 2018: 4).

Despite the frightening predictions about a world that has warmed more than 1.5 degrees Celsius (and these rather glaring omissions), the 2018 IPCC report does provide the international community with some good news: humanity still has time to avoid this awful fate. Although "it will require revolutionary changes in government and investment policies" that are unprecedented in history (Athanasiou, 2018: 4), global warming can still be limited to 1.5 degrees Celsius. To accomplish this critical but extremely difficult goal, however, the international political community has to recognize that the existing pledges to cut greenhouse gas emissions under the Paris Agreement do not come close to achieving the objective. Much steeper reductions must be made—and made very quickly. As Bob Berwyn (2018b: 3) of *Inside Climate News* contends, "Governments must get ready to commit to much more aggressive climate targets by 2020 at the latest, and they have to ditch coal." Not only must states commit to drastic cuts in carbon emissions, they must also negotiate a new climate treaty to replace the Paris Agreement, a stronger accord that makes these cuts legally binding under international law. Rising public condemnation of climate crimes must be linked to increasingly severe legal sanctions for those states and corporations that continue to engage in morally blameworthy climate harms.

The history of UNFCCC negotiations does not provide much assurance that these important goals will be realized. The 2018 1.5 degrees Celsius IPCC report was widely discussed at the Conference of the Parties (COP) 24 that December in Katowice, Poland, but the formal sessions focused more on developing a set of rules to guide countries' implementation of the inadequate Paris Agreement rather than creating a new, tougher, more adequate pact. As the "brutal news" that global carbon emissions had jumped to a record high in 2018 arrived on the eve of the Katowice conference (Carrington, 2018a), the Trump administration, along with some oil-friendly allies (Russia, Saudi Arabia, and Australia), was still trying to push the ludicrous idea that even more fossil fuels need to be extracted and simply burned more efficiently. COP 24 did approve a "rulebook" for implementing the Paris goals, but experts warned that the accord was "inadequate and lacked urgency" (Harvey, 2018b, 2018c). Katowice proved once again that despite popular campaigns for climate action over the years, meaningful global accords have repeatedly been "stifled by big power politics and diplomatic gridlock" (Tokar, 2014: 67). This has to change. The political will to create a new international agreement with the necessary and legally binding reductions in carbon emissions that can limit warming below 1.5 degrees Celsius must somehow be mustered.

There is no shortage of good ideas about what needs to be done, what specific elements could go into a new legally binding climate accord if the appropriate political will could be mustered. One of the most promising plans to phase out fossil fuels and make the great transition to renewables has come from Johan Rockström, executive director of the Stockholm Resilience Centre. He and his colleagues have developed "a roadmap for rapid decarbonization" that proposes global action over several decades. Their roadmap is based on a simple heuristic, a "carbon law" of cutting gross anthropogenic carbon dioxide emissions in half every decade from now to 2050. Rockström and his colleagues (2017: 1,270) suggest that future international climate talks could be based on their plan and that the UNFCCC "should transform into a vanguard forum where nations, businesses, nongovernmental organizations, and scientific communities meet to refine the roadmap." They argue that annual emissions from fossil fuels must start falling by 2020, proposing that "well-proven policy instruments" such as carbon tax schemes, cap-and-trade systems, feed-in tariffs, and quota approaches should "roll out at a wide scale" (Rockström et al., 2017: 1,270). They also propose that fossil fuel subsidies (currently at around $600 billion per year) should be eliminated by 2020 and a moratorium placed on investment in coal-based production facilities. The European researchers further suggest that all cities and major corporations should have "decarbonization strategies" in place by 2020 and go on to spell out additional "Herculean efforts" and "breakthrough" negative emissions technologies and renewable energy mechanisms that will be necessary to achieve net-zero emissions by 2050. Rockström and his colleagues (2017: 1,271) end with a plea for "climate stabilization" to be placed on par with economic development, human rights, democracy, and peace in the global governance process, arguing that "the design and implementation of the carbon roadmap should take center stage at the UN Security Council, as these quintessential objectives increasingly interact, influencing the stability and resilience of societies and the Earth System."

Rockström and colleagues' (2017) roadmap for rapid decarbonization, with its call for greater UN leadership, was a key component of a major report that resulted from the Global Climate Action Summit held in San Francisco in September 2018. The report, *Exponential Climate Action Roadmap*, was delivered by Rockström and former UNFCCC executive secretary Christiana Figueres, and it demonstrates the potential for all sectors of the global economy—energy, food and agriculture, industry, buildings, and transport—to halve greenhouse gas emissions by 2030. The trajectory leading to the report was borne of Rockström's roadmap, and the solutions database was drawn from Project Drawdown organized by American environmentalist Paul Hawken (2017), which describes itself as "the most comprehensive plan ever proposed to reverse global warming." The Global Climate Action Summit (2018) report also concludes that the transformation to renewable energy in the next decade could occur at a much faster pace than many forecast, as the price of solar and wind drops low enough to outcompete fossil fuels. That important conclusion is supported by a 2018 report from the International Renewable Energy Agency, which found that by 2020, electricity from renewables will be consistently cheaper than most fossil fuels (Dudley, 2018). In fact, American journalist Ryan Cooper (2018: 3) argues that wind and solar have become so cheap that, "in favorable locations, renewable energy is now able to stand without subsidies. The technology is largely where it needs to be, and it's getting better all the time." This conclusion about the potential for a transformation to clean energy is supported by a considerable amount of research (Klein, 2014; McKibben, 2016b, 2019a).

The speed and comprehensiveness of this global clean energy transformation is a major cause for hope. But the renewable energy transformation that is under way must continue to grow at an even more rapid pace. Market forces have driven much of this transition, but individual states and the international political community will have to play an increasingly larger role in this process (McKibben, 2019a). Lester Brown (2015: 135) points out several "basic policy instruments" that individual nation-states can use "to support the move to carbon-free renewable sources of energy." One instrument is the feed-in tariff, which guarantees renewable energy producers access to the grid and a long-term purchase price for the power they generate. Another government measure is called renewable portfolio standards, which mandates that a certain amount of electricity has to come from renewable sources. Tax credits are still another way that governments can support the deployment of wind and solar power. As Lester Brown (2015: 136) points out, "Such pro-renewable policies help level the playing field with artificially cheap fossil fuels that have been subsidized long past their debuts on the energy scene." State officials will have to be pressured to take these actions and others to support the transition to renewables.

In addition to national and subnational governmental activity, I still stress the critical importance of working at the global level. Even if tough decarbonization laws and public policies would be approved in individual nation-states, or by what Benjamin Barber (2017) calls "cool cities," strong international action remains essential to facilitating the shift from fossil fuels to clean energy. The 2018 IPCC report points out that the world will have to invest an average of around $3 trillion a year over the next three decades to transform its energy supply systems. The report does not call for a "new pot of money to be magically created" but suggests a "re-direction from investment in fossil fuels to efficiency and renewables" (McKenna, 2018: 2). Planetary cooperation and globally coordinated action will be necessary to redirect these investments, ramp up renewable energy and efficiency projects, and complete the Great Transition that is already under way. Stronger global governance and a new international climate treaty will still be required to decarbonize the world.

Not only must the international political community come up with legally binding targets for reducing greenhouse gas emissions and a global roadmap for transitioning to clean, renewable energy; it must also work to achieve climate justice. Issues such as global poverty and inequality, a just transition for fossil fuel workers, and fair, cooperative adaptation to climate disruptions must be incorporated into a new and stronger climate action treaty. A new global accord on climate change must make it a political and legal imperative for the nations of the Global North, who are primarily responsible for the vast majority of historical greenhouse gas emissions, to assist the Global South, particularly the smaller developing countries, in implementing mitigation and adaptation measures and in fostering sustainable development. As Mary Robinson (2018: 91), the former president of Ireland and past UN special envoy on climate change, argues, "the concept of climate justice must be broadened to ensure that smaller states are given a voice and a place at the negotiating table." The developed countries must provide both financial and technical assistance to the less developed nations that are often facing higher levels of poverty and economic inequality due to a long history of colonialism, neoliberal capitalist exploitation, and political and military domination.

Important as it is, the Green Climate Fund that was set up as a part of the Paris Agreement to aid the nations of the Global South with historically low carbon emissions is woefully inadequate. Climate activist and writer Oscar Reyes (2016: 1) estimates that mitigation finance transfers from the North to the South are currently around $15 billion annually but need to be upward of twenty times higher. He also points out that the "adaptation financing gap is even starker, with around $5 billion per year flowing from richer countries to poorer countries, compared with the $150 billion annually that may be needed by 2020—or the $500 billion needed by 2050 as the climate crisis worsens." The 2018 IPCC report, while omitting any mention of these types of financial transfers, does make frequent reference in its summary for policymakers to the critical issues of sustainable development, poverty alleviation, and the need to reduce economic inequality. As Mary Robinson (2018: 133) observes about the post-Paris climate negotiations, "As we pursue this new stage of bold action, we will succeed only if we recognize that the struggle to combat climate change is inextricably linked to tackling poverty, inequality, and exclusion. If we keep that in mind, our solutions will be more effective and more enduring."

This means that, in addition to dealing with structural inequalities of gender, race, and global region, the international community must also ensure that no fossil fuel workers or communities are economically disadvantaged by the transformation to renewable energy. This is the goal of a movement know as Just Transition. Mary Robinson (2018: 114) points out, "It is a people-first approach, arguing that workers should be given wage and benefit insurance, income support, and access to health care as they move from the fossil fuel sector into the clean-energy and other sectors." Climate activists have not always been as sensitive to the just transition issue as they should be. A consideration of this issue also highlights the fact that the transformation to clean energy can create millions of new jobs in what American economist Robert Pollin (2015) and others have called the Green New Deal. As Mary Robinson spells out in her book *Climate Justice: Hope, Resilience, and the Fight for a Sustainable Future* (2018), going forward after Paris, there must be a sustained and legally binding commitment to mitigation and adaptation solutions that concretely achieve social, economic, and political justice for smaller states in the Global South and displaced fossil fuel workers. Recall that women are also disproportionately victimized by climate disruption and must therefore also be the focus of special efforts to achieve climate justice, as a number of sociologists have so expertly spelled out (Nagel, 2016; Wonders and Danner, 2015).

Climate refugees are also an important concern. The problem, however, is that rather than striving for the kind of climate justice that Mary Robinson advocates for, many of the developed nations are responding to the climate crisis with the barbarous politics of the armed lifeboat. Instead of providing the needed technical and financial assistance and developing a formal agenda of economic redistribution on an international scale, the powerful states of the Global North are turning to the climate crime of militarized adaptation. This is especially true concerning the plight of internally displaced persons and climate refugees, and it has tragic consequences. As Ghosh (2016: 144) points out, "It goes without saying that if the world's most powerful nations adopt the 'politics of the armed lifeboat,' explicitly or otherwise, then millions of people in Asia, Africa, and elsewhere will face doom." Authoritarian and militaristic forms of adaptation to climate change are significantly related to the history of the American Empire and its culture of militarism. Ghosh (2016: 146) argues that "even if capitalism were to be magically transformed tomorrow, the imperatives of political and military dominance would remain a significant obstacle to progress on mitigatory action." This imperialism and militarism must be challenged.

In the current period, this means that the Trump administration's militarized border policies must be rolled back. As Parenti (2011: 226) argues, "we must find humane and just means of adaptation, or we face barbaric prospects." But this is a global problem, and planetary cooperation and worldwide action is necessary. The issue of how to move from "militarized borders to sustainable peace," as American sociologist Randall Amster (2015) puts it, has to be addressed and resolved through the international climate negotiation process. In *Peace Ecology*, Amster (2015: 193) reviews a substantial body of literature on the interrelated problems of environmental degradation and sociopolitical violence and attempts to "articulate a framework linking peace and nonviolence with ecology and sustainability, in the belief that this union is the crux of the matter." Amster describes emerging trends such as resource collaboration, conflict transformation, and sustainable relations, arguing that these practices will need to be expanded in a world facing increasing climate disruptions and conflicts. Peace ecology is possible. Humane and just means of climate adaptation do exist. As with most matters related to the climate crisis, the problem is not that we lack ideas or proposals for what to do; the problem is that those who are most concerned with these issues lack the political power to implement effective international or national policies that could result in mitigation and adaptation, effective policies that could resist the climate crimes of empire and produce climate justice.

THE SWERVE, CLIMATE MOBILIZATION, AND POLITICAL POWER

As my wife Jane and I marched in the "We Resist, We Build, We Rise" event organized by the Peoples Climate Movement in Washington, DC, on that hot April day back in 2017, we could sense the "climate swerve" described by Lifton (2017). The presence of several hundred thousand concerned citizens in our nation's capital (and hundreds of thousands of other marchers in various locations around the world that day), the political signs they carried, and the deeply felt messages they chanted demonstrated that a more informed awareness of the climate crisis was indeed evolving. To use Lifton's terms, the visual and audible displays of "species awareness" we experienced that day convinced us that a "mind set capable of constructive action" was truly emerging, and it did provide us with "a significant source of hope." At the same time, however, we were also struck by a grim awareness: the frightening reality that Trump was now president and that he and all the other climate change deniers in his administration seemed determined to go down in history as the greatest carbon criminals of all time. The emerging climate swerve was about to be confronted with the enormous obstacle of the Trump presidency.

Even before Trump became president, the Republican Party had devolved into one of the most "ideologically extreme" political parties in American history (T. Mann and Ornstein, 2016). The vast majority of Republicans have become extreme climate change deniers, and since they controlled both the House and the Senate during much of Obama's time in office, they were able to block or limit most of his efforts to deal with global warming that did not involve executive actions (Krugman, 2018a). This already dire situation was made worse by the election of Trump. As I have described, during the first three years of the Trump presidency, his administration engaged in all four of the major climate crimes analyzed in this book. First, Trump appointed climate change deniers to major cabinet positions such as secretary of state, secretary of energy, secretary of the interior, and head of the EPA. Second, at a time when oil, coal, and gas have to remain in the ground, Trump has been unwavering in his support for domestic fossil fuel production, which would permit and encourage the continuing extraction of these fuels and eventually result in more dangerous greenhouse gas emissions. Third, the Trump administration has attempted to roll back almost every element of the Climate Action Plan that Obama was able to accomplish, including automobile fuel-efficiency standards and the Clean Power Plan. In its environmental impact statement arguing against the Obama automobile mileage standards, the Trump administration made the alarming claim that since the planet is already on course to warm by a disastrous 7 degrees Fahrenheit (4 degrees Celsius), the increased carbon emissions that would be caused by this proposal do not really matter (Eilperin, Dennis, and Mooney, 2018). Climate activists were astounded by the cynical nature of the statement, and Noam Chomsky said that it "qualifies as a contender for the most evil document in history" (Horgan, 2018: 3). Trump has also announced that he will attempt to remove the United States from the Paris Agreement. Not only does he make it perfectly clear that he has no intention of doing anything to mitigate or adapt to climate change (the climate crime of political omission); he is actively seeking to destroy any previous government efforts to respond to the climate crisis. Finally, Trump has demonstrated deep hostility toward immigrants in general and climate refugees in particular and has ratcheted up the politics of the armed lifeboat through fear-mongering rhetoric and militaristic border policies.

Given the Trump administration's engagement in all four of the major forms of climate crimes, the Republican Party's blind adherence to a destructive neoliberal ideology, the party's public support of and organizational intersection with corporations in the fossil fuel industry, its extreme denial of climate change, and its political obstruction of any governmental efforts to mitigate or adapt to global warming, it is difficult to see how any elements of the climate action plan I outlined at the outset of this chapter could be enacted under the current government. Despite the fact that the Democratic Party won a majority of seats in the House of Representatives in the 2018 midterm elections, as long as Trump remains in the White House and an extremist Republican Party controls the Senate, climate action, including international treaties, will be blocked. Given this, the 2020 U.S. presidential and congressional elections may very well determine the climate future, not just for the United States but for the planet. Continued Republican control of the White House and the Senate will ensure, at a minimum, another four years of global warming denial and irresponsible inaction on the climate crisis; and at worst, the nightmare of what Naomi Klein (2019) calls "climate barbarism."

Even with a change in the White House and the Senate in 2021, it is difficult to see how climate actions—energy extraction infrastructure fights, fossil fuel divestment campaigns, disruption of the flow of money to coal, oil, and gas, climate litigation (whether directed at corporations or governments), international treaty negotiations, efforts to follow decarbonization roadmaps, or attempts to implement climate justice proposals—as separate political struggles will be able to hold global warming below 1.5 degrees Celsius. These separate political actions must be brought together to create a synergy that will enhance their effectiveness. Only a broad-based, integrated, worldwide mass social movement can mobilize the public sphere and international political community and bring about the "urgent and unprecedented" changes that the IPCC (2018) says are necessary, the "radical transformation" of the global economy and its energy systems that will allow us to meet this target.

I am not alone in drawing this conclusion. A number of environmental sociologists and climate activists have argued that only a mass social movement that is capable of disrupting existing power structures and reclaiming democratic politics will have a chance to decarbonize the world and achieve some measure of climate justice. In their study of shifting power relationships in international climate negotiations, for example, Ciplet, Roberts, and Khan (2015: 252) argue that only a transformed civil society (public sphere) approach that involves "linking movements" can bring about an

effective international climate treaty and advance climate justice. They conclude that "global climate justice necessitates a radical transformation from
isolated, fragmented, and top-down civil society efforts that conform to
dominant relations of power, to social movements that link grassroots activism to legislative efforts; unite unlikely, broad-based and diverse counterhe-
gemonic coalitions; and respond strategically to globalization's spatial
reorganization of environmental problems." Such an approach, Ciplet,
Roberts, and Khan (2015) argue, is the only way to end the continued
dominance of the fossil fuel corporations and their powerful state allies in
international climate negotiations.

In a series of studies featuring the comparative analysis of environmental movement organizations (EMOs), another team of sociologists has also
examined the evolution of the global climate movement, noting how it has
altered its strategies and tactics while attracting new actors to the effort.
Carl Cassegard, Linda Soneryd, Hakan Thorn, and Asa Wettergren (2017)
document how the development of global environmental networks has
increased contact between EMOs in the Global North and Global South,
and they also argue that the global climate movement's increasing reliance
on the normative frame of "climate justice" can help promote political
change. Their conclusion, too, is that a broad-based, integrated social movement will be necessary to deal with the climate crisis, noting that "managing climate change will require massive interventions in existing social
structures, necessitating discussions about burden sharing and responsibility, globally and nationally. To protect the rights and the interests of populations in the Global South, a diverse array of actors are today mobilizing
around the call for climate justice, including EMOs; women's, worker's and
youth movements; indigenous peoples and local communities" (Cassegard,
Soneryd, et al., 2017: 242).

Climate activists themselves have long recognized the need for greater
climate mobilization and movement building. As Naomi Klein (2014: 450)
has argued, "only mass social movements can save us now. Because we
know where the current system, left unchecked, is headed." She goes on to
point out that "the only remaining variable is whether some countervailing
power will emerge to block the road [to climate-related disasters], and
simultaneously clear some alternative pathways to destinations that are safer.
If that happens, well, it changes everything." Klein's call for the emergence
of some "countervailing power" has been echoed by other activists. In 2015,
prior to the Paris Climate Conference, key leaders from the global climate
justice movement came together and issued a joint statement titled "Stop
Climate Crimes," which affirmed their belief that only "mass popular mobi-
lizations across the planet" demanding a "drastic reckoning with the world's

fossil fuel paradigm will suffice" when it comes to confronting the threat of the climate crisis (350.org, 2015). In *Unprecedented Climate Mobilization: A Handbook for Citizens and Their Governments*, Elizabeth Woodworth and David Ray Griffin (2016: 16) argue that global society needs to shift into "emergency mode" and undertake "a full World War II style climate mobilization," an unprecedented "program of action" that will require "a leadership role for the United States." In a similar vein, Brecher (2017: 29) calls for the "rising up" of a "global nonviolent insurgency" to solve the problem of global warming. He notes that "the climate protection movement needs to weld the people of the world into an effective force capable of compelling corporations, governments, and institutions to shift from fossil fuels to clean energy."

In addition to calling for the creation of a larger, more engaged mass social movement that can become a greater consequential political force, these climate activists are also calling for more concerned citizens to become participants in a greater organized and effective politics of climate change. In "How the Active Many Can Overcome the Ruthless Few," Bill McKibben (2016a: 21) estimates that if just 10 percent of Americans "actually engaged in the work of politics," it "might well be sufficient to produce structural change of the size that would set us on a new course: a price on carbon, a commitment to massive subsidies for renewable energy, a legislative commitment to keep carbon in the ground." To speak of "politics" in this context means "addressing the key requisites of fundamental change: a well-defined alternative ideology, dynamic leadership and organization resources for mass mobilization, effective strategy, and an orientation toward winning governmental power" (Boggs, 2012: xvii). History shows that these key requisites have not always been present in environmental organizations or the broader climate movement. They will have to be developed more fully for a mass social movement to emerge as an effective political force. And, given the great power of the corporate state and the fossil fuel industry to resist change, mass civil disobedience will have to be used as a strategic tool in the struggle to save the planet. Both McKibben and Brecher discuss the important role that nonviolent resistance needs to play in the larger movement. In addition to the efforts of Blockadia, a recent example of a mass civil disobedience campaign comes from the United Kingdom where a group calling itself "Extinction Rebellion" rallied more than a thousand people to block Parliament Square in London on October 31, 2018. As Molly Scott Cato (2018: 2), a Green Party member of the European Parliament, explained, "Direct actions . . . have a long and proud history; it's time to carry them through in a systematic way to protect the climate, and be willing to be arrested for doing so." This action and other forms of nonviolent resistance that are taking place all over the world

(Extinction Rebellion chapters are now forming across the United States) are more indicators of the growing climate swerve.

In addition to mounting civil disobedience campaigns, a mass climate movement must take on a number of other political tasks in order to grow and be more effective. First, climate activists must counter the climate change denial countermovement and its false narratives. For far too long, conservative think tanks, the Republican Party, and other denialist organizations have been able to create doubts about climate science through their dishonest reports and tactics and therefore block state policies that would reduce greenhouse gas emissions. While research shows that some conservatives reject the scientific facts about climate change because it genuinely threatens their cultural worldview, ideological identity, and social group membership (Jost, Nosek, and Gosling, 2008; Kahan, Braman, et al., 2007; McCright and Dunlap, 2011), it is important to remember that social meanings and cultural beliefs are still rooted in historical social structures. As Klein (2011: 18) astutely points out, "The deniers are doing more than protecting their cultural worldview—they are protecting powerful interests that stand to gain from muddying the waters of the climate debate." Thus, socially organized climate change denial must be confronted, debated, and exposed as the propaganda that it is, as part of the larger political project.

A second political task for a mass climate movement is to continually challenge neoliberal ideology and its efforts to privatize the public sphere, deregulate the corporate sector, and lower taxes for the wealthy. The pursuit of these policies drastically reduces the role of the state in responding to serious social problems and regulating corporate harms at the very time that governmental action has become more imperative to deal with the climate crisis. As many scholars have made clear, climate change is first and foremost a political issue, and the state is central to any effort to mitigate carbon emissions or adapt to climate disruptions. Winning governmental power through the electoral process and using that power to decarbonize the economy, facilitate the great transition to renewable energy, and promote policies related to climate justice is critical. A mass climate movement will have to "come up with strategies that engage and attempt to transform the state" (Parenti, 2016b: 182).

A third and related task is to restructure the economic system. Since global warming is rooted in fossil capitalism and the destructive treadmill of production, some argue that the capitalist system itself must be confronted and changed, that an "ecological revolution" in the global political economy is necessary (Angus, 2016; Boggs, 2012; Foster, 2009; Moore, 2015). As the treadmill of production theorists have demonstrated, there is no doubt that capitalism causes environmental destruction in general and is the major social force that is responsible for the climate crisis in particular. No long-term solution to the problem of global warming is possible without a dramatic restructuring of the capitalist political economy. "System Change, Not Climate Change" is not an idle slogan. Parenti (2011: 241) argues, however, that we cannot afford to wait for an ecological or socialist revolution to occur. Time is not on our side. As he puts it, "The fact of the matter is time has run out on the climate issue. Either capitalism solves the crisis, or it destroys civilization." He goes on to assert that "we must begin immediately transforming the energy economy. Other necessary changes can and will follow." This means engaging the state.

A concrete example of how this process might work is currently taking place in the United States following the 2018 midterm elections, which brought a group of young, progressive activists to power. Led by Democrats Alexandria Ocasio-Cortez of New York, Rashida Tlaib of Michigan, and Ayanna Pressley of Massachusetts, a group of fourteen newly elected members of Congress, with strong support from the emerging Sunrise Movement (a youth-led climate activist group), called for the creation of a Select Committee for a Green New Deal (Lavelle, 2019). Under the proposal, this committee would be fully funded and empowered to draft climate legislation after consulting with a range of experts to map out a "detailed national, industrial, economic mobilization plan capable of making the U.S. economy carbon neutral while promoting economic and environmental justice and equality" (Klein, 2018b: 5). In February 2019, Representative Ocasio-Cortez and Senator Ed Markey followed up this proposal by introducing a nonbinding Green New Deal Resolution in Congress (Klein, 2019; Lavelle and Cushman, 2019). The resolution proposes a ten-year national mobilization to achieve net-zero greenhouse gas emissions, create millions of good, high-wage jobs, invest in infrastructure and industry, secure clean air and water, and promote justice and equity by addressing various forms of oppression (Brecher, 2019). While the Green New Deal has been attacked by conservative critics as "shocking" and "a call for enviro-socialism in America," environmental scholar Jedediah Britton-Purdy (2019: A21) argues that this proposal "is what realistic environmental policy looks like"; and Klein (2018b: 7) argues that "the Green New Deal has the capacity to mobilize a truly intersectional mass movement behind it—not despite its sweeping ambition, but precisely because of it."

As this ambitious Green New Deal proposal suggests, the climate movement must view the state as a site of social struggle and contend for governmental power to begin bringing about immediate regulations and reforms. Parenti (2016b: 182) argues that "To reform capitalism—and to move beyond it—the Left needs to place the state front and center in its strategic considerations." My belief is that the political goals of social reform (as illustrated by some elements of the Green New Deal) and the revolutionary transformation of the political economy are not necessarily mutually exclusive. As Ian Angus (2016: 222), a staunch advocate for ecosocialism, observes, "Fighting for immediate gains against capitalist destruction and fighting for the ecosocialist future are not separate activities, they are aspects of one integrated process." Progress toward significant social change, such as that required for dealing with the climate crisis, demands that we take multiple actions across multiple domains. As Rob White and I have written about societal efforts to prevent climate change ecocide, "History teaches us that momentum for revolutionary transformation must incorporate many different interest groups (in alliance formally or in united front), pursuing the struggle within existing state and civil institutions as well as fighting the power elite" (R. White and Kramer, 2015b: 7). It bears repeating, to save human civilization from climate change catastrophe, a broad-based, integrated, worldwide mass social movement must be brought together to quickly engage in the political struggle to effectively regulate, substantially reform, and ultimately restructure fossil capitalism to stop climate crimes. As Klein (2019: 261–262) insists, "The single largest determining factor in whether a Green New Deal mobilization pulls us back from the climate cliff will be the actions taken by social movements in the coming years." She goes on to argue that "as important as it is to elect politicians who are up for this fight, the decisive questions are not going to be settled through elections alone. At their core, they are about building political power—enough to change the calculus of what is possible."

The question is: Can such a mass social movement be mobilized in time to save the world from the unfolding climate catastrophe? That is, can such a movement actually win political power and use it effectively to implement a climate action plan or a Green New Deal style program? Can an unprecedented climate mobilization or insurgency stop climate crimes and hold carbon criminals accountable? Is Lifton (2017) correct in his assessment that the world's evolving awareness of our dire environmental predicament—the climate swerve—is, in fact, creating a mind-set capable of constructive actions? Does the global community have the political will to take the urgent and unprecedented actions in the next decade or two that the IPCC recommends to limit global warming to 1.5 degrees Celsius? Despite the mounting scientific evidence about the climate crisis that could very easily lead us to the brink of despair about the future of our children and grandchildren, is it still possible to have hope?

There is no way to really know the answers to these questions. It will depend on what we do as a global community, what social movements we join, what political actions we take. But, again, I choose optimism over despair. One reason to choose optimism, as Howard Zinn (1994: 207) points out in his memoir *You Can't Be Neutral on a Moving Train*, is that "pessimism becomes a self-fulfilling prophecy; it reproduces itself by crippling our willingness to act." A second reason that I hold out hope is that history is neither finite nor uniform. As Zinn (1994: 208) asserts, "To be hopeful in bad times is not just foolishly romantic. It is based on the fact that human history is a history not only of cruelty, but also of compassion, sacrifice, courage, kindness." When we look to history, we see that some past social movements, facing what appeared to be insurmountable obstacles, persevered and accomplished their significant goals. The U.S. abolitionist movement against slavery in the nineteenth century, the movement to secure human rights in the Soviet bloc that led to the fall of the Berlin Wall in 1989 and the eventual dissolution of the Soviet Union, and the anti-apartheid movement in South Africa that resulted in the election of Nelson Mandela in 1994 are prominent examples of historical swerves. As Zinn (1994: 208) observes, "If we remember those times and places—and there are so many where people have behaved magnificently, this gives us the energy to act, and at least the possibility of sending this spinning top of a world in a different direction."

The final reason that I have hope is that I sense that the climate swerve is real and building in an exponential manner. This book has provided a considerable amount of evidence that people, organizations, and broader movements all over the world are developing the species awareness that Lifton (2017) describes and starting to act on that awareness of the danger of global warming. As I was finishing this book, sixteen-year-old Greta Thunberg from Sweden inspired students from all over the world to "strike for climate," and her heroic protest has led to a nomination for the Nobel Peace Prize (Carrington, 2019). In both March and September 2019 millions of young people and others joined Greta in a global climate strike that appears to be building real momentum for a mass movement to demand action. The "climate kids" are coming "in massive and growing numbers," writes Mark Hertsgaard (2019: 25); they are rising up to take back their future, which they recognize is in grave danger. Naomi Klein (2019: 7) asserts that "young people around the world are cracking open the heart of the climate crisis, speaking of a deep longing for a future they thought that had but that is disappearing with each day that adults fail to act on the reality that we are in an emergency."

These climate kids, these young activists like Greta Thunberg inspire me. And as more and more of their elders join them and take whatever actions they can to bring about the necessary social changes, my hope is that we will reach a tipping point that leads to the emergence of a global mass movement and an avalanche of political action that mitigates carbon emissions, prevents the worst forms of climate disaster, and adapts in a just way to the harms that cannot be prevented. I also hope that a climate change criminology, speaking in the prophetic voice about state–corporate carbon criminals and their climate crimes, can make a meaningful contribution to that noble movement, help preserve a livable climate for our children and grandchildren, and safeguard the human prospect.

Acknowledgments

Writing a book is a social endeavor, and I have many people to thank for their assistance on this project.

First of all, I want to thank Rob White (University of Tasmania) for writing the foreword to the book and for his strong support of my work. Rob's writings on environmental harms and his exposition of a climate change criminology have been an inspiration to many criminologists. When I first started to study the field of green criminology, I encountered Rob's work at every turn. I remember remarking at the time to my wife, Jane, that "I have to meet this guy Rob White." Lucky for me, I did get to meet Rob shortly thereafter at a critical criminology conference, and subsequently we have done some writing together on the issue of climate change ecocide. I have learned much from his analysis of environmental crime in general and climate change criminology more specifically.

I started to work on this book in the Sonoran Desert near Goodyear, Arizona, while on sabbatical leave from Western Michigan University (WMU). I want to thank WMU for the opportunity to take that sabbatical and pursue this project; and I want to express my appreciation to the chair of the Department of Sociology, David Hartmann, and my departmental colleagues for their support as well.

In 2012, I was very fortunate to join with colleagues from across WMU to create an interdisciplinary study group on climate change under the auspices of the University Center for the Humanities. Our study group has evolved into the WMU Working Group on Climate Change, and I have learned much from my association with friends and colleagues in the group. Special thanks go to the chair of the Working Group, Denise Keele (Political Science and the Institute of the Environment and Sustainability), and Dave Karowe (Biological Sciences), Paul Clements (Political Science), Allen Webb (English), and Steve Bertman (Institute of the Environment and Sustainability).

I also want to thank Casey James Schotter, currently a doctoral candidate in the Department of Sociology at WMU, for his excellent research assistance, and for our many conversations about baseball.

Rob White, Denise Keele, Elizabeth (Lizzy) Stanley (Victoria University, Wellington, New Zealand), and Edo Weits (professor emeritus, Western Michigan University) all read the complete manuscript and gave me constructive feedback. Thanks to all of them.

I have learned a great deal about state–corporate crime from my former graduate students over the years. Special thanks go to Nancy Wonders (Northern Arizona University), Brian Smith (Henry Ford Community College), Dave Kauzlarich (University of North Carolina, Greensboro), Dawn Rothe (Florida Atlantic University), Rick Matthews (Carthage College), and Elizabeth Bradshaw (Central Michigan University).

Fellow criminologists and good friends David Friedrichs (University of Scranton), Gregg Barak (Eastern Michigan University), Mike Lynch (University of South Florida), Paul Stretesky (Northumbria University, United Kingdom), Kimberly Barrett (Eastern Michigan University), Peter Yeager (Boston University), Peter Iadicola (Indiana–Purdue University, Fort Wayne), Avi Brisman (Eastern Kentucky University), and Susan Carlson (Western Michigan University) have influenced my thinking about state–corporate crime and green criminology over the years. Special mention has to be made of the important influence that the late Bill Chambliss and Richard Quinney have had on my critical criminological imagination. And although I never got to meet him, I feel a special debt of gratitude to Edwin H. Sutherland, the sociologist who famously created the concept of white-collar crime.

Many of the ideas contained in this book were first presented at invited lectures and special conferences. These include the 2011 Annual Conference of the European Group for the Study of Deviance and Social Control, Universite de Savoie, Chambery, France (September 6); a keynote address to the 2012 conference Environmental Crime and Its Victims in Delft, the Netherlands (September 18); the 2015 International Conference on State Crime at the Free University of Berlin (February 11); the 2016 Ladhoff Lecture at Northern Arizona University (February 18); and the 2018 Dean's Lecture at the College of Justice and Safety, Eastern Kentucky University (March 22).

I discussed a number of the legal issues addressed in the book with my son, Andrew Kramer (an attorney in Wisconsin), and I thank him for his good counsel. I also want to express my special appreciation to Andrew and his wife, Dr. Sarah Endrizzi, for their love and support. This book is dedicated to their three sons (my grandsons): Truman, Malcolm, and Calvin Kramer.

My daughter, Sarah Kramer, a writer and editor with the Savannah College of Art and Design, was indispensable to this project. She skillfully edited the final draft, helped design the cover, and diligently prepared the index. She made this a much better book. Sarah and her husband, Chris

Cowgill, also provided love and support throughout the writing process, and I want to express my appreciation to them both.

My great friend and frequent writing partner Ray Michalowski (Northern Arizona University) was also indispensable to the production of this book. Ray and I created the concept and theory of state–corporate crime together, have collaborated on many research and writing projects over the years, and discussed every facet of this book from beginning to end. These discussions took place over the phone, through e-mail, and in person at criminology conferences (often over pancakes and coffee). Ray has shaped and influenced this book more than anyone else. He read every chapter and provided excellent editorial feedback and constructive criticism throughout. He also made this a much better book than it might have been. I cannot thank him enough.

I also cannot thank my wonderful wife, Jane, enough. She has lived with this book for many years, and she will never know how much she contributed to the project. Her assistance took many, many different forms, all of them important. All I can say is that I would not have been able to write this book without her. Her kindness, generosity, and genuine concern for others are constant inspirations for my work. My love and appreciation for her know no bounds.

REFERENCES

ABC News. 2016. "Obama Has Deported More People than Any Other President." *ABCNews.com* (August 29). Retrieved August 30, 2018 (https://abcnews.go.com /Politics/obamas-deportation-policy-numbers/story?id=41715661).

Agnew, Robert. 2011a. "Dire Forecast: A Theoretical Model of the Impact of Climate Change on Crime." *Theoretical Criminology* 16: 21–42.

———. 2011b. *Toward a Unified Criminology: Integrating Assumptions about Crime, People, and Society.* New York: New York University Press.

Aiken, Judge Ann. 2016. "Opinion and Order, Case No. 6;15-cv-01517-TC" (November 10), *Kelsey Cascadia Rose Juliana et al. v. United States of America, et al.* United States District Court for the District of Oregon, Eugene Division.

Amster, Randall. 2015. *Peace Ecology.* New York: Routledge.

Anderson, David, Mark Kasper, and David Pomerantz. 2017. *Utilities Knew: Documenting Electric Utilities' Early Knowledge and Ongoing Deception on Climate Change from 1968–2017.* San Francisco: Energy and Policy Institute.

Anderson, Fred, and Andrew Cayton. 2005. *The Dominion of War: Empire and Liberty in North America, 1500–2000.* New York: Penguin Books.

Anderson, Terry H. 2011. *Bush's Wars.* New York: Oxford University Press.

Angus, Ian. 2016. *Facing the Anthropocene: Fossil Capitalism and the Crisis of the Earth System.* New York: Monthly Review Press.

Antonio, Robert J., and Robert J. Brulle. 2011. "The Unbearable Lightness of Politics: Climate Change Denial and Political Polarization." *Sociological Quarterly* 52: 195–202.

Antonio, Robert J., and Brett Clark. 2015. "The Climate Change Divide in Social Theory." In *Climate Change and Society: Sociological Perspectives,* edited by R. Dunlap and R. J. Brulle, 333–368. New York: Oxford University Press.

Appy, Christian G. 2015. *American Reckoning: The Vietnam War and Our National Identity.* New York: Viking.

Armstrong, David. 2002. "Dick Cheney's Song of America: Drafting a Plan for Global Dominance." *Harper's Magazine* (October): 76–83.

Aronoff, Kate. 2018. "Denial by a Different Name: It's Time to Admit That Half-Measures Can't Stop Climate Change." *Intercept* (April 17). Retrieved May 4, 2018 (https://theintercept.com/2018/04/17/climate-change-denial-trump-germany/).

———. 2019. "It's Time to Try Fossil-Fuel Executives for Crimes against Humanity." *Jacobin* (February 5). Retrieved February 5, 2019 (https://jacobinmag.com/2019 /02/fossil-fuels-climate-change-crimes-against-humanity).

Associated Press. 2018. "Dutch Appeals Court Upholds Landmark Climate Case Ruling." *New York Times* (October 9). Retrieved October 17, 2018 (https://www .nytimes.com/aponline/2018/10/09/world/europe/ap-eu-netherlands-climate -change.html).

Athanasiou, Tom. 2018. "1.5 to Stay Alive, Says Landmark UN Climate Report." *Nation* (October 9). Retrieved October 9, 2018 (https://www.thenation.com /article/1-5-to-stay-alive-says-landmark-un-climate-report/).

Atkins, David. 2015. "Exxon Should be Prosecuted for Covering up Its Knowledge of Fossil Fuel-Induced Climate Change." *Washington Monthly* (October 18). Retrieved August 9, 2016 (https://washingtonmonthly.com/2015/10/18/exxon-should-be -prosecuted-for-covering-up-its-knowledge-of-fossil-fuel-induced-climate-change/).

Bacevich, Andrew J. 2002. *American Empire: The Realities and Consequences of U.S. Diplomacy.* Cambridge, MA: Harvard University Press.

———. 2005. *The New American Militarism: How Americans Are Seduced by War.* New York: Oxford University Press.

———. 2016. *America's War for the Greater Middle East: A Military History.* New York: Random House.

Bachrach, Peter, and Morton S. Baratz. 1970. *Power and Poverty: Theory and Practice.* New York: Oxford University Press.

Baker, Brandon. 2014. "Dr. David Suzuki Tells Bill Moyers Why It's Time to Get Real on Climate Change." *EcoWatch* (May 11). Retrieved May 12, 2014 (https:// www.ecowatch.com/dr-david-suzuki-tells-bill-moyers-why-its-time-to-get-real -on-climate—1881911418.html).

Baker, Peter. 2009. "The Mellowing of William Jefferson Clinton." *New York Times Magazine* (May 26). Retrieved June 10, 2018 (https://www.nytimes.com/2009/05 /31/magazine/31clintont.html?_r=1&ref=magazine&pagewanted=all).

Baker, Peter, and Coral Davenport. 2017. "Trump Revived Keystone Pipeline Rejected by Obama." *New York Times* (January 24). Retrieved November 12, 2017 (https:// www.nytimes.com/2017/01/24/us/politics/keystone-dakota-pipeline-trump .html).

Banerjee, Neela. 2015. "Exxon's Oil Industry Peers Knew about Climate Dangers in the 1970s, Too." *Inside Climate News* (December 22). Retrieved June 10, 2017 (https://insideclimatenews.org/news/22122015/exxon-mobil-oil-industry-peers -knew-about-climate-change-dangers-1970s-american-petroleum-institute-api -shell-chevron-texaco).

———. 2017. "How Big Oil Lost Control of Its Climate Misinformation Machine." *Inside Climate News* (December 22). Retrieved December 23, 2017 (https://inside climatenews.org/news/22122017/big-oil-heartland-climate-science-misinformation -campaign-koch-api-trump-infographic).

———. 2019. "How Much Would Trump's Climate Rule Rollbacks Worsen Health and Emissions?" *Inside Climate News* (March 6). Retrieved March 6, 2019 (https:// insideclimatenews.org/news/06032019/trump-climate-regulations-rollback-cost -health-emissions-clean-power-plan-cars-oil-gas-methane).

Banerjee, Neela, John H. Cushman, David Hasemyer, and Lisa Song. 2015. *Exxon: The Road Not Taken.* New York: Inside Climate News.

———. 2016. "CO2's Role in Global Warming Has Been on the Oil Industry's Radar since the 1960s." *Inside Climate News* (April 13). Retrieved November 7, 2017 (https://insideclimatenews.org/news/13042016/climate-change-global-warm ing-oil-industry-radar-1960s-exxon-api-co2-fossil-fuels).

Banerjee, Neela, Lisa Song, and David Hasemyer. 2015a. "Exxon Believed Deep Dive into Climate Research Would Protect Its Business." *Inside Climate News* (September 17). Retrieved September 21, 2015 (https://insideclimatenews.org/news /16092015/exxon-believed-deep-dive-into-climate-research-would-protect-its -business).

———. 2015b. "Exxon's Own Research Confirmed Fossil Fuels' Role in Global Warming Decades Ago." *Inside Climate News* (September 16). Retrieved September 21, 2015 (https://insideclimatenews.org/news/15092015/Exxons-own-research -confirmed-fossil-fuels-role-in-global-warming).

Barak, Gregg, ed. 1991. *Crimes by the Capitalist State: An Introduction to State Criminality.* Albany: State University of New York Press.

———. 2012. *Theft of a Nation: Wall Street Looting and Federal Regulatory Colluding.* Lanham, MD: Rowman & Littlefield.

———, ed. 2015. *The Routledge International Handbook of the Crimes of the Powerful.* New York: Routledge.

Barber, Benjamin R. 2017. *Cool Cities: Urban Sovereignty and the Fix for Global Warming.* New Haven, CT: Yale University Press.

Barnosky, Anthony D., Elizabeth A. Hadly, Jordi Bascompte, et. al., 2012. "Approaching a State Shift in Earth's Biosphere." *Nature* 486: 52–58.

Bastasch, Michael. 2016. "'Untapped Energy': Trump Promises a 750 Trillion Economic Stimulus." *Daily Caller* (September 23). Retrieved November 5, 2016 (http://dailycaller.com/2016/09/23/untapped-energy-trump-promises-a-50 -trillion-economic-stimulus).

Battistoni, Alyssa. 2018. "There's No Time for Gradualism." *Jacobin* (October 9). Retrieved October 9, 2018 (https://www.jacobinmag.com/2018/10/climate-change -united-nations-report-nordhaus-nobel).

Batstrand, Sondre. 2015. "More than Markets: A Comparative Study of Nine Conservative Parties on Climate Change." *Politics & Policy* 43: 538–561.

BBC. 2009. "Climate Talks Battle to Save Deal." *BBC News* (December 19). Retrieved September 19, 2011 (http://news.bbc.co.uk/2/hi/science/nature/8422031.stm).

Beder, Sharon. 1999. "Corporate Hijacking of the Greenhouse Debate." *Ecologist* 29: 119–122.

Benedictus, Leo. 2016. "Noam Chomsky on Donald Trump: 'Almost a Death Knell for the Human Species.'" *Guardian* (May 20). Retrieved February 2, 2017 (http:// www.theguardian.com/film/2016/may/20/noam-chomsky-on-donald-trump -almost-a-death-knell-for-the-human-species).

Bennett, Jeffrey. 2016. *A Global Warming Primer.* Boulder, CO: Big Kid Science.

Bennis, Phyllis. 2006. *Challenging Empire: How People, Governments, and the UN Defy U.S. Power.* Northampton, MA: Olive Branch Press.

Berwyn, Bob. 2017. "Climate Change Is Happening Faster than Expected, and It's More Extreme." *Inside Climate News* (December 26). Retrieved December 26, 2017 (https://insideclimatenews.org/news/26122017/climate-change-science-2017-year -review-evidence-impact-faster-more-extreme).

———. 2018a. "Global Warming Is Messing with the Jet Stream. That Means More Extreme Weather." *Inside Climate News* (October 31). Retrieved November 1, 2018 (https://insideclimatenews.org/news/31102018/jet-stream-climate-change -study-extreme-weather-arctic-amplification-temperature).

———. 2018b. "IPCC: Radical Energy Transformation Needed to Avoid 1.5 Degrees Global Warming." *Inside Climate News* (October 8). Retrieved October 8, 2018 (https://insideclimatenews.org/news/07102018/ipcc-climate-change-science-report -data-carbon-emissions-heat-waves-extreme-weather-oil-gas-agriculture).

Berwyn, Bob, and Zahra Hirji. 2016. "As Coral Bleaching Goes Global, Scientists Fear Worst Is Yet to Come." *Inside Climate News* (June 8). Retrieved July 20, 2018 (https://insideclimatenews.org/news/07062016/coral-bleaching-alarms-scientists -climate-change-global-warming-great-barrier-reef).

Betts, Richard A., Chris D. Jones, Jeff R. Knight, et al. 2016. "El Nino and a Record CO2 Rise." *Nature Climate Change* 6: 806–810.

Biello, David. 2015. "How Far Does Obama's Clean Power Plan Go in Slowing Climate Change?" *Scientific American* (August 6). Retrieved October 23, 2017 (https://www.scientificamerican.com/article/how-far-does-obama-s-clean-power-plan-go-in-slowing-climate-change/).

Blum, William. 2004. *Killing Hope: U.S. Military and C.I.A. Interventions since World War II*. Monroe, ME: Common Courage Press.

Blumm, Michael C., and Mary Christina Wood. 2017. "'No Ordinary Lawsuit': Climate Change, Due Process, and the Public Trust Doctrine." *American University Law Review* 67: 101–185.

Boggs, Carl. 2010. *The Crimes of Empire: Rogue Superpower and World Domination*. London: Pluto Press.

———. 2012. *Ecology and Revolution: Global Crisis and the Political Challenge*. New York: Palgrave Macmillan

———. 2017. *Origins of the Warfare State: World War II and the Transformation of American Politics*. New York: Routledge.

Bonds, Eric. 2015. "Challenging Global Warming's New 'Security Threat' Status." *Peace Review: A Journal of Social Justice* 27: 209–216.

———. 2016. "Upending Climate Violence Research: Fossil Fuel Corporations and the Structural Violence of Climate Change." *Human Ecology Review* 22: 3–23.

Borger, Julian. 2001. "Bush Kills Global Warming Treaty." *Guardian* (March 29). Retrieved February 20, 2018 (http://www.theguardian.com/environment/2001/mar/29/globalwarming.usnews).

———. 2017. "Trump Drops Climate Change from U.S. National Security Strategy." *Guardian* (December 18). Retrieved August 24, 2018 (https://www.theguardian.com/us-news/2017/dec/18/trump-drop-climate-change-national-security-strategy).

Bradshaw, Elizabeth A. 2014. "State–Corporate Environmental Cover-Up: The Response to the 2010 Gulf of Mexico Oil Spill." *State Crime: Journal of the International State Crime Initiative* 3: 163–181.

———. 2015a. "Blacking out the Gulf: State–Corporate Environmental Crime and the Response to the 2010 BP Oil Spill." In *The Routledge Handbook of the Crimes of the Powerful*, edited by G. Barak, 363–372. New York: Routledge.

———. 2015b. "'Obviously, We're All Oil Industry': The Criminogenic Structure of the Offshore Oil Industry." *Theoretical Criminology* 19: 376–395.

Braithwaite, John. 1989. *Crime, Shame and Reintegration*. Cambridge: Cambridge University Press.

Brecher, Jeremy. 2015. *Climate Insurgency: A Strategy for Survival*. Boulder, CO: Paradigm Press.

———. 2017. *Against Doom: A Climate Insurgency Manual*. Oakland, CA: PM Press.

———. 2019. *18 Strategies for a Green New Deal: How to Make the Climate Mobilization Work*. Labor Network for Sustainability Discussion Paper. Takoma Park, MD: Voices for a Sustainable Future.

Britton-Purdy, Jedediah. 2019. "The Green New Deal's Realism." *New York Times* (February 16): A21.

Bromwich, David. 2019. *American Breakdown: The Trump Years and How They Befall Us*. London: Verso.

Brown, Donald. 2010. "Is Climate Science Disinformation a Crime against Humanity?" *Guardian* (November 1). Retrieved March 26, 2011 (https://www.theguardian

.com/environment/cif-green/2010/nov/01/climate-science-disinformation
-crime).

Brown, Lester R. 2015. *The Great Transition: Shifting from Fossil Fuels to Solar and Wind Energy*. New York: W. W. Norton.

Brown, Patrick T., and Ken Caldeira. 2017. "Greater Future Global Warming Inferred from Earth's Recent Energy Budget." *Nature* 552: 45–50.

Brulle, Robert J. 2013. "Environmentalisms in the United States." In *Environmental Movements around the World Volume 1,* edited by T. Doyle and S. MacGregor, 163–194 Santa Barbara, CA: Praeger.

———. 2014. "Institutionalizing Delay: Foundation Funding and the Creation of U.S. Climate Change Counter-Movement Organizations." *Climatic Change* 122: 681–694.

Buck, Holly Jean. 2018. "The Need for Carbon Removal." *Jacobin* (July 24). Retrieved July 25, 2018 (https://www.jacobinmag.com/2018/07/carbon-removal-geoengineer ing-global-warming).

Burawoy, Michael. 2007. "For Public Sociology." In *Public Sociology: Fifteen Eminent Sociologists Debate Politics and the Profession in the Twenty-First Century,* edited by D. Clawson, R. Zussman, J. Misra, et al., 23–64. Berkeley: University of California Press.

Burns, Ronald G., Michael J. Lynch, and Paul Stretesky. 2008. *Environmental Law, Crime, and Justice*. New York: LFB Scholarly Publishing.

Butt, Ahsan I. 2019. "Why Did Bush Go to War in Iraq? The Answer Is More Sinister than You Think." *Common Dreams* (March 20). Retrieved March 21, 2019 (https:// www.commondreams.org/views/2019/03/20/why-did-bush-go-war-iraq-answer -more-sinister-you-think).

Buxton, Nick, and Ben Hayes, eds. 2016. *The Secure and the Dispossessed: How the Military and Corporations Are Shaping a Climate Changed World*. London: Pluto Press.

Callinicos, Alex. 2003. *The New Mandarins of American Power: The Bush Administration's Plans for the World*. Cambridge, UK: Polity Press.

Campbell, Kurt M., Jay Gulledge, J. R. McNeill, et al. 2007. *The Age of Consequences: The Foreign Policy and National Security Implications of Global Climate Change*. Washington, DC: Center for Strategic and International Studies.

Caney, Simon. 2010. "Climate Change, Human Rights, and Moral Thresholds." In *Climate Ethics: Essential Readings,* edited by S. Gardiner, S. Caney, D. Jamieson, and H. Shue, 163–176. New York: Oxford University Press.

Carmin, Joann, Kathleen Tierney, Eric Chu, et al. 2015. "Adaptation to Climate Change." In *Climate Change and Society: Sociological Perspectives,* edited by R. E. Dunlap and R. J. Brulle, 164–198. New York: Oxford University Press.

Carrington, Damian. 2016. "Climate Change Will Stir 'Unimaginable' Refugee Crisis, Says Military." *Guardian* (December 1). Retrieved December 2, 2016 (http:// www.theguardian.com/environment/2016/dec/01/climate-change-trigger -unimaginable-refugee-crisis-senior-military).

———. 2017. "Green Movement 'Greatest Threat to Freedom,' Says Trump Advisor." *Guardian* (January 30). Retrieved November 6, 2017 (http://www.theguardian .com/environment/2017/jan/30/green-movement-greatest-threat-freedom-says -trump-advisor-myron-ebell).

———. 2018a. "'Brutal News': Global Carbon Emissions Jump to All-Time High in 2018." *Guardian* (December 5). Retrieved December 6, 2018 (https://www .theguardian.com/environment/2018/dec/05/brutal-news-global-carbon -emissions-jump-to-all-time-high-in-2018).

————. 2018b. "Can Climate Litigation Save the World?" *Guardian* (March 20). Retrieved March 29, 2018 (https://www.theguardian.com/environment/2018/mar/20/can-climate-litigation-save-the-world).

————. 2018c. "Fossil Fuel Divestment Funds Rise to $6 tn." *Guardian* (September 10). Retrieved September 11, 2018 (https://www.theguardian.com/environment/2018/sep/10/fossil-fuel-divestment-funds-rise-to-6tn).

————. 2018d. "Humanity Has Wiped out 60% of Animal Populations since 1970, Report Shows." *Guardian* (October 29). Retrieved October 30, 2018 (https://www.theguardian.com/environment/2018/oct/30/humanity-wiped-out-animals-since-1970-major-report-finds).

————. 2019. "Greta Thunberg Nominated for Nobel Peace Prize." *Guardian* (March 14). Retrieved March 16, 2019 (https://www.theguardian.com/world/2019/mar/14/greta-thunberg-nominated-nobel-peace-prize).

Carrington, Damian, and Jelmer Mommers. 2017. "Shell's 1991 Warning: Climate Changing 'at Faster Rate than at Any Time since End of Ice Age.'" *Guardian* (February 28). Retrieved March 1, 2017 (https://www.theguardian.com/environment/2017/feb/28/shell-film-warning-climate-change-rate-faster-than-end-ice-age).

Carter, Peter D., and Elizabeth Woodworth. 2018. *Unprecedented Crime: Climate Science Denial and Game Changers for Survival.* Atlanta: Clarity Press.

Cassegard, Carl, Linda Soneryd, Haken Thorn, and Asa Wettergren, eds. 2017. *Climate Change in a Globalizing World: Comparative Perspectives on Environmental Movements in the Global North.* New York: Routledge.

Cato, Molly Scott. 2018. "Why I'm Turning from Law-Maker to Law-Breaker to Try to Save the Planet." *Guardian* (October 31). Retrieved November 1, 2018 (https://www.theguardian.com/commentisfree/2018/oct/31/law-breaker-save-planet-direct-action-civil-disobedience).

Ceccarelli, Leah. 2011. "Manufactured Scientific Controversy: Science, Rhetoric, and Public Debate." *Rhetoric and Public Affairs* 14: 195–228.

Center for Naval Analyses. 2007. *National Security and the Threat of Climate Change.* Retrieved August 23, 2018 (https://www.cna.org/cna_files/pdf/national%20security%20and%20the%20threat%20of%20climate%20change.pdf).

Chambliss, William. 1989. "State-Organized Crime: The American Society of Criminology, 1988 Presidential Address." *Criminology* 27: 183–208.

Charney, Jule. 1979. *Carbon Dioxide and Climate: A Scientific Assessment, Report of an Ad Hoc Study Group on Carbon Dioxide and Climate, Woods Hole, Massachusetts, July 23–27, 1979, to the Climate Research Board, National Research Council.* Washington, DC: National Academies Press.

Chernus, Ira. 2006. *Monsters to Destroy: The Neoconservative War on Terror and Sin.* Boulder, CO: Paradigm Publishers.

Chomsky, Aviva. 2018. "Talking Sense about Immigration: Rejecting the President's Manichaean Worldview." *TomDispatch* (March 13). Retrieved March 13, 2018 (http://www.tomdispatch.com/blog/176397).

Chomsky, Noam. 2000. *Rogue States: The Rule of Force in World Affairs.* Boston: South End Press.

————. 2003. *Hegemony or Survival: America's Quest for Global Dominance.* New York: Metropolitan Books.

————. 2013. *Nuclear War and Environmental Catastrophe* (with Larry Polk). New York: Seven Stories Press.

————. 2017. *Optimism over Despair: On Capitalism, Empire, and Social Change.* Chicago: Haymarket Books.

Ciplet, David, J. Timmons Roberts, and Mizan R. Khan. 2015. *Power in a Warming World: The New Global Politics of Climate Change and the Remaking of Environmental Inequality.* Cambridge, MA: The MIT Press.

Clark, Brett, and Andrew K. Jorgenson. 2012. "The Treadmill of Destruction and the Environmental Impacts of Militaries." *Sociology Compass* 6/7: 557–569.

Clark, Brett, Andrew K. Jorgenson, and Jeffrey Kentor. 2010. "Militarization and Energy Consumption." *International Journal of Sociology* 40: 23–43.

Clarke, Richard A. 2004. *Against All Enemies: Inside America's War on Terror.* New York: Free Press.

Clarke, Tony. 2008. *Tar Sands Showdown: Canada and the New Politics of Oil in an Age of Climate Change.* Toronto: James Lorimer.

Coady, David, Ian Parry, Louis Sears, and Baoping Shang. 2016. "How Large Are Global Fossil Fuel Subsidies?" *World Development* 91: 11–27

Cohan, William D. 2014. "To Save the Planet, We Need to Leave Fossil Fuels in the Ground—but Oil Companies Have Other Plans." *Nation* (December 10). Retrieved September 10, 2017 (https://www.thenation.com/article/save-planet -we-need-leave-fossil-fuels-ground-oil-companies-have-other-plans/).

Cohen, Stanley. 2001. *States of Denial: Knowing about Atrocities and Suffering.* Cambridge, UK: Polity Press.

Cole, Juan. 2015. "After the COP 21 Paris Climate Accord, What We Need Is an International Climate Court." *Truthdig* (December 13). Retrieved September 21, 2018 (https://www.truthdig.com/videos/after-the-cop21-paris-climate-accord -what-we-need-is-an-international-climate-court/).

———. 2019. "How Bush's War on Iraq Produced Trumpism, Instability, Refugees and Climate Catastrophe." *Informed Comment* (March 21). Retrieved March 22, 2019 (https://www.juancole.com/2019/03/trumpism-instability-catastrophe.html).

Coleman, Roy, Joe Sim, Steve Tombs, and David Whyte, eds. 2009. *State Power Crime.* London: Sage.

Coll, Steve. 2012. *Private Empire: ExxonMobil and American Power.* New York: Penguin Books.

Common Dreams. 2015. "Seething with Anger, Probe Demanded into Exxon's Unparalleled Climate Crime." *Common Dreams* (October 23). Retrieved October 23, 2015 (https://www.commondreams.org/news/2015/10/30/seething-anger-probe -demanded-exxons-unparalleled-climate-crime).

———. 2016. "SEC Probe Called Possible 'Moment of Reckoning' for Exxon's Climate Crimes." *Common Dreams* (September 20). Retrieved September 21, 2016 (https:// www.commondreams.org/news/2016/09/20/sec-probe-called-possible-moment -reckoning-exxons-climate-crimes).

———. 2019. "'In a Just World, It Would Be Treated as Crime against Humanity': New Report Exposes Big Oil's Real Agenda." *Common Dreams* (March 22). Retrieved March 22, 2019 (https://www.commondreams.org/news/2019/03/22 /just-world-it-would-be-treated-crime-against-humanity-new-report-exposes -big-oils).

Conason, Joe. 2016. *Man of the World: The Further Endeavors of Bill Clinton.* New York: Simon & Schuster.

Conley, Julia. 2018. "Urging Multi-Pronged Effort to Halt Climate Crisis, Scientists Say Protecting World's Forests as Vital as Cutting Emissions." *Common Dreams* (October 4). Retrieved October 4, 2018 (https://www.commondreams.org/news /2018/10/04/urging-multi-pronged-effort-halt-climate-crisis-scientists-say -protecting-worlds).

Cooper, Ryan. 2018. "The Case against Despair on Climate Change." *Week* (October 10). Retrieved October 11, 2018 (https://theweek.com/articles/800922/case-against-despair-climate-change).

Corbett, Jessica. 2018a. "As CO2 Levels Soar Past 'Troubling' 410 ppm Threshold, Trump Kills NASA Carbon Monitoring Program." *Common Dreams* (May 11). Retrieved May 13, 2018 (https://www.commondreams.org/news/2018/05/11/co2-levels-soar-past-troubling-410-ppm-threshold-trump-kills-nasa-carbon-monitoring).

———. 2018b. "'Watershed Moment for Climate Liability' as Rhode Island Files Historic Lawsuit against 21 Big Oil Companies." *Common Dreams* (July 2). Retrieved July 2, 2018 (https://www.commondreams.org/news/2018/07/02/watershed-moment-climate-liability-rhode-island-files-historic-lawsuit-against-21).

———. 2019. "From Premature Deaths to Planet-Heating Emissions, Analysis Reveals Costs of Trump's Fossil Fuel Giveaways." *Common Dreams* (January 28). Retrieved January 28, 2019 (https://www.commondreams.org/news/2019/01/28/premature-deaths-planet-heating-emissions-analysis-reveals-costs-trumps-fossil-fuel).

Coumou, Dim, Giorgia Di Capua, Steve Vavrus, et al. 2018. "The Influence of Artic Amplification on Mid-Latitude Summer Circulation." *Nature Communications* 9: 1–12.

Crank, John P., and Linda S. Jacoby. 2015. *Crime, Violence, and Global Warming.* London: Routledge.

Crawford, Neta C. 2019. *Pentagon Fuel Use, Climate Change, and the Costs of War.* Providence, RI: Watson Institute of International & Public Affairs, Brown University.

Currie, Elliott. 2007. "Against Marginality: Arguments for a Public Criminology." *Theoretical Criminology* 11: 175–190.

Cushman, John H. 1996. "In Shift, U.S. Will Seek Binding World Pact to Combat Global Warming." *New York Times* (July 17). Retrieved June 6, 2018 (https://www.nytimes.com/1996/07/17/world/in-shift-us-will-seek-binding-world-pact-to-combat-global-warming.html).

———. 1997. "Intense Lobbying against Global Warming Treaty." *New York Times* (December 7). Retrieved June 6, 2018 (https://www.nytimes.com/1997/12/07/us/intense-lobbying-against-global-warming-treaty.html).

———. 2018. "Shell Knew Fossil Fuels Created Climate Change Risks back in 1980s, Internal Documents Show." *Inside Climate News* (April 5). Retrieved April 5, 2018 (https://insideclimatenews.org/news/05042018/shell-knew-scientists-climate-change-risks-fossil-fuels-global-warming-company-documents-netherlands-lawsuits).

Cushman, John H., and David Hasemyer. 2018. "Judge Rejects Exxon's Attempt to Shut down Climate Fraud Investigations." *Inside Climate News* (March 29). Retrieved May 9, 2018 (https://insideclimatenews.org/news/29032018/exxon-climate-fraud-investigations-ruling-federal-judge-caproni-new-york-schneiderman-massachusetts-healey).

Cushman, John H., and Marianne Lavelle. 2017. "Inside the White House War over the Paris Climate Treaty." *Inside Climate News* (May 5). Retrieved May 31, 2017 (https://insideclimatenews.org/news/05052017/paris-agreement-climate-change-donald-trump-jared-kushner-steve-bannon-clean-power-plan).

Cutler, David, and Francesca Dominici. 2018. "A Breath of Bad Air: Cost of the Trump Environmental Agenda May Lead to 80,000 Extra Deaths per Decade." *Journal of the American Medical Association* 319: 2261–2262.

Davenport, Coral. 2014. "With Compromises, a Global Accord to Fight Climate Change Is in Sight." *New York Times* (December 9). Retrieved December 9, 2014

(https://www.nytimes.com/2014/12/10/world/with-compromises-a-global
-accord-to-fight-climate-change-is-in-sight.html).

———. 2015a. "Citing Climate Change, Obama Rejects Construction of Keystone
XL Oil Pipeline." *New York Times* (November 6). Retrieved November 7, 2015
(https://www.nytimes.com/2015/11/07/us/obama-expected-to-reject
-construction-of-keystone-xl-oil-pipeline.html).

———. 2015b. "Nations Approve of Landmark Climate Deal." *New York Times*
(December 13): A1

———. 2017a. "Climate Change Denialists in Charge." *New York Times* (March 27).
Retrieved March 27, 2017 (https://www.nytimes.com/2017/03/27/us/politics
/climate-change-denialists-in-charge.html).

———. 2017b. "E.P.A. Chief Doubts Consensus View of Climate Change." *New
York Times* (March 9). Retrieved March 9, 2017 (https://www.nytimes.com/2017
/03/09/us/politics/epa-scott-pruitt-global-warming.html).

———. 2017c. "Trump to Undo Vehicle Rules That Curb Global Warming." *New
York Times* (March 3). Retrieved March 3, 2017 (https://www.nytimes.com/2017
/03/03/us/politics/trump-vehicle-emissions-regulation.html).

———. 2018. "Major Climate Report Describes a Strong Risk of Crisis as Early as
2040." *New York Times* (October 7). Retrieved October 8, 2018 (https://www
.nytimes.com/2018/10/07/climate/ipcc-climate-report-2040.html).

———. 2019. "Trump to Scrap California's Role on Car Emissions" *New York Times*
(September 18): A1.

Davenport, Coral, and Mark Landler. 2019. "Trump Administration Hardens Its
Attack on Climate Science." *New York Times* (May 27). Retrieved May 27, 2019
(https://www.nytimes.com/2019/05/27/us/politics/trump-climate-science
.html).

Davenport, Coral, and Eric Lipton. 2017. "The Pruitt Emails: E.P.A. Chief Was Arm in
Arm with Industry." *New York Times* (February 22). Retrieved February 22, 2017
(https://www.nytimes.com/2017/02/22/us/politics/scott-pruitt-environmental
-protection-agency.html).

Davenport, Coral, and Kendra Pierre-Louis. 2018. "U.S. Climate Report Warns of
Damaged Environment and Shrinking Economy." *New York Times* (November 23). Retrieved November 24, 2018 (https://www.nytimes.com/2018/11/23
/climate/us-climate-report.html).

Davenport, Coral, and Alissa J. Rubin. 2017. "Trump Signs Executive Order Unwind-
ing Obama Climate Policies." *New York Times* (March 28). Retrieved March 28,
2017 (https://www.nytimes.com/2017/03/28/climate/trump-executive-order
-climate-change.html).

Davies, Jeremy. 2016. *The Birth of the Anthropocene*. Oakland: University of California
Press.

Deans, Bob. 2012. *Reckless: The Political Assault on the American Environment*. Lanham,
MD: Rowman & Littlefield.

DeMelle, Brendan, and Kevin Grandia. 2016. "'There Is No Doubt': Exxon Knew
CO2 Pollution Was a Global Threat by Late 1970s." *Desmog* (April 26). Retrieved
April 26, 2016 (https://www.desmogblog.com/2016/04/26/there-no-doubt
-exxon-knew-co2-pollution-was-global-threat-late-1970s).

Dennis, Brady, and Juliet Eilperin. 2017. "Fancy Dinners, Far-Flung Speeches: Cal-
endars Detail EPA Chief's Close Ties to Industry." *Washington Post* (October 3).
Retrieved October 3, 2017 (https://www.washingtonpost.com/news/energy
-environment/wp/2017/10/03/steakhouse-dinners-golf-resort-speeches
-calendars-detail-epa-chiefs-close-ties-to-industry/).

Derber, Charles. 2010. *Greed to Green: Solving Climate Change and Remaking the Economy*. Boulder, CO: Paradigm Publishers.

de Blasio, Bill, and Sadiq Khan. 2018. "As New York and London Mayors, We Call on All Cities to Divest from Fossil Fuels." *Guardian* (September 10). Retrieved September 28, 2019 (https://www.theguardian.com/commentisfree/2018/sep/10/london-new-york-cities-divest-fossil-fuels-bill-de-blasio-sadiq-khan).

de Graaf, K. J., and J. H. Jans. 2015. "The Urgenda Decision: Netherlands Liable for Role in Causing Dangerous Global Climate Change." *Journal of Environmental Law* 27: 517–527.

Diethelm, Pascal, and Martin McKee. 2009. "Denialism: What Is It and How Should Scientists Respond?" *European Journal of Public Health* 19: 2–4.

Dillion, Lindsey. 2018. "The Environmental Protection Agency in the Early Trump Administration: Prelude to Regulatory Capture." *American Journal of Public Health* 108: S89–S94.

Dobson, Michael. 2018. "The Radical Paris Agreement." *Jacobin* (January 5). Retrieved January 5, 2018 (https://www.jacobinmag.com/2018/01/paris-climate-agreement-global-warming-trump).

Dorrien, Gary. 2004. *Imperial Designs: Neo-Conservatism and the New Pax Americana*. New York: Routledge.

Dorroh, Jennifer. 2018a. "Baltimore Becomes Latest City to Sue Fossil Fuel Companies for Climate Damages." *Climate Liability News* (July 20). Retrieved July 20, 2018 (https://www.climateliabilitynews.org/2018/07/20/baltimore-climate-damages-liability-lawsuit/).

———. 2018b. "Safe Climate, Environment Should Be Held as a Human Right, UN Expert Says." *Climate Liability News* (March 13). Retrieved March 13, 2018 (https://www.climateliabilitynews.org/2018/03/13/un-climate-environment-human-right-john-knox/).

Dow, Kirstin, and Thomas E. Downing. 2011. *The Atlas of Climate Change: Mapping the World's Greatest Challenge*. Third edition. Berkeley: University of California Press.

Dower, John W. 2017. *The Violent American Century: War and Terror since World War II*. Chicago: Haymarket Books.

Downey, Liam. 2015. *Inequality, Democracy, and the Environment*. New York: New York University Press.

Doyon, Jacquelynn A., and Elizabeth A. Bradshaw. 2015. "Unfettered Fracking: A Critical Examination of Hydraulic Fracturing in the United States." In *The Routledge Handbook of the Crimes of the Powerful*, edited by G. Barak, 235–246. New York: Routledge.

Dreyfuss, Robert. 2005. *Devil's Game: How the United States Helped Unleash Fundamentalist Islam*. New York: Metropolitan Books.

Drugmand, Dana. 2018. "Despite Two Dismissals, Climate Liability Lawsuits Only Just Getting Started." *Climate Liability News* (August 2). Retrieved August 2, 2018 (https://www.climateliabilitynews.org/2018/08/02/climate-liability-lawsuits-nyc-san-francisco-oakland/).

———. 2019. "Exxon Climate History on Trial: Oil Giant's Legal Challenges Reach Critical Mass This Fall." *Climate Liability News* (September 10). Retrieved September 11, 2019 (https://www.climateliabilitynews.org/2019/09/10/exxon-climate-fraud-courts-lawsuits/).

Dryzek, John S., Richard B. Norgaard, and David Schlosberg, eds. 2011. *The Oxford Handbook of Climate Change and Society*. Oxford: Oxford University Press.

Dudley, Dominic. 2018. "Renewable Energy Will Be Consistently Cheaper Than Fossil Fuels By 2020, Report Claims." *Forbes* (January 13). Retrieved September 28, 2018 (https://www.forbes.com/sites/dominicdudley/2018/01/13/renewable-energy-cost-effective-fossil-fuels-2020/#6ba6ed9b4ff2).

Dunbar-Ortiz, Roxanne. 2014. *An Indigenous Peoples' History of the United States*. Boston: Beacon Press.

Dunlap, Riley E., and Aaron M. McCright. 2015. "Challenging Climate Change: The Denial Countermovement." In *Climate Change and Society: Sociological Perspectives*, edited by R. E. Dunlap and R. J. Brulle, 300–332. New York: Oxford University Press.

Dunlap, Riley E., Aaron M. McCright, and Jerrod H. Yarosh. 2016. "The Political Divide on Climate Change: Partisan Polarization Widens in the U.S." *Environment Science and Policy for Sustainable Development* 58: 4–23.

Dunn, Timothy J. 1996. *Militarization of the U.S.–Mexico Border: Low Intensity Warfare Comes Home*. Austin: University of Texas Press.

Dyer, Gwynne. 2010. *Climate Wars: The Fight for Survival as the World Overheats*. Oxford: Oneworld Publications.

Ehrhardt-Martinez, Karen, Juliet B. Schor, Wokje Abrahamse, et al. 2015. "Consumption and Climate Change." In *Climate Change and Society: Sociological Perspectives*, edited by R. E. Dunlap and R. J. Brulle, 93–126. New York: Oxford University Press.

Eilperin, Juliet. 2017. "EPA Now Requires Political Aide's Sign-off for Agency Awards, Grant Applications." *Washington Post* (September 4). Retrieved September 4, 2017 (https://www.washingtonpost.com/politics/epa-now-requires-political-aides-sign-off-for-agency-awards-grant-applications/2017/09/04/2fd707a0-88fd-11e7-a94f-3139abce39f5_story.html?utm_term=.4ae4e75e9c8f).

Eilperin, Juliet, Brady Dennis, and Chris Mooney. 2018. "Trump Administration Sees a 7-Degree Rise in Global Temperatures by 2100." *Washington Post* (September 28). Retrieved October 2, 2018 (https://www.washingtonpost.com/national/health-science/trump-administration-sees-a-7-degree-rise-in-global-temperatures-by-2100/2018/09/27/b9c6fada-bb45-11e8-bdc0-90f81cc58c5d_story.html?utm_term=.6466d8b32b1e).

Eilperin, Juliet, Steven Mufson, and Philip Rucker. 2016. "The Oil and Gas Industry Is Quickly Amassing Power in Trump's Washington." *Washington Post* (December 14). Retrieved December 14, 2016 (https://www.washingtonpost.com/politics/the-oil-and-gas-industry-is-quickly-amassing-power-in-trumps-washington/2016/12/14/0d4b26e2-c21c-11e6-9578-0054287507db_story.html).

Ekwurzel, Brenda, James Boneham, M. W. Dalton, et al. 2017. "The Rise in Global Atmospheric CO_2, Surface Temperature, and Sea Level from Emissions Traced to Major Carbon Producers." *Climatic Change* 144: 579–590.

Energy Information Administration. 2016. *International Energy Outlook 2016*. Washington, DC: U.S. Department of Energy.

Engelhardt, Tom. 2014. *Shadow Government: Surveillance, Secret Wars, and a Global Security State in a Single Superpower World*. Chicago: Haymarket Books.

———. 2018. *A Nation Unmade by War*. Chicago: Haymarket Books.

Environmental Data & Governance Initiative. 2017. "Pursuing a Toxic Agenda: Environmental Injustice in the Early Trump Administration." *EDGI* (September). Retrieved November 15, 2017 (https://100days.envirodatagov.org).

———. 2018. "Changing the Digital Climate: How Climate Change Web Content is Being Censored Under the Trump Administration." *EDGI* (January). Retrieved January 15, 2018 (https://envirodatagov.org/publication/changing-digital-climate).

Environmental Justice Foundation. 2017. *Beyond Borders: Our Changing Climate—Its Role in Conflict and Displacement.* London: Environmental Justice Foundation.

Environmental Protection Agency. 2015. *Factsheet: The Clean Power Plan.* U.S. EPA Archive Document. Washington, DC: Author.

Everest, Larry. 2004. *Oil, Power and Empire: Iraq and the U.S. Global Agenda.* Monroe, ME: Common Courage Press.

Expert Group on Global Climate Change Obligations. 2015. *Oslo Principles on Global Climate Change Obligations.* The Hague: Eleven International Publishing.

Falk, Richard. 2004. *The Declining World Order: America's Imperial Geopolitics.* New York: Routledge.

Fallows, James. 2004. "Bush's Lost Year." *Atlantic Monthly* (October): 68–84.

Farrell, Justin. 2015. "Network Structure and Influence of the Climate Change Counter-Movement." *Nature Climate Change* 6: 370–374.

———. 2016. "Corporate Funding and Ideological Polarization about Climate Change." *Proceedings of the National Academy of Sciences* 113: 92–97.

Ferguson, Niall. 2004. *Colossus: The Price of America's Empire.* New York: Penguin Press.

Fiala, Andrew G. 2008. *The Just War Myth: The Moral Illusions of War.* Lanham, MD: Rowman & Littlefield.

Fischer, Hubertus, Katrin J. Meissner, and Liping Zhou. 2018. "Palaeoclimate Constraints on the Impact of 2 Degrees C Anthropogenic Warming and Beyond." *Nature Geoscience* 11: 474–485.

Flannery, Tim. 2015. *Atmosphere of Hope: Searching for Solutions to the Climate Crisis.* New York: Atlantic Monthly Press.

Fong, Benjamin Y. 2017. "The Climate Crisis? It's Capitalism, Stupid." *New York Times* (November 20). Retrieved November 20, 2017 (https://www.nytimes.com /2017/11/20/opinion/climate-capitalism-crisis.html/).

Foster, John Bellamy. 2009. *The Ecological Revolution: Making Peace with the Planet.* New York: Monthly Review Press.

———. 2017. *Trump in the White House: Tragedy and Farce.* New York: Monthly Review Press.

Foster, John Bellamy, Brett Clark, and Richard York. 2010. *The Ecological Rift: Capitalism's War on the Earth.* New York: Monthly Review Press.

Fountain, Henry, and Brad Plumer. 2017. "Damaging Floods, Filthier Air and Political Instability." *New York Times* (November 4): A11.

Frank, Andre Gunder. 1969. *Capitalism and the Underdevelopment of Latin America.* New York: Monthly Review Press.

Franta, Benjamin. 2018. "On Its 100th birthday in 1959, Edward Teller Warned the Oil Industry about Global Warming." *Guardian* (January 1). Retrieved January 2, 2018 (https://www.theguardian.com/environment/climate-consensus-97-per-cent/2018 /jan/01/on-its-hundredth-birthday-in-1959-edward-teller-warned-the-oil -industry-about-global-warming).

Freedman, Andrew. 2019. "A Tipping Point on Climate Change." *Axios* (February 21). Retrieved February 21, 2019 (https://www.axios.com/climate-change -action-tipping-point-056c7163-3258-4f18-ab19-8efc289ef956.html).

Fremstad, Anders, and Mark Paul. 2018. "Why Socialists Should Back a Carbon Tax." *Jacobin* (September 19). Retrieved September 20, 2018 (https://jacobinmag .com/2018/09/carbon-tax-divided-peoples-policy-project).

Friedman, Lisa. 2017a. "E.P.A. Cancels Talk on Climate Change by Agency Scientists." *New York Times* (October 22) Retrieved October 22, 2017 (https://www.nytimes .com/2017/10/22/climate/epa-scientists).

———. 2017b. "E.P.A. Scrubs a Climate Website of 'Climate Change.'" *New York Times* (October 20). Retrieved October 20, 2017 (https://www.nytimes.com /2017/20/20/climate/epa-climate-change).

———. 2017c. "Pruitt Bars Some Scientists from Advising E.P.A." *New York Times* (October 31). Retrieved November 11, 2017 (https://www.nytimes.com/2017/10 /31/climate/pruitt-epa-science-advisory-boards.html).

Friedman, Lisa, and Brad Plumer. 2017. "E.P.A. Announces Bid to Roll Back Emissions Policy." *New York Times* (October 10): A1.

Friedrichs, David O. 2010. *Trusted Criminals: White Collar Crime in Contemporary Society.* Fourth edition. Belmont, CA: Wadsworth.

Friends of the Earth. 2017. "Leaving Paris Agreement Makes U.S. Foremost Climate Villain." *Friends of the Earth* News Release (May 31), Washington D.C.

Freeman, Joshua B. 2012. *American Empire: The Rise of a Global Power, the Democratic Revolution at Home, 1945–2000.* New York: Viking.

Frumhoff, Peter C., Richard Heede, and Naomi Oreskes. 2015. "The Climate Responsibilities of Industrial Carbon Producers." *Climatic Change* 132: 157–171.

Galliher, John. 1991. *Deviant Behavior and Human Rights.* Englewood Cliffs, NJ: Prentice Hall.

Galtung, Johan. 1969. "Violence, Peace, and Peace Research." *Journal of Peace Research* 6: 167–191.

Gasser, Thomas, Meddi Kechlar, Phillippe Ciais, et al. 2018. "Path-Dependent Reductions in CO_2 Emission Budgets Caused by Permafrost Carbon Release." *Nature Geoscience* 11: 830–835.

Geis, Gilbert, and Colin Goff. 1982. "Edwin H. Sutherland: A Biographical and Analytical Commentary." In *White-Collar and Economic Crime: Multidisciplinary and Cross-National* Perspectives, edited by P. Wickman and T. Dailey, 1–18. Lexington, MA: Lexington Books.

Gelbspan, Ross. 1998. *The Heat Is On: The Climate Crisis, the Cover-up, the Prescription.* Updated edition. Reading, MA: Perseus Books.

———. 2004. *Boiling Point: How Politicians, Big Oil and Coal, Journalists, and Activists Are Fueling the Climate Crisis—and What We Can Do to Avert Disaster.* New York: Basic Books.

George, Susan. 2015. *Shadow Sovereigns: How Global Corporations Are Seizing Power.* Cambridge, UK: Polity Press.

Germanos, Andrea. 2018a. "'Crime against Humanity' and 'International Embarrassment': Trump Refuses to 'Believe' Climate Report." *Common Dreams* (November 26). Retrieved November 26, 2018 (https://www.commondreams.org/news /2018/11/26/crime-against-humanity-trump-denounced-international -embarrassment-refusing-believe).

———. 2018b. "Judge Drops Charges against 13 Who Argued Pipeline Civil Disobedience Action Was 'Necessary' to Save Planet." *Common Dreams* (March 28). Retrieved March 28, 2018 (https://www.commondreams.org/news/2018/03/28 /judge-drops-charges-against-13-who-argued-pipeline-civil-disobedience -action-was).

Ghosh, Amitav. 2016. *The Great Derangement: Climate Change and the Unthinkable.* Chicago: University of Chicago Press.

Giddens, Anthony. 2011. *The Politics of Climate Change.* Second edition. Cambridge, UK: Polity Press.

Gilens, Martin, and Benjamin Page. 2014. "Testing Theories of American Politics: Elites, Interest Groups, and Average Citizens." *Perspectives on Politics* 12: 564–581.

Gillis, Justin, and Clifford Krauss. 2015. "Inquiry Weighs Whether Exxon Lied on Climate." *New York Times* (November 6): A1.

Givens, Jennifer E. 2014. "Global Climate Change Negotiations, the Treadmill of Destruction, and World Society: An Analysis of Kyoto Protocol Ratification." *International Journal of Sociology* 44: 7–36.

Global Climate Action Summit. 2018. *Exponential Climate Action Roadmap.* (September). Retrieved October 22, 2018 (https://exponentialroadmap.org/wp-content/uploads/2018/09/Exponential-Climate-Action-Roadmap-September-2018.pdf).

Goldenberg, Suzanne. 2013a. "Climate Change Inaction the Fault of Environmental Groups, Report Says." *Guardian* (January 14). Retrieved September 25, 2017 (https://www.theguardian.com/environment/2013/jan/14/environmental-groups-climate-change-inaction).

———. 2013b. "Just 90 Companies Caused Two-Thirds of Man-Made Global Warming Emissions." *Guardian* (November 23). Retrieved August 23, 2015 (https://www.theguardian.com/environment/2013/nov/20/90-companies-man-made-global-warming-emissions-climate-change).

Goodell, Jeff. 2017a. "Scott Pruitt's Crimes against Nature." *Rolling Stone* (August 10): 44–51.

———. 2017b. *The Water Will Come: Rising Seas, Sinking Cities, and the Remaking of the Civilized World.* New York: Little, Brown.

———. 2019. "Can We Survive Extreme Heat?" *Rolling Stone* (September): 76–84.

Gore, Al. 1992. *Earth in the Balance.* Boston: Houghton Mifflin.

———. 2006. *An Inconvenient Truth: The Planetary Emergency of Global Warming and What We Can Do about It.* Emmaus, PA: Rodale.

Gould, Kenneth A. 2007. "The Ecological Costs of Militarization." *Peace Review: A Journal of Social Justice* 19: 331–334.

Gould, Kenneth A., David N. Pellow and Allan Schnaiberg. 2008. *The Treadmill of Production: Injustice and Unsustainability in the Global Economy.* Boulder, CO: Paradigm Publishers.

Gramsci, Antonio. 1971. *Selections from the Prison Notebooks.* London: Lawrence and Wishart.

Green, Duncan. 2016. *How Change Happens.* Oxford: Oxford University Press.

Green, Penny, and Tony Ward. 2004. *State Crime: Governments, Violence and Corruption.* London: Pluto Press.

Greenpeace. 2011. *Who's Holding Us Back? How Carbon Intensive Industry Is Preventing Effective Climate Legislation.* Amsterdam: Author.

———. 2013a. *Dealing in Doubt: The Climate Denial Machine vs Climate Science.* Amsterdam: Author.

———. 2013b. *Point of No Return: The Massive Climate Threats We Must Avoid.* Amsterdam: Author.

Griffin, David Ray. 2015. *Unprecedented: Can Civilization Survive the CO_2 Crisis?* Atlanta: Clarity Press.

Hacker, Jacob, and Paul Pierson. 2010. *Winner-Take-All Politics: How Washington Made the Rich Richer—and Turned Its Back on the Middle Class.* New York: Simon & Schuster.

Hadden, Jennifer. 2017. "Learning from Defeat: The Strategic Reorientation of the U.S. Climate Movement." In *Climate Change in a Globalizing World: Comparative Perspectives on Environmental Movements in the Global North,* edited by C. Cassegard, L. Soneryd, H. Thorn, and A. Wettergren, 127–148. New York: Routledge.

Hahn, Steven. 2016. *A Nation without Borders: The United States and Its World in an Age of Civil Wars, 1830–1910.* New York: Viking.

Halper, Stefan, and Jonathan Clarke. 2004. *America Alone: The Neoconservatives and the Global Order.* Cambridge: Cambridge University Press.

Hamilton, Clive. 2010. *Requiem for a Species: Why We Resist the Truth about Climate Change.* London: Earthscan.

Hansen, James. 2009. *Storms of My Grandchildren: The Truth about the Coming Climate Catastrophe and Our Last Chance to Save Humanity.* New York: Bloomsbury.

Hansen, James, Makiko Sato, Paul Hearty, et al. 2016. "Ice Melt, Sea Level Rise and Superstorms: Evidence from Paleoclimate Data, Climate Modeling, and Modern Observations that 2 Degrees C Global Warming Could Be Dangerous." *Atmospheric Chemistry and Physics* 16: 3761–3812.

Hartjen, Clayton. 1978. *Crime and Criminalization.* Second edition. New York: Holt, Rinehart and Winston/Praeger.

Hartung, William D. 2004. *How Much Are You Making on the War Daddy? A Quick and Dirty Guide to War Profiteering in the Bush Administration.* New York: Nation Books.

———. 2011. *Prophets of War: Lockheed Martin and the Making of the Military–Industrial Complex.* New York: Nation Books.

Harvey, David. 2010. *The Enigma of Capital and the Crises of Capitalism.* New York: Oxford University Press.

Harvey, Fiona. 2018a. "Climate Change Soon to Cause Movement of 140 M People, World Bank Warns." *Guardian* (March 19). Retrieved March 20, 2018 (https://www.theguardian.com/environment/2018/mar/19/climate-change-soon-to-cause-mass-movement-world-bank-warns).

———. 2018b. "UN Climate Accord 'Inadequate' and Lacks Urgency, Experts Warn." *Guardian* (December 16). Retrieved December 17, 2018 (https://www.theguardian.com/environment/2018/dec/16/un-climate-accord-inadequate-and-lacks-urgency-experts-warn).

———. 2018c. "'We Can Move Forward Now': UN Climate Talks Take Significant Step." *Guardian* (December 16). Retrieved December 17, 2018 (https://www.theguardian.com/environment/2018/dec/16/katowice-we-can-move-forward-now-un-climate-talks-take-significant-step).

Hasemyer, David. 2016a. "Class-Action Lawsuit Adds to ExxonMobil's Climate Change Woes." *Inside Climate News* (November 21). Retrieved November 21, 2016 (https://insideclimatenews.org/news/18112016/exxon-climate-change-research-oil-reserves-stranded-assets-lawsuit).

———. 2016b. "Justice Department Refers Exxon Investigation Request to FBI." *Inside Climate News* (March 2). Retrieved March 3, 2016 (https://insideclimatenews.org/news/02032016/justice-department-refers-exxon-investigation-request-fbi-climate-change-research-denial).

———. 2017a. "Exxon Concealed Tillerson's 'Alias' Emails from NY Climate Fraud Probe, AG Claims." *Inside Climate News* (March 14). Retrieved May 20, 2018 (https://insideclimatenews.org/news/14032017/rex-tillerson-exxonmobil-climate-change-scandal-eric-schneiderman-wayne-tracker).

———. 2017b. "Exxon May Have Erased 7 Years of Tillerson's 'Wayne Tracker' Emails, Witness Says." *Inside Climate News* (June 8). Retrieved May 20, 2018 (https://insideclimatenews.org/news/08062017/exxon-climate-fraud-rex-tillerson-erased-wayne-tracker-emails-witness).

———. 2017c. "With Bare Knuckles and Big Dollars, Exxon Fights Climate Probe to a Legal Stalemate." *Inside Climate News* (June 5). Retrieved May 20, 2018 (https://insideclimatenews.org/news/05062017/exxon-climate-change-fraud-investigation-eric-schneiderman-rex-tillerson-exxonmobil).

————. 2018a. "Fossil Fuels on Trial: Where the Major Climate Change Lawsuits Stand Today." *Inside Climate News* (November 2). Retrieved November 8, 2018 (https://insideclimatenews.org/news/04042018/climate-change-fossil-fuel-company-lawsuits-timeline-exxon-children-california-cities-attorney-general).

————. 2018b. "New York AG Sues Exxon, Says Oil Giant Defrauded Investors over Climate Change." *Inside Climate News* (October 24). Retrieved October 24, 2018 (https://insideclimatenews.org/news/24102018/exxon-climate-fraud-lawsuit-new-york-attorney-general-investigation-tillerson).

Hasemyer, David, and John H. Cushman. 2015. "Exxon Sowed Doubt about Climate Science for Decades by Stressing Uncertainty." *Inside Climate News* (October 22). Retrieved November 7, 2016 (https://insideclimatenews.org/news/22102015/Exxon-Sowed-Doubt-about-Climate-Science-for-Decades-by-Stressing-Uncertainty).

Hasemyer, David, and Sabrina Shankman. 2016. "Climate Fraud Investigation of Exxon Draws Attention of 17 Attorneys General." *Inside Climate News* (March 30). Retrieved June 10, 2017 (https://insideclimatenews.org/news/30032016/climate-change-fraud-investigation-exxon-eric-shneiderman-18-attorneys-general).

Hauter, Wenonah. 2016. *Frackopoly: The Battle for the Future of Energy and the Environment*. New York: New Press.

Hawken, Paul, ed. 2017. *Drawdown: The Most Comprehensive Plan Ever Proposed to Reverse Global Warming*. New York: Penguin Books.

Hayden, Tom. 2009. *The Long Sixties: From 1960 to Barack Obama*. Boulder, CO: Paradigm Publishers.

Hayes, Ben, Steve Wright, and April Humble. 2016. "From Refugee Protection to Militarised Exclusion: What Future for 'Climate Refugees'?" In *The Secure and the Dispossessed: How the Military and Corporations Are Shaping a Climate Changed World*, edited by N. Buxton and B. Hayes, 111–132. London: Pluto Press.

Hayes, Chris. 2014. "The New Abolitionism." *Nation* (April 22). Retrieved December 25, 2018 (https://www.thenation.com/article/new-abolitionism/).

Hedges, Chris. 2009. *Empire of Illusion: The End of Literacy and the Triumph of Spectacle*. New York: Nation Books

————. 2010. *The Death of the Liberal Class*. New York: Nation Books.

————. 2018. *America: The Farewell Tour*. New York: Simon & Schuster.

Heede, Richard. 2014. "Tracing Anthropogenic Carbon Dioxide and Methane Emissions to Fossil Fuel and Cement Producers, 1854–2010." *Climatic Change* 122: 229–241.

Heilbroner, Robert. 1974. *An Inquiry into the Human Prospect*. New York: W. W. Norton.

Helvarg, David. 2017. "Defending the Earth from Donald Trump." *Progressive* (February): 21–23.

Henson, Robert. 2019. *The Thinking Person's Guide to Climate Change*. Second edition. Boston: American Meteorological Society.

Hertsgaard, Mark. 2011a. "Durban: Where the Climate Deniers-in-Chief Run the Show." *Nation* (December 14). Retrieved February 2, 2012 (https://www.thenation.com/article/durban-where-climate-deniers-chief-ran-show/).

————. 2011b. *Hot: Living through the Next Fifty Years on Earth*. Boston: Houghton Mifflin Harcourt.

————. 2012. "Déjà Vu at Rio+20." *Nation* (June 14). Retrieved June 14, 2012 (http://www.thenation.com/article/168387/deja-vu-rio20).

————. 2017. "Enemy of Humanity: That's Trump, Our Climate Denier in Chief." *Nation* (July 3/10): 10–11.

————. 2018. "Why L.A. Burns." *Nation* (January 1/8): 4.

————. 2019. "The Climate Kids Are Coming: The March 15 Strikes Are Game-Changing." *Nation* (March 25): 18–19, 26.

Hess, David J. 2014. "When Green Became Blue: Epistemic Rift and the Corralling of Climate Science." *Political Power and Social Theory* 27: 127–153.

Higgins, Polly. 2010. *Eradicating Ecocide: Exposing the Corporate and Political Practices Destroying the Planet and Proposing the Laws Needed to Eradicate Ecocide.* London: Shepheard-Walwyn Publishers.

————. 2012. *Earth Is Our Business: Changing the Rules of the Game.* London: Shepheard-Walwyn Publishers.

Higgins, Polly, Damien Short, and Nigel South. 2013. "Protecting the Planet: A Proposal for a Law of Ecocide." *Crime, Law and Social Change* 59: 251–266.

Hodgson, Godfrey. 2009. *The Myth of American Exceptionalism.* New Haven, CT: Yale University Press.

Hoffman, Andrew J. 2015. *How Culture Shapes the Climate Debate.* Stanford, CA: Stanford University Press.

Hoggan, James. 2009. *Climate Cover-Up.* Vancouver: Greystone Books.

Hooks, Gregory, and Chad L. Smith. 2004. "The Treadmill of Destruction: National Sacrifice Areas and Native Americans." *American Sociological Review* 69: 558–575.

————. 2005. "Treadmills of Production and Destruction: Threats to the Environment Posed by Militarism." *Organizations & Environment* 18: 19–37.

Horgan, John. 2018. "Noam Chomsky Calls Trump and Republican Allies 'Criminally Insane.'" *Scientific American* (November 3). Retrieved November 3, 2018 (https://blogs.scientificamerican.com/cross-check/noam-chomsky-calls-trump-and-republican-allies-criminally-insane/).

Horton, Radley, Katharine Hayhoe, Robert Kopp, and Sarah Doherty. 2017. "The Climate Risks We Face." *New York Times* (November 6). Retrieved November 6, 2017 (https://www.nytimes.com/2017/11/06/opinion/climate-report-global-warming.html).

Howe, Joshua P. 2014. *Behind the Curve: Science and the Politics of Global Warming.* Seattle: University of Washington Press.

Iadicola, Peter. 2010. "The Centrality of Empire in the Study of State Crime and Violence." In *State Crime in the Global Age*, edited by W. Chambliss, R. Michalowski, and R. Kramer, 31–44. Cullompton, UK: Willan Publishing.

Immerman, Richard H. 2010. *Empire for Liberty: A History of American Imperialism from Benjamin Franklin to Paul Wolfowitz.* Princeton, NJ: Princeton University Press.

InfluenceMap. 2019. "Big Oil's Real Agenda on Climate Change." *InfluenceMap* (March). Retrieved March 22, 2019 (https://influencemap.org/report/How-Big-Oil-Continues-to-Oppose-the-Paris-Agreement-38212275958aa21196dae3b76 220bddc).

Inside Climate News. 2018. "Six of the G7 Commit to Climate Action. Trump Wouldn't Even Join Conversation." *Inside Climate News* (June 10). Retrieved June 11, 2018 (https://insideclimatenews.org/news/10062018/g7-summit-climate-change-communique-trump-allies-estranged-germany-france-canada).

International Bar Association. 2014. *Achieving Justice and Human Rights in an Era of Climate Disruption.* London: Author.

Intergovernmental Panel on Climate Change. 1990. *Climate Change: The IPCC Impacts Assessment Report.* Canberra: Australian Government Publishing Service. Prepared for IPCC by Working Group II, Canberra.

———. 2013. *Fifth Assessment Report. Climate Change 2013: The Physical Science Basis.* Geneva: Author.

———. 2018. *Global Warming of 1.5 Degree C.* Geneva: Author.

Irfan, Umair. 2018. "Climate Change Is a Global Injustice. A New Study Shows Why." *Vox* (September 26). Retrieved October 3, 2018 (https://www.vox.com/2018/9/26/17897614/climate-change-social-cost-carbon).

———. 2019. "Pay Attention to the Growing Wave of Climate Change Lawsuits." *Vox* (February 22). Retrieved February 22, 2019 (https://www.vox.com/energy-and-environment/2019/2/22/17140166/climate-change-lawsuit-exxon-juliana-liability-kids).

Jacques, Peter J. 2009. *Environmental Skepticism: Ecology, Power and Public Life.* London: Routledge.

Jacques, Peter J., Riley Dunlap, and Mark Freeman. 2008. "The Organization of Denial: Conservative Think Tanks and Environmental Skepticism." *Environmental Politics* 17: 349–385.

Jamail, Dahr. 2016. "Atmospheric Carbon Dioxide Concentration Has Passed the Point of No Return." *Truthout* (May 23). Retrieved May 23, 2016 (https://truthout.org/articles/atmospheric-carbon-dioxide-concentration-has-passed-the-point-of-no-return/).

———. 2017. "Trump Signs Executive Order Obliterating Regulations on Carbon Emissions and Pollution." *Truthout* (March 28). Retrieved March 28, 2017 (https://truthout.org/articles/trump-s-environmental-shock-and-awe/).

———. 2019. *The End of Ice: Bearing Witness and Finding Meaning in the Path of Climate Disruption.* New York: New Press.

Jamieson, Dale. 2014. *Reason in a Dark Time: Why the Struggle against Climate Change Failed—and What It Means for Our Future.* New York: Oxford University Press.

Jennings, Katie, Dino Grandoni, and Susanne Rust. 2015. "How Exxon Went from Leader to Skeptic on Climate Change Research." *Los Angeles Times* (October 23). Retrieved October 24, 2015 (http://graphics.latimes.com/exxon-research/).

Jensen, Robert. 2009. *All My Bones Shake: Seeking a Progressive Path to the Prophetic Voice.* Berkeley, CA: Soft Skull Press.

Jerving, Sara, Katie Jennings, Masako Melissa Hirsch, and Susanne Rust. 2015. "What Exxon Knew about the Earth's Melting Arctic." *Los Angeles Times* (October 9). Retrieved November 9, 2016 (http://graphics.latimes.com/exxon-arctic/).

Johnson, Chalmers. 2000. *Blowback: The Costs and Consequences of American Empire.* New York: Henry Holt.

———. 2004. *The Sorrows of Empire: Militarism, Secrecy, and the End of the Republic.* New York: Metropolitan Books.

Johnson, Lyndon B. 1965. "President Lyndon B. Johnson's Special Message to the Congress on Conservation and Restoration of National Beauty, February 8, 1965." In *Public Papers of the Presidents of the United States: Lyndon B. Johnson, 1965. Volume I, Entry 54,* 155–165. Washington, DC: Government Printing Office.

Johnston, Ian. 2016. "Climate Change May Be Escalating So Fast It Could Be 'Game Over,' Scientists Say." *Independent* (November 9). Retrieved September 20, 2017 (https://www.independent.co.uk/news/science/climate-change-game-over-global-warming-climate-sensitivity-seven-degrees-a7407881.html).

Jorgenson, Andrew K., and Brett Clark. 2009. "The Economy, Military, and Ecologically Unequal Exchange Relationships: A Panel Study of the Ecological Footprints of Nations, 1975–2000." *Social Problems* 56: 621–646.

———. 2016. "The Temporal Stability and Developmental Differences in the Environmental Impacts of Militarism: The Treadmill of Destruction and Consumption-Based Carbon Emissions." *Sustainability Science* 11: 505–514.

Jorgenson, Andrew K., Brett Clark, and Jennifer E. Givens. 2012. "The Environmental Impacts of Militarization in Comparative Perspective: An Overlooked Relationship." *Nature and Culture* 7: 314–337.

Jorgenson, Andrew K., Brett Clark, and Jeffrey Kentor. 2010. "Militarization and the Environment: A Panel Study of Carbon Dioxide Emissions and the Ecological Footprints of Nations, 1970–2000." *Global Environmental Politics* 10: 7–29.

Jost, John T., Brian A. Nosek, and Samuel D. Gosling. 2008. "Ideology: Its Resurgence in Social, Personality, and Political Psychology." *Perspectives on Psychological Science* 3: 126–136.

Judis, John B. 2004. *The Folly of Empire: What George W. Bush Could Learn from Theodore Roosevelt and Woodrow Wilson.* New York: A Lisa Drew Book/Scribner.

Juhasz, Antonia. 2006. *The Bu$h Agenda: Invading the World One Economy at a Time.* New York: Regan Books (HarperCollins).

Kabaservice, Geoffrey M. 2012. *Rule and Ruin: The Downfall of Moderation and the Destruction of the Republican Party, from Eisenhower to the Tea Party.* New York: Oxford University Press.

Kahn, Brian. 2016a. "Antarctic CO2 Hit 400 PPM for First Time in 4 Million Years." *Climate Central.* (September 27). Retrieved July 10, 2017 (http://www.climatecentral.org/news/antarctica-co2-400-ppm-million-years-2045).

———. 2016b. "Earth's CO2 Passes the 400 PPM Threshold—Maybe Permanently." *Scientific American* (September 27). Retrieved July 10, 2018 (https://www.scientificamerican.com/article/earth-s-co2-passes-the-400-ppm-threshold-maybe-permanently/).

Kahan, Dan M., Donald Braman, John Gastil, et al. 2007. "Culture and Identity-Protective Cognition: Explaining the White-Male Effect in Risk Perception." *Journal of Empirical Legal Studies* 4: 465–505.

Kahneman, Daniel. 2011. *Thinking, Fast and Slow.* New York: Farrar, Straus and Giroux.

Kaiser, David, and Lee Wasserman. 2016. "The Rockefeller Family Fund vs. Exxon." *New York Review of Books* (December 8 and 22). Retrieved October 10, 2017 (https://www.nybooks.com/articles/2016/12/08/the-rockefeller-family-fund-vs-exxon/).

Karlin, Mark. 2014. "The Numbing Down of America about How Big Oil Is Bringing Us to the Precipice of Extinction." *Buzzflash at Truthout* (March 25). Retrieved March 26, 2014 (http://buzzflash.com/commentary/the-numbing-down-of-america-about-how-big-oil-is-bringing-us-to-the-precipice-of-extinction).

Kaufman, Alexander C. 2017. "9 Times Donald Trump Showed He Doesn't Understand the Climate Pact He Just Quit." *Huffington Post* (June 1). Retrieved June 2, 2017 (https://www.huffingtonpost.com/entry/trump-paris-speech-fact-check_us_59309ab8e4b02478cb99f151).

Kauzlarich, David, and Ronald C. Kramer. 1998. *Crimes of the American Nuclear State: At Home and Abroad.* Boston: Northeastern University Press.

Kauzlarich, David, Christopher Mullins, and Rick Matthews. 2003. "A Complicity Continuum of State Crime." *Contemporary Justice Review* 6: 241–254.

Keenan, Neil. 2008. "Global Warming due to Greenhouse Gas Emissions: The Success of State Solutions as a Model for a Federal Solution; Note." *Journal of Legislation* 34: 168–189.

Kelly, Colin P., Mohtadi Shahrzad, Mark A. Cane, et al. 2015. "Climate Change in
the Fertile Crescent and Implications of the Recent Syrian Drought." *Proceedings of
the National Academy of Sciences* 112: 3241–3246.

Kennedy, Robert F., Jr. 2005. *Crimes against Nature: How George W. Bush & His Corporate Pals Are Plundering the Country & Hijacking Our Democracy.* New York: Harper
Perennial.

Kerry, John. 2018. *Every Day Is Extra.* New York: Simon & Schuster.

Kessler, Glenn, and Michelle Ye Hee Lee. 2017. "Fact-Checking President Trump's
Claims on the Paris Climate Change Deal." *Washington Post* (June 1). Retrieved
June 1, 2017 (https://www.washingtonpost.com/news/fact-checker/wp/2017/06
/01/fact-checking-president-trumps-claims-on-the-paris-climate-change-deal
/?utm_term=.dc3da97785de).

King, Marcus. 2011. "World Resources Report: Climate Change Adaptation and National
Security." *World Resources Institute.* Retrieved August 28, 2018 (http://www.wri.org
/our-work/project/world-resources-report/climate-change-adaptation-and
-national-security).

Kinzer, Stephen. 2006. *Overthrow: America's Century of Regime Change from Hawaii to
Iraq.* New York: Times Books/Henry Holt.

Klare, Michael. 2001. *Resource Wars: The New Landscape of Global Conflict.* New York:
Metropolitan Books.

———. 2004. *Blood and Oil: The Dangers and Consequences of America's Growing Dependency on Imported Petroleum.* New York: Metropolitan Books.

———. 2012. *The Race for What's Left: The Global Scramble for the World's Last
Resources.* New York: Metropolitan Books.

———. 2016a. "Drowning the World in Oil: Trump's Carbon-Obsessed Energy
Policy and the Planetary Nightmare to Come." *TomDispatch* (December 15).
Retrieved December 15, 2016 (http://www.tomdispatch.com/blog/176222).

———. 2016b. "Hooked! The Unyielding Grip of Fossil Fuels on Global Life." *Tom-
Dispatch* (July 14). Retrieved July 14, 2016 (http://www.tomdispatch.com/blog
/176164).

———. 2017a. "America's Carbon-Pusher in Chief: Trump's Fossil-Fueled Foreign
Policy." *TomDispatch* (July 30). Retrieved July 30, 2017 (http://www.tomdispatch
.com/blog/176313).

———. 2017b. "Is Trump Launching a New World Order? The Petro-Powers vs. the
Greens." *TomDispatch* (June 11). Retrieved June 11, 2017 (http://www.tomdispatch
.com/blog/176294/tomgram%3A_michael_klare%2C_%22the_battle_lines_of
_the_future%22).

Klein, Naomi. 2011. "Capitalism vs. the Climate." *Nation* (November 28): 11–21.

———. 2014. *This Changes Everything: Capitalism vs the Climate.* New York: Simon &
Schuster.

———. 2017. *No Is Not Enough: Resisting Trump's Shock Politics and Winning the World
We Need.* Chicago: Haymarket Books.

———. 2018a. "Capitalism Killed Our Climate Momentum, Not Human Nature."
Intercept (August 3). Retrieved August 4, 2018 (https://theintercept.com/2018/08
/03/climate-change-new-york-times-magazine/).

————. 2018b. "The Game-Changing Promise of a Green New Deal." *Intercept* (November 27). Retrieved November 28, 2018 (https://theintercept.com/2018 /11/27/green-new-deal-congress-climate-change/).

————. 2019. *On Fire: The (Burning) Case for a Green New Deal.* New York: Simon & Schuster.

Koh, Harold H. 2003. "Foreword: On American Exceptionalism." *Stanford Law Review* 55: 1479–1527.

Kolbert, Elizabeth. 2014. *The Sixth Extinction: An Unnatural History.* New York: Henry Holt.

————. 2017. "Going Negative: Can Carbon-Dioxide Removal Save the World?" *New Yorker* (November 20): 64–73.

Kopp, Robert E., Andrew C. Kemp, Klaus Bittermann, et al. 2016. "Temperature-Driven Global Sea-Level Variability in the Common Era." *Proceedings of the National Academy of Science* 113: E1434-E1441.

Kotchen, Matthew J. 2017. "Trump Will Stop Paying into the Green Climate Fund. He Has No Idea What It Is." *Washington Post* (June 2). Retrieved June 2, 2017 (https://www .washingtonpost.com/posteverything/wp/2017/06/02/trump-will-stop-paying-into -the-green-climate-fund-he-has-no-idea-what-it-is/?utm_term=.16d928703633).

Kramer, Ronald C. 1982. "The Debate over the Definition of Crime: Paradigms, Value Judgments, and Criminological Work." In *Ethics, Public Policy and Criminal Justice* edited by F. Elliston and N. Bowie, 33–58. Cambridge, MA: Oelgeschlager, Gunn, and Hain.

————. 2012. "Public Criminology and the Responsibility to Speak in the Prophetic Voice Concerning Global Warming." In *State Crime and Resistance,* edited by E. Stanley and J. McCulloch, 41–53. London: Routledge.

————. 2013a. "Carbon in the Atmosphere and Power in America: Climate Change as State–Corporate Crime." *Journal of Crime & Justice* 36: 153–170.

————. 2013b. "Expanding the Core: Blameworthy Harms, International Law and State–Corporate Crimes." American Society of Criminology 2013 Annual Meeting Presidential Papers (https://www.asc41.com/resources.html).

————. 2018. "Curbing State Crime by Challenging the U.S. Empire." In *Routledge Handbook of Critical Criminology,* second edition, edited by W. S. DeKeseredy and M. Dragiewicz, 431–443. Abingdon, UK: Routledge.

Kramer, Ronald C., David Kauzlarich, and Brian Smith. 1992. "Toward the Study of Governmental Crime: Nuclear Weapons, Foreign Intervention, and International Law." *Humanity and Society* 16: 543–563.

Kramer, Ronald C., and Sam Marullo. 1985. "Toward a Sociology of Nuclear Weapons." *Sociological Quarterly* 26: 277–292.

Kramer, Ronald C., and Raymond Michalowski. 2012. "Is Global Warming a State–Corporate Crime?" In *Climate Change from a Criminological Perspective,* edited by R. White, 71–88. New York: Springer.

Krauthammer, Charles. 1991. "The Unipolar Moment." *Foreign Affairs* 70: 23–33.

Krugman, Paul. 2018a. "Climate Denial as the Crucible for Trumpism." *New York Times* (December 3): A25.

————. 2018b. "The Depravity of Climate-Change Denial." *New York Times* (November 27): A22.

Kusnetz, Nicholas. 2017a. "Exxon's Climate Accounting a 'Sham' under Rex Tillerson, New York's AG Says." *Inside Climate News* (June 2). Retrieved June 3, 2017 (https:// insideclimatenews.org/news/02062017/exxon-climate-change-greenhouse-gas -accounting-sham-rex-tillerson-new-york-attorney-general-court).

———. 2017b. "200+ Investors Tell Trump: Don't Abandon Paris Climate Accord." *Inside Climate News* (May 8). Retrieved May 31, 2017 (https://insideclimatenews .org/news/08052017/investors-trump-paris-climate-accord-G7-G20).

———. 2018a. "New York City Sues Oil Companies over Climate Change, Says It Plans to Divest." *Inside Climate News* (January 11). Retrieved January 11, 2018 (https://insideclimatenews.org/news/10012018/new-york-city-divest-sued-big -oil-climate-change-costs-exxon-chevron-bp-shell-mayor-deblasio).

———. 2018b. "2017 One of Hottest Years on Record, and Without El Nino." *Inside Climate News* (January 19). Retrieved January 19, 2018 (https://insideclimatenews .org/news/18012018/2017-third-hottest-year-record-wildfires-hurricanes-noaa -nasa-annual-climate-change-report).

———. 2019. "A Surge of Climate Lawsuits Targets Human Rights, Damages from Fossil Fuels." *Inside Climate News* (January 4). Retrieved January 4, 2019 (https:// insideclimatenews.org/news/04012019/climate-change-lawsuits-2018-year -review-exxon-fossil-fuel-companies-human-rights-children-government).

Landler, Mark. 2014. "U.S. and China Reach Climate Accord after Months of Talks." *New York Times* (November 11). Retrieved August 29, 2017 (https://www.nytimes .com/2014/11/12/world/asia/china-us-xi-obama-apec.html).

Lavelle, Marianne. 2015. "A 50th Anniversary Few Remember: LBJ's Warning on Carbon Dioxide." *Daily Climate* (February 2). Retrieved December 20, 2016 (https://reneweconomy.com.au/a-50th-anniversary-few-remember-lbjs-warning -on-carbon-dioxide-37099/).

———. 2017. "Trump's Executive Order: More Fossil Fuels, Regardless of Climate Change." *Inside Climate News* (March 28). Retrieved June 14, 2018 (https:// insideclimatenews.org/news/28032017/trump-executive-order-climate-change -paris-climate-agreement-clean-power-plan-pruit).

———. 2018a. "Mike Pompeo, Climate Policy Foe, Picked to Replace Tillerson as Secretary of State." *Inside Climate News* (March 13). Retrieved July 9, 2018 (https:// insideclimatenews.org/news/13032018/mike-pompeo-koch-brothers-secretary-state -climate-change-rex-tillerson-exxon).

———. 2018b. "6 Ways EPA's New Leader, a Former Coal Lobbyist, Could Shape Climate Policy." *Inside Climate News* (July 7). Retrieved July 9, 2018 (https:// insideclimatenews.org/news/07072018/epa-andrew-wheeler-trump-climate -policy-coal-lobbyist-auto-emissions-methane-science-clean-power-plan-pruitt).

———. 2018c. "With Democratic Majority, Climate Change Is back on U.S. House Agenda." *Inside Climate News* (November 7). Retrieved November 8, 2018 (https:// insideclimatenews.org/news/07112018/election-2018-climate-change-democratic -majority-house-congress-energy-senate-governors-initiatives).

———. 2018d. "House Votes to Denounce Carbon Taxes." *Inside Climate News* (July 19). Retrieved October 10, 2018 (https://insideclimatenews.org/news/19072018/anti -carbon-tax-resolution-house-vote-climate-solutions-caucus-curbelo-scalise-koch -influence-congress).

———. 2019. "New Congress Members See Climate Solutions and Jobs in a Green New Deal." *Inside Climate News* (January 3). Retrieved January 4, 2019 (https:// insideclimatenews.org/news/03012019/green-new-deal-climate-solutions-jobs -2018-year-review-ocasio-cortez-castor-sunrise-movement-congress).

Lavelle, Marianne, and John H. Cushman, Jr. 2019. "The Green New Deal Lands in Congress." *Inside Climate News* (February 7). Retrieved February 18, 2019 (https:// insideclimatenews.org/news/07022019/green-new-deal-alexandria-ocasio-cortez -ed-markey-climate-change).

Leaf, Dennis. 2001. "Managing Global Atmospheric Change: A U.S. Policy Perspective." *Human and Ecological Risk Assessment* 7: 1211–1226.

Leber, Rebecca. 2019. "How a Revolution in Climate Science Is Putting Big Oil on Trial." *Mother Jones* (September). Retrieved September 16, 2019 (https://www.motherjones.com/politics/2019/09/how-a-revolution-in-climate-science-is-putting-big-oil-back-on-trial/).

Leichenko, Robin, and Karen O'Brien. 2019. *Climate and Society: Transforming the Future*. Cambridge, UK: Polity Press.

Lens, Sidney. 1971. *The Forging of the American Empire: From the Revolution to Vietnam: A History of Imperialism*. New York: Thomas Y. Crowell Company (2003 edition published by Pluto Press [London] and Haymarket Books [Chicago]).

Lessig, Lawrence. 2011. *Republic, Lost: How Money Corrupts Congress—and a Plan to Stop It*. New York: Grand Central Publishing.

Lewis, Simon L., and Mark A. Maslin. 2018. *The Human Planet: How We Created the Anthropocene*. New Haven, CT: Yale University Press.

Lieberman, Amy, and Susanne Rust. 2015. "Big Oil Braced for Global Warming While It Fought Regulations." *Los Angeles Times* (December 31). Retrieved January 2, 2016 (http://graphics.latimes.com/oil-operations/).

Lifton, Robert Jay. 2017. *The Climate Swerve: Reflections on Mind, Hope, and Survival*. New York: New Press.

Lofoten Declaration. 2017. "Climate Leadership Requires a Managed Decline of Fossil Fuel Production." Retrieved October 10, 2018 (http://www.lofotendeclaration.org/).

Loki, Reynard. 2016. "Rockefeller Family Fund to Divest from ExxonMobil, Says Oil Giant Is 'Morally Reprehensible.'" *Alternet* (March 25). Retrieved May 1, 2016 (https://www.alternet.org/environment/rockefeller-family-fund-divest-exxonmobil-says-oil-giant-morally-reprehensible).

Lukes, Steven. 2005. *Power: A Radical View*. Second edition. Houndsmill, UK: Palgrave Macmillan.

Lutts, Ralph H. 1985. "Chemical Fallout: Rachel Carson's *Silent Spring*, Radioactive Fallout, and the Environmental Movement." *Environmental Review* 9: 210–225.

Lynas, Mark. 2008. *Six Degrees: Our Future on a Hotter Planet*. Washington, DC: National Geographic.

Lynch, Michael. 1990. "The Greening of Criminology: A Perspective for the 1990s." *Critical Criminologist* 2: 3–4, 11–12.

Lynch, Michael J., Ronald Burns, and Paul B. Stretesky, 2010. "Global Warming and State–Corporate Crime: The Politicization of Global Warming under the Bush Administration." *Crime, Law and Social Change* 54: 213–239.

Lynch, Michael J., Michael A. Long, Paul B. Stretesky, and Kimberly L. Barrett. 2017. *Green Criminology: Crime, Justice, and the Environment*. Oakland: University of California Press.

Lynch, Michael J., and Paul B. Stretesky. 2003. "The Meaning of Green: Contrasting Criminological Perspectives." *Theoretical Criminology* 7: 217–238.

———. 2010. "Global Warming, Global Crime: A Green Criminological Perspective." In *Global Environmental Harm: Criminological Perspectives*, edited by R. White, 62–84. Cullompton, UK: Willan Publishing.

Lynch, Michael J., Paul B. Stretesky, and Michael Long. 2016. "A Proposal for the Political Economy of Green Criminology: Capitalism and the Case of the Alberta Tar Sands." *Canadian Journal of Criminology and Criminal Justice* 58: 137–160.

Lynch, Michael J., Paul B. Stretesky, Michael A. Long, and Kimberly L. Barrett. 2017. "Social Justice, Environmental Destruction, and the Trump Presidency: A Criminological View." *Social Justice: A Journal of Crime, Conflict and World Order* (January 12). Retrieved January 13, 2017 (http://www.socialjusticejournal.org/social-justice-environmental-destruction-and-the-trump-presidency-a-criminological-view/).

MacLean, Nancy. 2017. *Democracy in Chains: The Deep History of the Radical Right's Stealth Plan for America*. New York: Viking.

Madrick, Jeff. 2011. *Age of Greed: The Triumph of Finance and the Decline of America 1970 to the Present*. New York: Alfred A. Knopf.

Malm, Andreas. 2016. *Fossil Capital: The Rise of Steam Power and the Roots of Global Warming*. London: Verso.

———. 2018. *The Progress of This Storm: Nature and Society in a Warming World*. London: Verso.

Mann, James. 2004. *Rise of the Vulcans: The History of Bush's War Cabinet*. New York: Viking.

Mann, Michael. 2003. *Incoherent Empire*. London: Verso.

Mann, Michael E. 2017. "Fear Won't Save Us: Putting a Check on Climate Doom." *Common Dreams* (July 10). Retrieved July 10, 2017 (https://www.commondreams.org/views/2017/07/10/fear-wont-save-us-putting-check-climate-doom).

Mann, Michael E., and Lee R. Kump. 2015. *Dire Predictions: Understanding Climate Change*. New York: Dorling Kindersley.

Mann, Michael E., and Tom Toles. 2016. *The Madhouse Effect: How Climate Change Denial Is Threatening Our Planet, Destroying Our Politics, and Driving Us Crazy*. New York: Columbia University Press.

Mann, Thomas, and Norman Ornstein. 2016. *It's Even Worse than It (Looks) Was: How the American Constitutional System Collided with the New Politics of Extremism*. New and expanded edition. New York: Basic Books.

Mark, Jason. 2018. "The Climate-Wrecking Industry . . . and How to Beat It." *Nation* (September 24/October 1): 14–17.

Marlon, Jennifer, Peter Howe, Matto Mildenberger, et al. 2018. "Yale Climate Opinion Maps 2018." Yale Program on Climate Change Communication (August 7). Retrieved September 7, 2018 (http://climatecommunication.yale.edu/visualizations-data/ycom-us-2018/?est=happening&type=value&geo=county).

Marshall, George. 2014. *Don't Even Think about It: Why Our Brains Are Wired to Ignore Climate Change*. New York: Bloomsbury.

Matthews, Dylan. 2017. "Donald Trump Has Tweeted Climate Change Skepticism 115 Times." *Vox* (June 1). Retrieved November 5, 2017 (https://www.vox.com/policy-and-politics/2017/6/1/15726472/trump-tweets-global-warming-paris-climate-agreement).

Mayer, Jane. 2017. *Dark Money: The Hidden History of the Billionaires behind the Rise of the Radical Right*. New York: Anchor Books.

McCauley, Lauren. 2016. "Refusing to Name Names, DOE Fearful of 'Climate Purge' under Trump." *Common Dreams* (December 13). Retrieved December 14, 2016 (https://www.commondreams.org/news/2016/12/13/refusing-name-names-doe-fearful-climate-purge-under-trump).

McCoy, Alfred W. 2017. *In the Shadows of the American Century: The Rise and Decline of U.S. Global Power*. Chicago: Haymarket Books.

McCright, Aaron, and Riley Dunlap. 2000. "Challenging Global Warming as a Social Problem: An Analysis of the Conservative Movement's Counter-Claims." *Social Problems* 47: 499–522.

———. 2003. "Defeating Kyoto: The Conservative Movement's Impact on U.S. Climate Change Policy." *Social Problems* 50: 348–373.

———. 2010. "Anti-Reflexivity: The American Conservative Movement's Success in Undermining Climate Science and Policy." *Theory, Culture & Society* 27: 100–133.

———. 2011a. "Cool Dudes: The Denial of Climate Change among Conservative White Males in the United States." *Global Environmental Change* 21: 1163–1172.

———. 2011b. "The Politicization of Climate Change and Polarization in the American Public's Views of Global Warming, 2001–2010." *Sociological Quarterly* 52: 155–194.

McKenna, Phil. 2017. "Judge Allows 'Necessity' Defense by Climate Activists in Oil Pipeline Protest." *Inside Climate News* (October 16). Retrieved October 16, 2017 (https://insideclimatenews.org/news/16102017/climate-change-activists-arrest-pipeline-shutdown-necessity-defense).

———. 2018. "That $3 Trillion-a-Year Clean Energy Transformation? It's Already Underway." *Inside Climate News* (October 11). Retrieved October 11, 2018 (https://insideclimatenews.org/news/11102018/ipcc-clean-energy-transformation-cost-trillion-climate-change-global-warming-renewable-coal-fossil-fuels).

McKibben, Bill. 2010a. "Copenhagen: Rallying for Next Steps." *Mother Jones* (January 19). Retrieved March 15, 2012 (http://motherjones.com/politics/2010/01/copenhagen-next-steps).

———. 2010b. *Eaarth: Making Life on a Tough New Planet.* New York: Times Books/Henry Holt.

———, ed. 2011. *The Global Warming Reader: A Century of Writing about Climate Change.* New York: Penguin Books.

———. 2012a. "Global Warming's Terrifying New Math: Three Simple Numbers that Add up to Global Catastrophe—and that Make Clear Who the Real Enemy Is." *Rolling Stone* (August 2). Retrieved July 19, 2013 (http://www.rollingstone.com/politics/news/global-warming-terrifying-new-math).

———. 2012b. "The Ultimate Corporation." *New York Review of Books* (June 7). Retrieved October 10, 2017 (https://www.nybooks.com/articles/2012/06/07/ultimate-corporation/).

———. 2013. *Oil and Honey: The Education of an Unlikely Activist.* New York: Times Books/Henry Holt.

———. 2014. "Exxon Mobil's Response to Climate Change Is Consummate Arrogance." *Guardian* (April 3). Retrieved December 19, 2018 (https://www.theguardian.com/environment/2014/apr/03/exxon-mobil-climate-change-oil-gas-fossil-fuels).

———. 2015a. "Exxon's Climate Lie: No Corporation Has Ever Done Anything This Big or Bad." *Guardian* (October 14). Retrieved August 23, 2017 (https://www.theguardian.com/environment/2015/oct/14/exxons-climate-lie-change-global-warming).

———. 2015b. "Obama's Catastrophic Climate-Change Denial." *New York Times* (May 12). Retrieved April 17, 2017 (https://www.nytimes.com/2015/05/13/opinion/obamas-catastrophic-climate-change-denial.html).

———. 2015c. "What Exxon Knew about Climate Change." *New Yorker* (September 18). Retrieved August 23, 2017 (https://www.newyorker.com/news/daily-comment/what-exxon-knew-about-climate-change).

———. 2016a. "The Active Many Can Overcome the Ruthless Few." *Nation* (December 19/26): 10–21.

————. 2016b. "Exxon's Never-Ending Big Dig: Flooding the Earth with Fossil Fuels." *TomDispatch* (February 18). Retrieved February 19, 2016 (http://www .tomdispatch.com/blog/176105/tomgram:_bill_mckibben,_it's_not_just_what _exxon_did,_it's_what_it's_doing/).

————. 2016c. "Global Warming's Terrifying New Chemistry: Our Leaders Thought Fracking Would Save Our Climate. They Were Wrong. Very Wrong." *Nation* (April 11/18): 12–18.

————. 2016d. "Recalculating the Climate Math: The Numbers on Global Warming Are Even Scarier than We Thought." *New Republic* (September 22). Retrieved September 27, 2016 (http://newrepublic.com/article/136987/recalculating -climate-math).

————. 2017a. "Keep It 100: The Unimaginable Is Now Possible: 100% Renewable Energy." *In These Times* (August 22). Retrieved August 23, 2017 (http://inthesetimes .com/features/bill_mckibben_renewable_energy_100_percent_solution.html).

————. 2017b. "The New Battle Plan for the Planet's Climate Crisis." *Rolling Stone* (January 24). Retrieved January 25, 2017 (https://www.rollingstone.com/politics /politics-features/the-new-battle-plan-for-the-planets-climate-crisis-114410/).

————. 2017c. "With the Rise of Trump, Is It Game over for the Climate Fight?" *Yale Environment360* (January 23). Retrieved January 24, 2017 (https://e360.yale .edu/features/with-the-ascent-of-trump-is-it-game-over-for-the-climate-fight).

————. 2018a. "Hit Fossil Fuels Where It Hurts—the Bottom Line." *Rolling Stone* (May 21). Retrieved May 21, 2018 (https://www.rollingstone.com/politics /politics-news/hit-fossil-fuels-where-it-hurts-the-bottom-line-627746/).

————. 2018b. "A Very Grim Forecast." *New York Review of Books* (November 22). Retrieved November 28, 2018 (https://www.nybooks.com/articles/2018/11/22 /global-warming-very-grim-forecast/).

————. 2019a. "A Future without Fossil Fuels." *New York Review of Books* (April 4). Retrieved April 4, 2019 (https://www.nybooks.com/articles/2019/04/04/future -without-fossil-fuels/).

————. 2019b. *Falter: Has the Human Game Begun to Play Itself Out?* New York: Henry Holt and Company.

————. 2019c. "Money Is the Oxygen on Which the Fire of Global Warming Burns." *The New Yorker* (September 17). Retrieved September 18, 2019 (https://www .newyorker.com/news/daily-comment/money-is-the-oxygen-on-which-the-fire -of-global-warming-burns).

McNall, Scott. 2011. *Rapid Climate Change: Causes, Consequences, and Solutions.* New York: Routledge.

Meyer, Robinson. 2016a. "How Obama Could Lose His Big Climate Case." *Atlantic* (September 29). Retrieved October 23, 2017 (https://www.atlantic.com/science /archive/2016/09/Obama-clean-power-dc-circuit-legal/502115).

————. 2016b. "The Supreme Court Didn't Block These Obama Climate Policies." *Atlantic* (February 16). Retrieved October 23, 2017 (https://www.theatlantic.com /technology/archive/2016/02/the-supreme-court-didnt-block-these-climate -policies/462618/).

Michalowski, Raymond J. 1985. *Order, Law and Crime.* New York: Random House.

————. 2010. "In Search of 'State and Crime' in State Crime Studies." In *State Crime in the Global Age*, edited by W. Chambliss, R. Michalowski, and R. C. Kramer, 13–30. Cullompton, UK: Willan Publishing.

Michalowski, Raymond, and Ronald C. Kramer. 2006. *State-Corporate Crime: Wrongdoing at the Intersection of Business and Government.* New Brunswick, NJ: Rutgers University Press.

———. 2007. "State-Corporate Crime and Criminological Inquiry." In *International Handbook of White-Collar and Corporate Crime*, edited by H. Pontell and G. Geis, 200–219. New York: Springer.

———. 2014. "Transnational Environmental Crime." In *Transnational Crime and Justice Handbook*, second edition, edited by P. Reichel and J. Albanese, 189–212. Thousand Oaks, CA: Sage.

Miller, Todd. 2017. *Storming the Wall: Climate Change, Migration, and Homeland Security*. San Francisco: City Lights.

———. 2018a. "The Border Fetish: The U.S. Frontier as a Zone of Profit and Sacrifice." *TomDispatch* (April 24). Retrieved April 24, 2018 (http://www.tomdispatch.com/blog/176414).

———. 2018b. "Why the Migrant Caravan Story Is a Climate Change Story." *Yes Magazine* (November 27). Retrieved November 29, 2018 (https://www.yesmagazine.org/peace-justice/why-the-migrant-caravan-story-is-a-climate-change-story-20181127).

———2019. *Empire of Borders: The Expansion of the US Border around the World*. London: Verso.

Mills, C. Wright. 1956. *The Power Elite*. New York: Oxford University Press.

———. 1963. "The Structure of Power in American Society." In *Power, Politics & People: The Collected Essays of C. Wright Mills,* edited by I. L. Horowitz, 23–38. New York: Oxford University Press.

Milman, Oliver. 2017. "Hurricanes and Heatwaves: Stark Signs of Climate Change 'New Normal.'" *Guardian* (December 28). Retrieved January 2, 2018 (https://www.theguardian.com/environment/2017/dec/28/climate-change-2017-warmest-year-extreme-weather).

———. 2018. "Ex-NASA Scientist: 30 Years on, World Is Failing 'Miserably' to Address Climate Change." *Guardian* (June 19). Retrieved June 19, 2018 (https://www.theguardian.com/environment/2018/jun/19/james-hansen-nasa-scientist-climate-change-warning).

———. 2019. "Americans are Waking Up: Two Thirds Say Climate Crisis Must be Adressed." *Guardian* (September 15). Retrieved September 15, 2019 (https://www.theguardian.com/science/2019/sep/15/americans-climate-change-crisis-cbs-poll).

Mitchell, Timothy. 2013. *Carbon Democracy*. London: Verso.

Molotch, Harvey. 1970. "Oil in Santa Barbara and Power in America." *Sociological Inquiry* 40: 131–144.

Monbiot, George. 2010. "The Process Is Dead." *Guardian* (September 20). Retrieved June 19, 2018 (http://www.monbiot.com/2010/09/20/the-process-is-dead).

———. 2015. "Grand Promises of Paris Climate Deal Undermined by Squalid Retrenchments." *Guardian* (December 12). Retrieved August 2, 2017 (https://www.theguardian.com/environment/georgemonbiot/2015/dec/12/paris-climate-deal-governments-fossil-fuels).

Mooney, Chris. 2005. *The Republican War on Science*. New York: Basic Books.

———. 2012. *The Republican Brain: The Science of Why They Deny Science—and Reality*. Hoboken, NJ: John Wiley & Sons.

———. 2016. "30 Years ago Scientists Warned Congress on Global Warming. What They Said Sounds Eerily Familiar." *Washington Post* (June 11). Retrieved September 20, 2016 (https://www.washingtonpost.com/news/energy-environment/wp/2016/06/11/30-years-ago-scientists-warned-congress-on-global-warming-what-they-said-sounds-eerily-familiar/).

————. 2018. "Earth's Atmosphere Just Crossed Another Troubling Climate Change Threshold." *Washington Post* (May 3). Retrieved May 3, 2018 (https://www .washingtonpost.com/news/energy-environment/wp/2018/05/03/earths -atmosphere-just-crossed-another-troubling-climate-change-threshold/?utm _term=.87fb376ecc44).

Mooney, Chris, and Brady Dennis. 2018. "The World Has Just over a Decade to Get Climate Change under Control, UN Scientists Say." *Washington Post* (October 7). Retrieved October 8, 2018 (https://www.washingtonpost.com/energy-environ ment/2018/10/08/world-has-only-years-get-climate-change-under-control-un -scientists-say/?noredirect=on&utm_term=.75119b4fbb9e).

Moore, Jason. 2015. *Capitalism in the Web of Life: Ecology and the Accumulation of Capital.* London: Verso.

————, ed. 2016. *Anthropocene or Capitalocene? Nature, History, and the Crisis of Capitalism.* Oakland, CA: PM Press.

Mora, Camilo, Daniele Spirandelli, Erik Franklin, et al. 2018. "Broad Threat to Humanity from Cumulative Climate Hazards Intensified by Greenhouse Gas Emissions." *Nature Climate Change* 8: 1062–1071.

Morgan, Jennifer, and Kevin Kennedy. 2013. "First Take: Looking at President Obama's Climate Action Plan." *World Resources Institute* (June 25). Retrieved February 4, 2018 (https://www.wri.org/blog/2013/06/first-take-looking-president -obama-s-climate-action-plan).

Nagel, Joanne. 2016. *Gender and Climate Change: Impacts, Science, Policy.* New York: Routledge.

National Oceanic and Atmospheric Administration. 2017. "International Report Confirms 2016 Was Warmest Year on Record for the Globe" (August 10). Retrieved July 10, 2018 (https://www.noaa.gov/news/international-report -confirms-2016-was-warmest-year-on-record-for-globe).

————. 2018. "Billion-Dollar Weather and Climate Disasters: Overview." National Centers for Environmental Information (NCEI), National Oceanic and Atmospheric Administration. Retrieved January 9, 2018 (https://www.ncdc.noaa.gov/billions/).

Negin, Elliott. 2016. "Broadcast Networks Largely Ignored Climate Change This Election Year." *Huffington Post* (November 16). Retrieved December 6, 2016 (https://www.huffingtonpost.com/elliott-negin/broadcast-networks-largel_b _13002776.html).

Nelson, Arthur. 2015. "Pentagon to Lose Emissions Exemption under Paris Climate Deal." *Guardian* (December 14). Retrieved March 18, 2017 (https://www .theguardian.com/environment/2015/dec/14/pentagon-to-lose-emissions -exemption-under-paris-climate-deal).

————. 2018. "Dutch Appeals Court Upholds Landmark Climate Change Ruling: Netherlands Ordered to Increase Emissions Cuts in Historic Ruling that Puts All World Governments on Notice." *Guardian* (October 9). Retrieved October 17, 2018 (https://www.theguardian.com/environment/2018/oct/09/dutch-appeals -court-upholds-landmark-climate-change-ruling).

Nesbit, Jeff. 2018. *This Is the Way the World Ends: How Droughts and Die-Offs, Heat Waves and Hurricanes Are Converging on America.* New York: Thomas Dunne Books.

Neukom, Raphael, Luis A. Barboza, Michael P. Erb, et al. 2019. "Consistent Multi-decadal Variability in Global Temperature Reconstructions and Simulations over the Common Era." *Nature Geoscience* 12: 643–649.

New York Times. 1989. "The White House and the Greenhouse." May 9, 1989. Retrieved February 20, 2018 (https://www.nytimes.com/1989/05/09/opinion /the-white-house-and-the-greenhouse.html).

———. 2008. "Barack Obama's Remarks in St. Paul." June 3, 2008. Retrieved September 25, 2017 (https://www.nytimes.com/2008/06/03/us/politics/03text-obama.html).

———. 2009. "The Endangerment Finding." December 7, 2009, A36.

———. 2017a. "President Trump Risks the Planet." March 28, 2017, A26.

———. 2017b. "President Trump's War on Science." September 10, 2017, 10SR.

Nixon, Rob. 2011. *Slow Violence and the Environmentalism of the Poor.* Cambridge, MA: Harvard University Press.

Norgaard, Kari Marie. 2011. *Living in Denial: Climate Change, Emotions, and Everyday Life.* Cambridge, MA: MIT Press.

Nugent, Walter. 2008. *Habits of Empire: A History of American Expansion.* New York: Vintage Books.

O'Harrow, Robert, Jr. 2017. "A Two-Decade Crusade by Conservative Charities Fueled Trump's Exit from Paris Climate Accord." *Washington Post* (September 5). Retrieved September 6, 2017 (https://www.washingtonpost.com/investigations/a-two-decade-crusade-by-conservative-charities-fueled-trumps-exit-from-paris-climate-accord/2017/09/05/fcb8d9fe-6726-11e7-9928-22d00a47778f_story.html).

Olson, Bradley, and Aruna Viswanatha. 2016. "SEC Probes Exxon over Accounting for Climate Change." *Wall Street Journal* (September 20). Retrieved November 6, 2016 (https://www.wsj.com/articles/sec-investigating-exxon-on-valuing-of-assets-accounting-practices-1474393593).

Oppenheimer, Michael, Naomi Oreskes, Dale Jamieson, et al. 2019. *Discerning Experts: The Practices of Scientific Assessment for Environmental Policy.* Chicago: University of Chicago Press.

Oreskes, Naomi. 2014. "Wishful Thinking about Natural Gas: Why Fossil Fuels Can't Solve the Problems Created by Fossil Fuels." *TomDispatch* (July 27). Retrieved July 28, 2014 (http://www.tomdispatch.com/post/175873/tomgram%3A_naomi_oreskes%2C_a_%22green%22_bridge_to_hell/).

———. 2015. "Exxon's Climate Concealment." *New York Times* (October 10): A21.

Oreskes, Naomi, and Erik Conway. 2010. *Merchants of Doubt: How a Handful of Scientists Obscured the Truth on Issues from Tobacco to Global Warming.* New York: Bloomsbury Press.

Otto, Shawn. 2016. *The War on Science: Who's Waging It, Why It Matters, What We Can Do about It.* Minneapolis: Milkweed Editions.

Parenti, Christian. 2011. *Tropic of Chaos: Climate Change and the New Geography of Violence.* New York: Nation Books.

———. 2012. "Why Climate Change Will Make You Love Big Government." *Nation* (January 26). Retrieved July 5, 2018 (https://www.thenation.com/article/why-climate-change-will-make-you-love-big-government/).

———. 2016a. "The Catastrophic Convergence: Militarism, Neoliberalism and Climate Change." In *The Secure and the Dispossessed: How the Military and Corporations are Shaping a Climate Changed World,* edited by N. Buxton and B. Hayes, 23–38. London: Pluto Press.

———. 2016b. "Environment-Making in the Capitalocene: Political Ecology of the State." In *Anthropocene or Capitalocene? Nature, History and the Crisis of Capitalism,* edited by J. Moore, 166–184. Oakland, CA: PM Press.

Park, Chang-Eui, Su-Jong Jeong, Manoj Joshi, et al. 2018. "Keeping Global Warming within 1.5 Degree C Constrains Emergence of Aridification." *Nature Climate Change* 8: 70–74.

Parr, Adrian. 2012. *The Wrath of Capital: Neoliberalism and Climate Change Politics.* New York: Columbia University Press.

Penn, Ivan. 2016. "California to Investigate Whether Exxon Mobil Lied about Climate-Change Risks." *Los Angeles Times* (January 20). Retrieved January 20, 2016 (https://www.latimes.com/business/la-fi-exxon-global-warming-20160120 -story.html).

Perrow, Charles, and Simone Pulver. 2015. "Organizations and Markets." In *Climate Change and Society: Sociological Perspectives*, edited by R. E. Dunlap and R. J. Brulle, 61–92. New York: Oxford University Press.

Phillips, Tom. 2016. "Climate Change a Chinese Hoax? Beijing Gives Donald Trump a Lesson in History." *Guardian* (November 17). Retrieved November 5, 2017 (https://www.theguardian.com/us-news/2016/nov/17/climate-change-a-chinese -plot-beijing-gives-donald-trump-a-history-lesson).

Pielke, Roger A., Jr., 2000. "Policy History of U.S. Global Change Research Program: Part I. Administrative Development." *Global Environmental Change* 10: 9–25; "Part II. Legislative Process." *Global Environmental Change* 10: 133–144.

Pinko, Nicole. 2018. "What Is ExxonMobil's Present Position and Role on Climate Change?" *Union of Concerned Scientists* (May). Retrieved September 12, 2018 (https://www.ucsusa.org/our-work/ucs-publications/Exxon#.XBnWos9Kj-Y).

Platt, Tony. 1974. "Prospects for a Radical Criminology in the United States." *Crime and Social Justice* 1: 2–10.

Plumer, Brad. 2018a. "New UN Climate Report Says Put a High Price on Carbon." *New York Times* (October 8): A7.

———. 2018b. "Trump Put a Low Cost on Carbon Emissions. Here's Why It Matters." *New York Times* (August 24): A16.

———. 2019. "Wildlife Facing Extinction Risk All Over the Globe." *New York Times* (May 7): A1.

Plumer, Brad, and Lisa Friedman. 2018. "Trump Has Allies on Fossil Fuels." *New York Times* (December 11): A1.

Polanyi, Karl. 1944. *The Great Transformation.* New York: Rinehart & Company.

Pollin, Robert. 2015. *Greening the Global Economy.* Cambridge, MA: MIT Press.

Popovich, Nadia, Livia Albeck-Ripka, and Kendra Pierre-Louis. 2019. "85 Environmental Rules Being Rolled Back under Trump." *New York Times* (September 12). Retrieved September 27, 2019 (https://www.nytimes.com/interactive/2019 /climate/trump-environment-rollbacks.html).

Powell, James Lawrence. 2011. *The Inquisition of Climate Science.* New York: Columbia University Press.

Project for the New American Century. 2000. *Rebuilding America's Defenses: Strategy, Forces and Resources for a New Century.* Washington, DC: Author.

Prupis, Nadia. 2015a. "Sanders Joins Call for DOJ Investigation into Exxon's Climate Coverup." *Common Dreams* (October 21). Retrieved October 30, 2016 (https:// www.commondreams.org/news/2015/10/21/sanders-joins-call-doj-investigation -exxons-climate-coverup).

———. 2015b. "UN Report Shows World's Pledges for Paris Are Recipe for Climate 'Disaster.'" *Common Dreams* (October 30). Retrieved October 30, 2015 (https:// www.commondreams.org/news/2015/10/30/un-report-shows-worlds-pledges -paris-are-recipe-climate-disaster).

Queally, Jon. 2015. "'It Can Be Done': New Report Details Path to 100% Renewables by 2050." *Common Dreams* (September 21). Retrieved October 10, 2016 (https://www.commondreans.org/news/2015/09/21/it-can-be-done-new-report-details-path-100-renewables-2050).

Rabe, Barry G. 2018. *Can We Price Carbon?* Cambridge, MA: MIT Press.

Reiner, Robert. 2016. *Crime: The Mystery of the Common-Sense Concept.* Cambridge, UK: Polity Press.

Renner, Michael. 1991. "Assessing the Military's War on the Environment." In *State of the World*, edited by L. Stark, 132–152. New York: W. W. Norton.

Resplandy, Laure, Ralph F. Keeling, Yassir A. Eddebbar, et al. 2018. "Quantification of Ocean Heat Uptake from Changes in Atmospheric O_2 and CO_2 Composition." *Nature* 563: 105–108.

Reuveny, Rafael. 2007. "Climate Change-Induced Migration and Violent Conflict." *Political Geography* 26: 656–673.

Reyes, Oscar. 2016. "The Little-Known Fund at the Heart of the Paris Climate Agreement." *Foreign Policy in Focus* (July 14). Retrieved July 15, 2016 (https://fpif.org/little-known-fund-heart-paris-climate-agreement/).

Rich, Nathaniel. 2018a. "Losing Earth." *New York Times Magazine* (August 5): 1–70.

———. 2018b. "The Most Honest Book about Climate Change Yet." *Atlantic* (October). Retrieved September 9, 2018 (https://www.theatlantic.com/magazine/archive/2018/10/william-vollmann-carbon-ideologies/568309/).

———. 2019. *Losing Earth: A Recent History.* New York: MCD/Farrar, Straus and Giroux.

Richardson, Katherine, Will Steffan, and Diana Liverman. 2011. *Climate Change: Global Risks, Challenges and Decisions.* New York: Cambridge University Press.

Ricke, Katharine, Laurent Drouet, Ken Caldeira, and Massimo Tavoni. 2018. "Country-Level Social Cost of Carbon." *Nature and Climate Change* 8: 895–900.

Rieff, David. 2003. "Were Sanctions Right?" *New York Times Magazine* (July 23). Retrieved August 2, 2018 (https://www.nytimes.com/2003/07/27/magazine/were-sanctions-right.html).

Risen, James. 2006. *State of War: The Secret History of the CIA and the Bush Administration.* New York: Free Press.

Robinson, Eugene. 2016. "A Vote for Trump Is a Vote to Undo Vital Progress on Climate Change." *Real Clear Politics* (September 20). Retrieved September 20, 2016 (https://www.realclearpolitics.com/articles/2016/09/20/a_vote_for_trump_is_a_vote_to_undo_vital_progress_on_climate_change_131829.html?mobile_redirect=true).

Robinson, Mary. 2018. *Climate Justice: Hope, Resilience, and the Fight for a Sustainable Future.* New York: Bloomsbury.

Rockström, Johan, Owen Gaffney, Joeri Rogelj, et al. 2017. "A Roadmap for Rapid Decarbonization." *Science* 355: 1269–1271.

Romm, Joseph. 2019. *Climate Change: What Everyone Needs to Know.* Second edition. New York: Oxford University Press.

Rosa, Eugene A., Thomas K. Rudel, Richard York, et al. 2015. "The Human (Anthropogenic) Driving Forces of Global Climate Change." In *Climate Change and Society: Sociological Perspectives*, edited by R. E. Dunlap and R. J. Brulle, 32–60. New York: Oxford University Press.

Roston, Eric. 2016. "Trying to Put a Price on Big Oil's 'Climate Obstruction' Efforts." *Bloomberg* (April 7). Retrieved June 30, 2016 (https://www.bloomberg.com/news/articles/2016-04-07/trying-to-put-a-price-on-big-oil-s-climate-obstruction-efforts).

Rothe, Dawn, and David Kauzlarich. 2016. *Crimes of the Powerful: An Introduction.* New York: Routledge.

Roy, Arundhati. 2004. *An Ordinary Person's Guide to Empire.* Cambridge, MA: South End Press.

Royal Society and U.S. National Academy of Sciences. 2014. *Climate Change: Evidence and Causes.* Washington, DC: National Academies Press.

Royden, Amy. 2002. "U.S. Climate Change Policy under President Clinton: A Look Back." *Golden Gate University Law Review* 32: 415–475.

Ryan, David. 2007. *Frustrated Empire: U.S. Foreign Policy, 9/11 to Iraq.* London: Pluto Press.

Salamon, Margaret Klein. 2017. *The Transformative Power of Climate Truth: Ecological Awakening in the Age of Trump.* New York: Climate Mobilization.

Sanders, Barry. 2009. *The Green Zone: The Environmental Costs of Militarism.* Oakland, CA: AK Press.

———. 2017. "Fueling the Engines of Empire." In *The War and Environment Reader,* edited by G. Smith, 199–205. Charlottesville, VA: Just World Books.

Sands, Philippe. 2005. *Lawless World.* New York: Penguin Books.

Santana, Deborah. 2002. "Resisting Toxic Militarism: Vieques versus the U.S. Navy." *Social Justice* (Spring–Summer): 37–48.

Sargent, Niall. 2017. "'Historic' High Court Decision Recognizes Constitutional Right to Environmental Protection." *Shell to Sea* (November 21). Retrieved October 17, 2018 (http://www.shelltosea.com/content/%E2%80%98historic%E2%80%99-high -court-decision-recognises-constitutional-right-environmental-protection).

Savage, Karen. 2017. "Ireland Recognizes Constitutional Right to a Safe Climate and Environment." *Climate Liability News* (December 11). Retrieved October 14, 2018 (https://www.climateliabilitynews.org/2017/12/11/ireland-constitutional-right -climate-environment-fie/).

———. 2018a. "Exxon Funds Push for a Carbon Tax that Ends Climate Liability Suits." *Climate Liability News* (October 9). Retrieved October 10, 2018 (https://www .climateliabilitynews.org/2018/10/09/exxon-carbon-tax-climate-liability/).

———. 2018b. "Hidden Gem for Big Oil in Carbon Tax Plan: Ending Climate Liability Suits." *Climate Liability News* (July 17). Retrieved October 10, 2018 (https:// www.climateliabilitynews.org/2018/07/17/carbon-tax-climate-liability-waiver/).

Schattschneider, E. E. 1960. *The Semi-Sovereign People: A Realist's View of Democracy in America.* New York: Holt, Rinehart & Winston.

Scheer, Robert. 2010. *The Great American Stickup.* New York: Nation Books.

Scheuer, James H. 1990. "Bush's 'Whitewash Effect' on Warming." *New York Times* (March 3). Retrieved February 2, 2018 (https://www.nytimes.com/1990/03/03 /opinion/bush-s-whitewash-effect-on-warming.html).

Schnaiberg, Allan. 1980. *The Environment: From Surplus to Scarcity.* New York: Oxford University Press.

Schwartz, John. 2015. "Studies Look for Signs of Climate Change in 2014's Extreme Weather Events." *New York Times* (November 6): A16.

———. 2016. "Exxon Mobil Accuses the Rockefellers of a Climate Conspiracy." *New York Times* (November 22): D1.

Schwartz, Michael. 2008. *War without End: The Iraq War in Context.* Chicago: Haymarket Books.

Schwartz, Peter, and Doug Randall. 2003. *An Abrupt Climate Change Scenario and Its Implications for United States Security.* Retrieved August 23, 2018 (https://eesc .columbia.edu/courses/v1003/readings/Pentagon.pdf).

Schwendinger, Herman, and Julia Schwendinger. 1970. "Defenders of Order or Guardians of Human Rights?" *Issues in Criminology* 5: 123–157.

———. 1977. "Social Class and the Definition of Crime." *Crime and Social Justice* 7: 4–13.

Shabecoff, Philip. 1988. "Global Warming Has Begun, Expert Tells Senate." *New York Times* (June 24): A1

Shankman, Sabrina. 2018. "Capturing CO2 from Air: To Keep Global Warming under 1.5C, Emissions Must Go Negative, IPCC Says." *Inside Climate News* (October 12). Retrieved October 13, 2018 (https://insideclimatenews.org/news/12102018/global-warming-solutions-negative-emissions-carbon-capture-technology-ipcc-climate-change-report).

Shankman, Sabrina, and Bob Berwyn. 2017. "Polar Ice Is Disappearing, Setting off Climate Alarms." *Inside Climate News* (December 27). Retrieved December 27, 2017 (https://insideclimatenews.org/news/27122017/arctic-antarctic-sea-ice-sheets-2017-year-review-glaciers-disappearing-polar-records).

Shankman, Sabrina, and Paul Horn. 2017. "The Most Powerful Evidence Climate Scientists Have of Global Warming." *Inside Climate News* (October 3). Retrieved October 5, 2017 (https://insideclimatenews.org/news/03102017/infographic-ocean-heat-powerful-climate-change-evidence-global-warming).

Shiva, Vandana. 2012. "Rio+20: An Undesirable U-Turn." *Asian Age* (June 25). Retrieved July 3, 2012 (http://www.asianage.com/columnists/rio20-undesirable-u-turn).

Shwom, Rachael L., Aaron M. McCright, and Steven R. Brechin. 2015. "Public Opinion on Climate Change." In *Climate Change and Society: Sociological Perspectives*, edited by R. E. Dunlap and R. J. Brulle, 269–300. New York: Oxford University Press.

Silvano, Alessandro. 2018. "Freshening by Glacial Meltwater Enhances Melting of Ice Shelves and Reduces Formation of Antarctic Bottom Water." *Science Advances* 4: 1–11.

Simison, Bob. 2015. "New York Attorney General Subpoenas Exxon on Climate Research." *Inside Climate News* (November 5). Retrieved May 31, 2017 (https://insideclimatenews.org/news/05112015/new-york-attorney-general-eric-schneiderman-subpoena-Exxon-climate-documents).

Skocpol, Theda. 2013. *Naming the Problem: What It Will Take to Counter Extremism and Engage Americans in the Fight against Global Warming.* Cambridge, MA: Scholars Strategy Network. Retrieved October 17, 2016 (www.scholarsstrategynetwork.org/sites/default/files/skocpol_captrade_report_january_2013_0.pdf).

Skocpol, Theda, and Vanessa Williamson. 2012. *The Tea Party and the Remaking of Republican Conservatism.* New York: Oxford University Press.

Smandych, Russell, and Rodney Kueneman. 2010. "The Canadian-Alberta Tar Sands: A Case Study of State–Corporate Environmental Crime." In *Global Environmental Harm: Criminological Perspectives*, edited by R. White, 87–109. Cullompton, UK: Willan Publishing.

Smith, Gar, ed. 2017. *The War and Environment Reader.* Charlottesville, VA: Just World Books.

Somaiya, Ravi. 2015. "Columbia Disputes Exxon Mobil on Climate Risk Articles." *New York Times* (December 1). Retrieved May 20, 2018 (https://www.nytimes.com/2015/12/02/business/media/columbia-disputes-exxon-mobil-on-climate-risk-articles.html).

Steffen, Alex. 2016. "Predatory Delay and the Rights of Future Generations." *Medium* (April 29). Retrieved November 11, 2018 (https://medium.com/@AlexSteffen/predatory-delay-and-the-rights-of-future-generations-69b06094a16).

Steffen, Will, Paul J. Crutzen, and John R. McNeill. 2007. "The Anthropocene: Are Humans Now Overwhelming the Great Forces of Nature?" *Ambio* 36: 614–621.

Steffen, Will, Johan Rockström, Katherine Richardson, et al. 2018. "Trajectories of the Earth System in the Anthropocene." *Proceedings of the National Academy of Sciences* 115: 8252–8259. Retrieved September 18, 2018 (https://www.pnas.org/content/115/33/8252/tab-article-info).

Stephenson, Wen. 2015. *What We're Fighting for Now Is Each Other: Dispatches from the Front Lines of Climate Justice.* Boston: Beacon Press.

Stiglitz, Joseph E., and Linda J. Bilmes. 2008. *The Three Trillion Dollar War: The True Cost of the Iraq Conflict.* New York: W. W. Norton.

Stretesky, Paul B., Michael A. Long, and Michael J. Lynch. 2014. *The Treadmill of Crime: Political Economy and Green Criminology.* London: Routledge.

Sullivan, Kaitlin. 2019. "Utilities Knew: Report Shows Power Industry Studied Climate Change, Stuck with Coal. *Climate Liability News* (January 16). Retrieved January 17, 2019 (https://www.climateliabilitynews.org/2019/01/16/utilities-climate-change-liability/).

Suskind, Ron. 2004. *The Price of Loyalty: George W. Bush, the White House, and the Education of Paul O'Neill.* New York: Simon & Schuster.

———. 2006. *The One Percent Doctrine: Deep inside America's Pursuit of Its Enemies since 9/11.* New York: Simon & Schuster.

Sutherland, Edwin H. 1940. "White Collar Criminality." *American Sociological Review* 5: 1–12.

———. 1949. *White Collar Crime.* New York: Holt, Rinehart and Winston.

Swanson, David. 2018. *Curing Exceptionalism: What's Wrong with How We Think about the United States.* Charlottesville, VA: David Swanson Publisher.

Talbot, Margaret. 2018. "Dirty Politics: Scott Pruitt's E.P.A. Is Giving Even Ostentatious Polluters a Reprieve." *New Yorker* (April 2): 38–51.

Tappan, Paul. 1947. "Who Is the Criminal?" *American Sociological Review* 12: 96–102.

Teske, Sven. 2015. *Energy [R]evolution: A Sustainable World Energy Outlook 2015.* Amsterdam: Greenpeace International.

350.org. 2015. "Freeze Fossil Fuel Extraction to Stop Climate Crimes." Retrieved October 10, 2016 (https://350.org/climate-crimes/).

———. 2017. "Make Exxon Pay for Harvey and Irma." Retrieved September 8, 2017 (https://350.org/press-release/350-org-make-exxon-pay-for-harvey-and-irma/).

Tifft, Larry, and Dennis Sullivan. 1980. *The Struggle to Be Human: Crime, Criminology and Anarchism.* Over the Water, UK: Cienfuegos Press.

———. 2001. "A Needs-Based, Social Harms Approach to Defining Crime." In *What Is Crime? Controversies over the Nature of Crime and What to Do about It*, edited by S. Henry and M. Lanier, 179–203. Lanham, MD: Rowman & Littlefield.

Tokar, Brian. 2014. *Towards Climate Justice: Perspectives on the Climate Crisis and Social Change.* Porsgrunn, Norway: New Compass Press.

Tombs, Steve, ed. 2014. "Special Issue on State–Corporate Crime." *State Crime: Journal of the International State Crime Initiative* 3 (2).

Tooze, Adam. 2014. *The Deluge: The Great War, America and the Remaking of the Global Order, 1916–1931.* New York: Viking.

Trout, Kelly. 2019. *Drilling towards Disaster: Why U.S. Oil and Gas Expansion Is Incompatible with Climate Limits.* Washington, DC: Oil Change International.

Turse, Nick. 2008. *The Complex: How the Military Invades Our Everyday Lives.* New York: Metropolitan Books.

———. 2013. *Kill Anything that Moves: The Real American War in Vietnam.* New York: Metropolitan Books.

United Nations Environment Programme. 2017. *The Status of Climate Change Litigation: A Global Review*. Nairobi, Kenya: Law Division, United Nations Environment Programme.

Union of Concerned Scientists. 2011. "Outcome at Durban Climate Negotiations Offers Limited Progress." *Common Dreams* (December 13). Retrieved March 15, 2012 (https://www.commondreams.org/newswire/2011/12/13-8).

———. 2012a. *A Climate of Corporate Control: How Corporations Have Influenced the U.S. Dialogue on Climate Science and Policy*. Cambridge, MA: UCS Publications.

———. 2012b. *Heads They Win, Tails We Lose: How Corporations Corrupt Science at the Public's Expense*. Cambridge, MA: UCS Publications.

———. 2013. "Global Warming Skeptic Organizations" (February 9). Retrieved February 9, 2013. (https://www.ucsusa.org/global-warming/solutions/fight-misinformation /global-warming-skeptic.html#.WOT209hKj-y).

———. 2014. "The U.S. Military and Oil." Retrieved December 12, 2018 (https:// www.ucsusa.org/clean_vehicles/smart-transportation-solutions/us-military-oil -use.html#.XBGZ8BNKj-a).

———. 2018. "Is Global Warming Fueling Increased Wildfire Risks?" (July 24). Retrieved November 22, 2018 (https://www.ucsusa.org/global-warming/science -and-impacts/impacts/global-warming-and-wildfire.html#.XAQF0RNKjq0).

Urry, John. 2011. *Climate Change & Society*. Cambridge, UK: Polity Press.

U.S. Department of Defense. 2016. *DOD Directive 4715.21: Climate Change Adaptation and Resilience* (January 14). Retrieved August 23, 2018 (https://dod.defense.gov /Portals/1/Documents/pubs/471521p.pdf).

U.S. Global Change Research Program. 2017. *Climate Science Special Report: Fourth National Climate Assessment*. Volume 1. Washington, DC: Author.

———. 2018. *Climate Science Special Report: Fourth National Climate Assessment*. Volume 2. Washington, DC: Author.

U.S. Supreme Court. 2007. *Opinion of the Court in Massachusetts v. Environmental Protection Agency*, 549 U.S. 497.

Vaughn, Diane. 1996. *The Challenger Launch Decision: Risky Technology, Culture, and Deviance at NASA*. Chicago: University of Chicago Press.

Vidal, John. 2005. "Revealed: How Oil Giant Influenced Bush." *Guardian* (June 8). Retrieved July 5, 2018 (http://www.theguardian.com/news/2005/jun/08/usnews .climatechange).

Vidal, John, and Owen Bowcott. 2016. "ICC Widens Remit to Include Environmental Destruction Cases." *Guardian* (September 15). Retrieved September 16, 2016 (https://www.theguardian.com/global/2016/sep/15/hague-court-widens-remit -to-include-environmental-destruction-cases).

Vollmann, William T. 2018. *No Immediate Danger*. Vol. 1 of *Carbon Ideologies*; *No Good Alternative*. Vol. 2 of *Carbon Ideologies*. New York: Viking.

Walia, Harsha. 2013. *Undoing Border Imperialism (Anarchist Interventions)*. Chico, CA: AK Press.

Wallace-Wells, David. 2017. "The Uninhabitable Earth." *New York Magazine* (July 9). Retrieved July 10, 2017 (http://nymag.com/intelligencer/2017/07/climate-change -earth-too-hot-for-humans.html).

———. 2019. *The Uninhabitable Earth: Life after Warming*. New York: Tim Duggan Books.

Wallerstein, Immanuel. 1989. *The Modern World System*. New York: Academic Press.

Wang, Ucilia. 2018. "What Oil Companies Knew about Climate Change and When: A Timeline. *Climate Liability News* (April 5). Retrieved January 31, 2019 (https://www.climateliabilitynews.org/2018/04/05/climate-change-oil-companies-knew-shell-exxon/).

Wapner, Paul. 2001. "Clinton's Environmental Legacy." *Tikkun* 16: 11–14.

Ward, Bob. 2012. "Heartland Institute Leak Exposes Strategies of Climate Attack Machine." *Common Dreams* (February 21). Retrieved February 21, 2012 (http://www.commondreams.org/view/2012/02/21-9).

Washington, Hayden, and John Cook. 2011. *Climate Change Denial: Heads in the Sand.* London: Earthscan.

Watts, Jonathon. 2017a. "Alarm as Study Reveals World's Tropical Forests Are Huge Carbon Emission Source." *Guardian* (September 28). Retrieved September 29, 2017 (https://www.theguardian.com/environment/2017/sep/28/alarm-as-study-reveals-worlds-tropical-forests-are-huge-carbon-emission-source).

———. 2017b. "Global Atmospheric CO2 Levels Hit Record High." *Guardian* (October 30). Retrieved October 30, 2017 (https://www.theguardian.com/environment/2017/oct/30/global-atmospheric-co2-levels-hit-record-high).

———. 2017c. "'We Should Be on the Offensive'—James Hansen Calls for Wave of Climate Lawsuits." *Guardian* (November 17). Retrieved April 4, 2018 (https://www.theguardian.com/environment/2017/nov/17/we-should-be-on-the-offensive-james-hansen-calls-for-wave-of-climate-lawsuits).

———. 2018. "We Have 12 Years to Limit Climate Change Catastrophe Warns UN." *Guardian* (October 8). Retrieved October 8, 2018 (https://www.theguardian.com/environment/2018/oct/08/global-warming-must-not-exceed-15c-warns-landmark-un-report).

Watts, Nick, Markus Amann, Nigel Arneil, et al. 2018. "The Lancet Countdown on Health and Climate Change." *The Lancet* 392: 2479–2514.

Watts, Rob. 2016. *States of Violence and the Civilising Process: On Criminology and State Crime.* London: Palgrave Macmillan.

Weiss, Daniel J. 2010. "Anatomy of a Senate Climate Bill Death." *Center for American Progress* (October 12). Retrieved September 9, 2017 (https://www.americanprogress.org/issues/green/news/2010/10/12/8569/anatomy-of-a-senate-climate-bill-death/).

Weisse, Mikaela, and Liz Goldman. 2018. "2017 Was the Second-Worst Year on Record for Tropical Tree Cover Loss." *Global Forest Watch* (June 27). Retrieved November 30, 2018 (https://blog.globalforestwatch.org/data/2017-was-the-second-worst-year-on-record-for-tropical-tree-cover-loss).

White, Damian F., Alan P. Rudy, and Brian J. Gareau. 2016. *Environments, Natures and Social Theory: Towards a Critical Hybridity.* London: Palgrave.

White, Gregory. 2011. *Climate Change and Migration: Security and Borders in a Warming World.* New York: Oxford University Press.

White, Rob, ed. 2010. *Global Environmental Harm: Criminological Perspectives.* Cullompton, UK: Willan Publishing.

———. 2011. *Transnational Environmental Crime: Toward an Eco-Global Criminology.* London: Routledge.

———. ed. 2012. *Climate Change from a Criminological Perspective.* New York: Springer.

———. 2018. *Climate Change Criminology.* London: Bristol University Press.

White, Rob, and Diane Heckenberg. 2014. *Green Criminology: An Introduction to the Study of Environmental Harm.* London: Routledge.

White, Rob, and Ronald C. Kramer, eds. 2015a. *Climate Change from a Criminological Perspective.* Special issue of *Critical Criminology: An International Journal* 23, no. 4 (November).

———. 2015b. "Critical Criminology and the Struggle against Climate Change Ecocide." *Critical Criminology: An International Journal* 23: 383–399.

Whyte, David. 2014. "Regimes of Permission and State–Corporate Crime." *State Crime: Journal of the International State Crime Initiative* 3: 237–246.

Williams, William Appleman. 1959. *The Tragedy of American Diplomacy*. New York: Norton.

———. 1980. *Empire as a Way of Life*. New York: Oxford University Press.

Wilson, William Julius. 2009. *More than Just Race: Being Black and Poor in the Inner City*. New York: W. W. Norton.

Wolin, Sheldon. 2008. *Democracy Incorporated: Managed Democracy and the Specter of Inverted Totalitarianism*. Princeton, NJ: Princeton University Press.

Wonders, Nancy A., and Mona Danner. 2015. "Gendering Climate Change: A Feminist Criminological Perspective." *Critical Criminology: An International Journal* 23: 401–416.

Wong, Edward. 2016. "Trump Has Called Climate Change a Chinese Hoax. Beijing Says It Is Anything But." *New York Times* (November 18). Retrieved November 19, 2016 (https://www.nytimes.com/2016/11/19/world/asia/china-trump-climate-change.html).

Wood, Mary Christina. 2014. *Nature's Trust: Environmental Law for a New Ecological Age*. New York: Cambridge University Press.

Woodbury, Zhiva. 2015. "After Paris: Making the Case for a People-Powered Transition to a New Climate Culture." *Truthout* (December 15). Retrieved December 15, 2015 (https://truthout.org/articles/after-paris-making-the-case-for-a-people-powered-transition-to-a-new-climate-culture/).

Woodward, Bob. 2004. *Plan of Attack*. New York: Simon & Shuster.

Woodworth, Elizabeth, and David Ray Griffin. 2016. *Unprecedented Climate Mobilization: A Handbook for Citizens and Their Governments*. Atlanta: Clarity Press.

Wright, Christopher, and Daniel Nyberg. 2015. *Climate Change, Capitalism, and Corporations: Processes of Creative Destruction*. Cambridge: Cambridge University Press.

Xu, Yangyang, and Veerabhadran Ramanathan. 2017. "Well below 2 Degree C: Mitigation Strategies for Avoiding Dangerous to Catastrophic Climate Changes." *Proceedings of the National Academy of Sciences* 114: 10315–10323.

Yamin, Farhana. 1998. "The Kyoto Protocol: Origins, Assessment, and Future Challenges." *Review of European, Comparative & International Law* 7: 113–127.

Ye, Connie. 2017. "Irish High Court Recognizes Personal Constitutional Right to Environment." *Human Rights Law Centre* (November 21). Retrieved October 17, 2018 (https://www.hrlc.org.au/human-rights-case-summaries/2018/6/18/irish-high-court-recognises-personal-constitutional-right-to-environment).

Young, Marilyn B. 1991. *The Vietnam Wars, 1945–1990*. New York: HarperCollins.

Zimet, Abby 2017. "A Spectacle Both Awe-Inducing and Horrifying." *Common Dreams* (June 1). Retrieved June 2, 2017 (https://www.commondreams.org/further/2017/06/01/spectacle-both-awe-inducing-and-horrifying).

Zinn, Howard. 1980. *A People's History of the United States: 1492–Present*. New York: HarperCollins.

———. 1994. *You Can't Be Neutral on a Moving Train: A Personal History of Our Times*. Boston: Beacon Press.

Zukowski, Dan. 2017. "Climate Destabilization Is Causing Thousands of New Species Migrations: Plant, Animal, Insect, Bird." *Truthout* (June 26). Retrieved July 15, 2018 (https://truthout.org/articles/climate-destabilization-causing-thousands-of-new-species-migrations-plant-animal-insect-bird/).

Index

legislation of, 28, 30–31, 34, 101–102, 106, 126, 133, 210; Republican Party and, 110, 113, 116–117, 121, 131–133, 200

Urgenda Foundation, 212–213

Urgenda Foundation v. the State of the Netherlands, 212

U.S. Border Patrol, 186. *See also* border enforcement; militarized border policies

U.S. Climate Action Partnership (USCAP), 127, 130

U.S. Department of Commerce, 31

U.S. Department of Defense, 51, 155–156, 171, 175, 177, 185. *See also* Pentagon

U.S. Department of Energy, 24, 31, 48, 69–70, 142–143

U.S. Department of Justice, 75–79, 87–88

U.S. Department of State, 24, 31

U.S. Department of the Interior, 24, 103, 109

U.S. Department of Transportation, 147

U.S. Global Change Research Program, 30, 34, 151

U.S. House of Representatives, 30, 110, 114, 116, 122, 127–128, 133, 200, 210, 222–223. *See also* United States Congress

U.S.–Mexico border, 186–188

U.S. Securities and Exchange Commission (SEC), 79, 207

U.S. Senate: climate change denial and, 130, 222; James Hansen testimony for, 27–28, 32, 48, 71, 89, 94, 102; legislation of, 115, 127–128; Republican Party and, 110, 114, 130, 133, 138, 200, 214, 222–223. *See also* United States Congress

U.S. Supreme Court, 78, 88, 96, 117, 122, 133, 147, 170, 206

Venezuela, 64

Vietnam, 1, 89, 95, 160, 162, 165–166, 168–169

war crimes, 89, 159–160, 209

warfare state, 25, 155, 157, 159, 174

weapons of mass destruction, 167–168, 172

White, Rob, 6–8, 19, 42, 128, 193, 199, 206, 209–210, 214, 228

white-collar crime, 1, 8, 13, 15–17

Whitman, Christine Todd, 81, 118–119, 143

wildfires, 35, 39, 215

Wilson, Woodrow, 157–158, 164

Wirth, Timothy E., 27–28, 114

Wood, Mary Christina, 86, 88, 206

World Bank, 42, 158

World Conference on the Changing Atmosphere, 32, 102

World Meteorological Organization (WMO), 32, 102

world system theory, 54, 159

World Trade Center, 170

World Trade Organization, 111–112

World War I, 157–158

World War II, 157–159, 161, 177, 225. *See also* post–World War II era

xenophobia, 141, 186–187

Zinn, Howard, 25, 158, 192, 228–229

About the Author

Ronald C. Kramer is professor of sociology at Western Michigan University. He received his doctorate in sociology from the Ohio State University, specializing in criminology and law. His previous books are *Crimes of the American Nuclear State: At Home and Abroad* (with Dave Kauzlarich); *State-Corporate Crime: Wrongdoing at the Intersection of Business and Government* (with Raymond J. Michalowski); and *State Crime in the Global Age* (edited with William J. Chambliss and Raymond J. Michalowski). Dr. Kramer is the recipient of the Lifetime Achievement Award from the Division of Critical Criminology of the American Society of Criminology. He lives in Kalamazoo, Michigan, with his wife, Jane.

Available titles in the Critical Issues in Crime and Society series: